Functional Molecular Gels

RSC Soft Matter Series

Series Editors:
Hans-Jürgen Butt, *Max Planck Institute for Polymer Research, Germany*
Ian W. Hamley, *University of Reading, UK*
Howard A. Stone, *Princeton University, USA*
Chi Wu, *The Chinese University of Hong Kong, China*

Titles in this Series:
1: Functional Molecular Gels

How to obtain future titles on publication:
A standing order plan is available for this series. A standing order will bring delivery of each new volume immediately on publication.

For further information please contact:
Book Sales Department, Royal Society of Chemistry, Thomas Graham House, Science Park, Milton Road, Cambridge, CB4 0WF, UK
Telephone: +44 (0)1223 420066, Fax: +44 (0)1223 420247
Email: booksales@rsc.org
Visit our website at www.rsc.org/books

Functional Molecular Gels

Edited by

Beatriu Escuder and Juan F. Miravet
Universitat Jaume I, Spain
Email: escuder@uji.es; miravet@uji.es

ROYAL SOCIETY
OF **CHEMISTRY**

RSC Soft Matter No. 1

ISBN: 978-1-84973-665-7
ISSN: 2048-7681

A catalogue record for this book is available from the British Library

Published by The Royal Society of Chemistry,
Thomas Graham House, Science Park, Milton Road,
Cambridge CB4 0WF, UK

Registered Charity Number 207890

For further information see our web site at www.rsc.org

Preface

The field of Molecular Gels – gels formed by low-molecular weight compounds – has been latent beneath diverse aspects of Science for more than a century but it has experienced a blossoming only in the last two decades. The number of papers related to the topic is growing exponentially and in the last few years there has been a gradual shift from considering these materials from an academic point of view towards their application in a wide variety of fields. In this new context, the design of functional molecular gelators that will confer a specific function to the supramolecular material takes on a relevant role, but also the emergence of new supramolecular properties as a direct consequence of the self-assembly process. This book proposes a tour starting from the fundamentals of design and characterisation of molecular gels (Chapters 1 and 2) followed by one of their most relevant properties, responsiveness to external stimuli (Chapters 3 and 4) and ending with a panoramic view of the most exciting applications in Chemistry and (Bio)Materials Science (Chapters 5 to 8). The book is not meant to be an exhaustive description but to provide with a few strokes of the brush a global view of the current state-of-the-art in the field of Molecular Gels. The editors are most grateful to all the authors contributing to this monograph for their willingness to spend their valuable time to participate in this project and to take the editors suggestions into account.

Juan F. Miravet
Beatriu Escuder

Castelló de la Plana, Spain
July 2013

RSC Soft Matter No. 1
Functional Molecular Gels
Edited by Beatriu Escuder and Juan F. Miravet
© The Royal Society of Chemistry 2014
Published by the Royal Society of Chemistry, www.rsc.org

Contents

RSC Soft Matter No. 1
Functional Molecular Gels
Edited by Beatriu Escuder and Juan F. Miravet
© The Royal Society of Chemistry 2014
Published by the Royal Society of Chemistry, www.rsc.org

CHAPTER 1

The Design of Molecular Gelators

NIEK ZWEEP[a] AND JAN H. VAN ESCH*[b]

[a] Stahl International bv, Sluisweg 10, 5145 PE, Waalwijk, The Netherlands;
[b] Department of Chemical Engineering, Delft University of Technology, Julianalaan 136, 2628BL Delft, The Netherlands
*Email: j.h.vanesch@tudelft.nl

1.1 Introduction

Supramolecular gels are hot! Since 1990 the number of publications on supramolecular gels has increased from less than five to at least five hundred papers per year.[1] The current interest in supramolecular gels has followed a long period of silence after their original discovery in the 1930s, when it was found that certain low molecular weight organic compounds could turn organic solvents into a gelly-like substance.[2] Because of this property they found widespread use as thickeners and lubricants, but at that time they did not raise much scientific interest. It took until the rise of supramolecular chemistry in the 1970s and 1980s, to realise that gel formation by these low molecular weight compounds is an example of a supramolecular system par excellence.

But what are gels? In daily life most people encounter gels, often without realising it. From a scientific point of view, the gel state has been recognised already for over 150 years. Nevertheless, the definition of a gel has long been under debate, mainly the term *"gel"* may point to chemically and physically very diverse systems.[3] In 1861 Thomas Graham gave the following description: *"while the rigidity of the crystalline structure shuts out external expressions, the*

RSC Soft Matter No. 1
Functional Molecular Gels
Edited by Beatriu Escuder and Juan F. Miravet
© The Royal Society of Chemistry 2014
Published by the Royal Society of Chemistry, www.rsc.org

softness of the gelatinous colloid partakes of fluidity, and enables the colloid to become a medium for liquid diffusion, like water itself."[4]

In the years that followed scientists attempted to define the "gel" state in a more explicit manner. Dorothy Jordon Lloyd wrote in 1926 that the colloidal condition, the "gel" state, is easier to recognise than to define[5] and proposed: "*only one rule seems to hold for all gels and that is that they must be built up from two components, one which is a liquid at the temperature under consideration and the other which, the gelling substance proper, often spoken of as the gelator, is a solid. The gel itself has the mechanical properties of a solid, i.e. it can maintain its form under stress of its own weight and under any mechanical stress it shows the phenomenon of strain*". Ever since, more rigorous definitions were proposed in attempts to link the microscopic and macroscopic properties of a gel.[3] Based on these definitions a substance is a gel if it (1) has a continuous microscopic structure with macroscopic dimensions that is permanent on the time scale of analytical experiments and (2) is solid-like in its rheological behaviour despite being mostly liquid.

In general, a gaseous or liquid system consisting of two or more components turns into a gel when one of the components forms a 3-dimensional (3D) entangled solid network within the bulk gas or liquid phase. The presence of this solid network restricts flow of the remaining gas or liquid bulk phase, and as a result the whole system appears macroscopically as a solid. Following this description, there are several possibilities to classify gels, *e.g.* according to the chemical nature of the components, or the physical state of the bulk phase.

A very common and useful classification is to distinguish between chemical and physical gels (Figure 1.1).[6] Chemical gels are gels in which formation of the

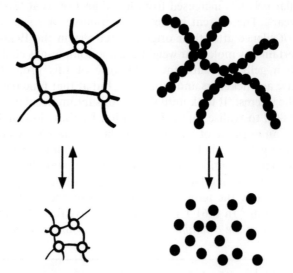

Figure 1.1   A change in the environment of a chemical gel (left) may cause it to shrink (or swell) but all network connections remain intact, whereas with a physical gel (right) the network connections may break (or form), eventually leading to complete loss of the gel state.

3D network has occurred through covalent crosslinking of the network components, resulting in a permanent network structure unless the covalent crosslinks are broken. Examples of such chemical gels include many crosslinked polymeric systems used in separation technology, and inorganic aerogels used as thermal insulators. In physical gels, formation of the 3D network has occurred through the formation of noncovalent interactions between the network components, and hence, the formation of physical gels is usually thermo-reversible. Physical gels can be formed by many, very different combinations of substances, for instance with clays, polymers, proteins, colloids, and certain small organic compounds as network-forming component in, for instance, organic solvents or water as the liquid component.

Supramolecular gels are thus a type of physical gels that are formed by small organic compounds as network-forming component.[7] These small organic compounds are called low molecular weight gelators (LMWGs). LMWGs are gelators consisting of organic compounds with a molecular weight of less than 2000 Da, that are capable of gelling organic solvents or water. Depending on whether the liquid phase is an organic solvent or water, they are also known as organogelators or hydrogelators, respectively. Gels from LMGWs are most commonly prepared by cooling a solution of the LMWG in the solvent to be gelled, leading to a supersaturated solution. The supersaturation causes a rapid assembly of the gelator molecules into elongated fibres of typically 5–100 nm diameter, and subsequently these fibres aggregate into a fibrous 3D entangled network, thereby turning the liquid into a supramolecular gel. Such a supramolecular gel consists of an entangled fibrous network of gelator molecules, which is held together by highly specific intermolecular interactions between the gelator molecules.

Hence, supramolecular gels are a clear manifestation of supramolecular chemistry in action. A central paradigm in supramolecular chemistry is that one can design supramolecular devices and materials with a desired function, by programming the assembly properties of their molecular building blocks *via* molecular shape and intermolecular interactions.[8] Therefore, over and over the question arose whether it would also be possible to design new supramolecular gels with tailor-made properties, by following the guidelines and principles of supramolecular chemistry. This chapter deals with the design of supramolecular gels. First, a brief introduction to the different types of LMW gelators as well as the main characteristics of supramolecular gels will be given, and then we will discuss the current status in the design of supramolecular gels.

1.2 Intermolecular Interactions in Molecular Gels

There are many different LMWGs with very different structures. However, despite the structural diversity, they have in common that their self-assembly into fibrous networks is driven by noncovalent interactions, like van der Waals interactions, π-interactions, dipolar interactions, hydrogen bonding, and coulomb interactions. Also, solvophobic effects play an important role. These solvophobic effects originate from moieties or functional groups in the gelator

molecule that are poorly soluble in the solvent to be gelled, and contribute to
the gelating ability by reducing the overall solubility of the gelator in that
solvent. Some examples of well-known LMWGs are given in Figure 1.2.

A large group of gelators is related to amphiphiles. Amongst the oldest of
gelators known are metallic soaps, which are used in cosmetic applications. An
example of a metallic soap gelator[2,9] is the lithium salt of 3-hydroxy stearic acid
1. Gelator **1** is based on 12-hydroxystearic acid, which is obtained from castor
oil. This gelator aggregates *via* several intermolecular interactions: ionic
interactions between the metal ions, van der Waals interactions between the
alkyl tails and hydrogen-bonding interactions between hydroxyl groups.
Cationic surfactants are structurally related to the metallic soaps. Quaternary
ammonium salts with long alkyl chains, like compound **2** show gelation behaviour due to aggregation driven by Coulomb interactions, as well as the poor
solubility of the quarternary ammonium moiety in many organic solvents.[10]

Less obvious cases of amphiphile-type gelators are **3** and **4**. Gelator **3** is a
perfluorcarbon-hydrocarbon block-compound, of which in hydrocarbon
solvents the perfluoroblock is insoluble and hence solvophobic, whereas the

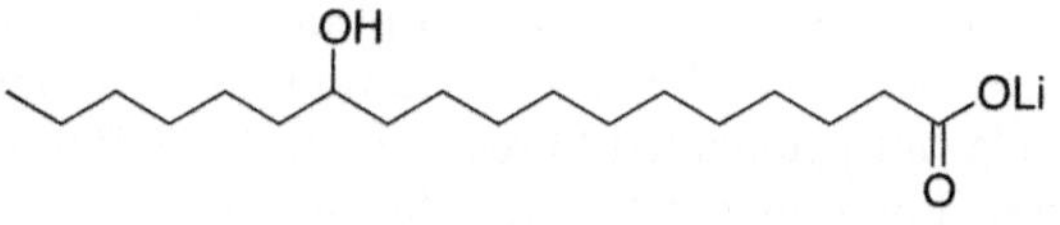

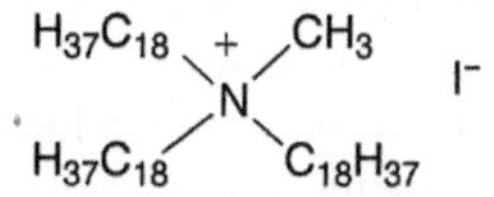

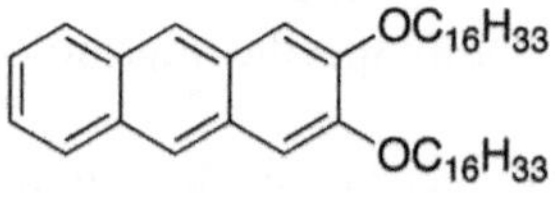

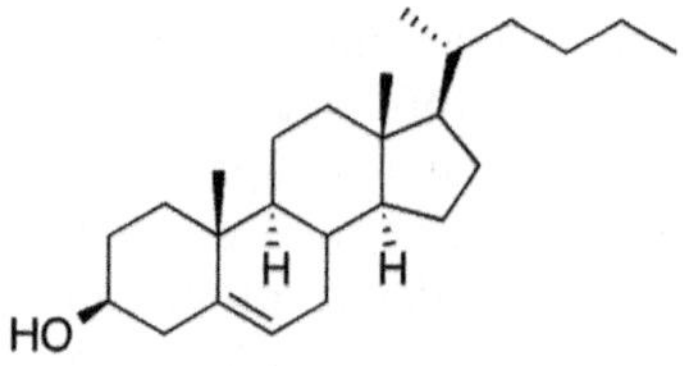

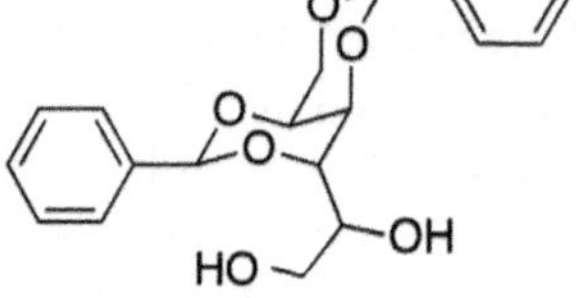

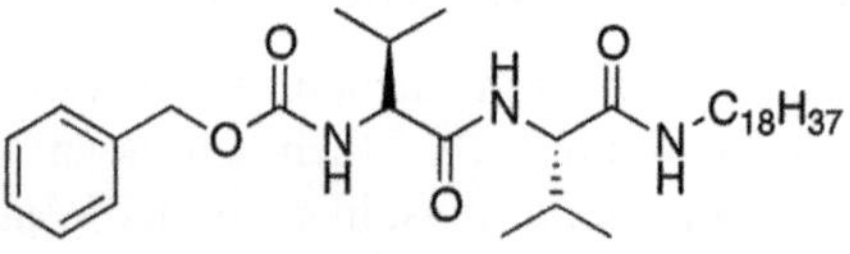

Figure 1.2   Structure of various low molecular weight gelators.

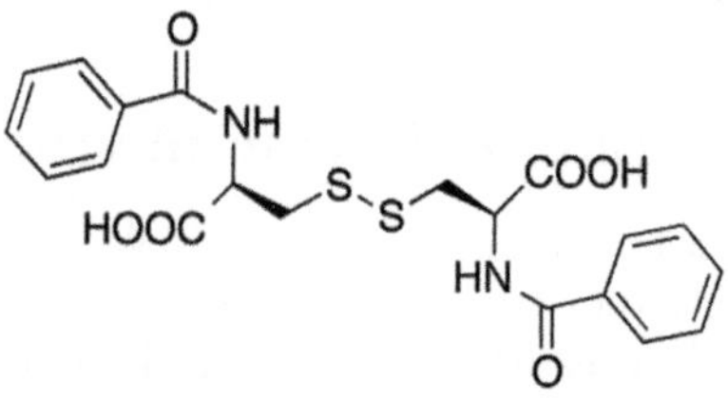

aliphatic part is very soluble, and hence solvophylic.[11] At room temperature compound **3** is immiscible with aliphatic hydrocarbons, leading to phase separation and gel formation at concentrations as low as 2 (w/w)%. Also, bis(alkyloxy)anthracene **4** can be considered as an amphiphile for hydrocarbon solvents, with the aliphatic hydrocarbon and anthracene moieties as solvophylic and solvophobic parts, respectively.[12] Apart from solvophobic effects aggregation of **4** is also controlled by π-interactions and van der Waals interactions.

In addition to solvophobic effects and specific intermolecular interactions, conformational rigidity may also contribute to the gelation ability. For instance, many steroids like cholesterol **5**, but also cholic acids are efficient gelators for various organic solvents.[13,14] The strong aggregation behaviour of these compounds is thought to originate in their molecular rigidity, which reduces entropic losses upon aggregation. Also, dibenzilydene-D-sorbitol (DBS) **6**, a simple sugar derivative and known since 1942 as an efficient gelator for both organic solvents and water, is a fairly rigid molecule with little conformational freedom.[15,16]

In the above examples hydrogen bonds are mostly absent or contribute only little to their aggregation behaviour. For many gelator molecules, however, hydrogen-bonding interactions are essential for their gelation ability. For instance, simple peptides like **7** or even without alkyl chains have been reported to gelate a range of different organic solvents, due to the formation of strong hydrogen bonds between the peptide amide groups.[17] Another well-known example of an amino acid based gelator is dibenzoylcysteine **8**, which is a very potent hydrogelator capable of gelling water at concentrations as low as 2 mM or (0.1% by weight).[18] With this compound again solvophobic effects, or more precisely hydrophobic effects because of the special character of water, and π-interactions between the pendant benzyl groups make a major contribution to the stability of its aggregates.

1.3 Structure and Properties of Molecular Gels

The gelators systems discussed above have been known for over 20 years. To be able to design new gelators with tailor-made properties, or even to predict gelation behaviour from the molecular structure requires a basic insight into the physiochemical basis for their gelation behaviour, however, such studies have been complicated by the large structural diversity of these systems as well as the structure of the gels itself. Many different techniques have been applied to establish the intermolecular interactions that are involved in self-assembly leading to gelation, to elucidate the structure and morphology of the gel and gel fibres, and to determine the thermotropic and viscoelastic properties.[7,19]

One important characteristic of gelators is the minimum required amount of gelator to form a gel in a specific solvent, also called the critical gelation concentration, cgc. The cgc consist of two contributions, that is the concentration of material required for the formation of a 3D network (C_{agg}), and the concentration of gelator molecules that remain in solution (C_{sol}) (Figure 1.3).[20,21]

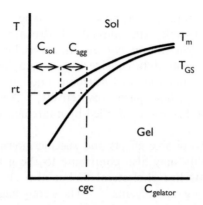

Figure 1.3 Example of a phase diagram for a gelator in a solvent.

Gel fibres are formed *via* noncovalent interactions and with increasing temperature these aggregates gradually dissolve of the increase of the solubility (C_{sol}). The temperature at which the gel loses its structural integrity is called the gel–sol phase transition (T_{gs}), and depends on the structure of the gelator, the nature of the solvent and the total concentration of gelator. The T_{gs} can be determined by various visual inspection techniques, *e.g.* by the "dropping ball" technique, bubble motion, or the inverted test tube method. The T_{gs} corresponds to the temperature at which the gel loses its structural integrity and some of the compound may be aggregated still, however, these aggregates are too small to sustain a network. The T_{gs} should not be confused with the dissolution temperature or melting temperature (T_m), the temperature at which all aggregates are completely dissolved. At T_m the total gelator concentration is equal to C_{sol}. The solubility curve has to be determined by alternative techniques that either determines T_m for a known total gelator concentration, or that determines the concentration of gelator in solution, *e.g.* by NMR, or chromatographic methods. From measurements of T_{gs} and T_m in a solvent over a range of concentrations one can draw the phase diagram of that gelator–solvent combination, from which one can read the cgc, C_{agg} and C_{sol} at each temperature (Figure 1.3).

These phase diagrams can also be used to determine the strength or enthalpy (ΔH_m) of the intermolecular interactions in the gel. If it is assumed that the gel–sol transition can be interpreted as dissolution of crystals in ideal solutions, one can determine the melting enthalpy of the gel *via* the van't Hoff equation (eqn (1.1)):[20–22]

$$\frac{d(\ln C_{sol})}{d\left(\dfrac{1}{T_m}\right)} = -\frac{\Delta H_m}{R} \qquad (1.1)$$

This method is based on the relationship between the logarithm of the concentration and $1/T_m$. However, in many studies only T_{gs} has been determined, which overestimates the dissolved amount of gelator, and leads to too low values for ΔH_m. A direct method to determine the melting enthalpy is differential scanning calorimetry (DSC). Unfortunately, the sensitivity of ordinary DSC is often too low for accurate measurements of ΔH_m, especially at low gelator concentrations and slow scan rates that are required to maintain dissolution equilibrium during the measurement. When this technique is applied for gels the energy required for the dissolution of the gel fibre is measured and ΔH_m can be determined directly.

The dropping ball, bubble motion and inverted test tube methods are also called "table-top rheology methods", and they already give a clear indication if a gel has been formed or not.[23] However, definite proof has to come from more quantitative studies on the viscoelastic properties of the supposed gels, *e.g.* by oscillatory rheology experiments.[24] In an oscillatory rheology experiment the sample is subjected to oscillating stress, and the viscoelastic response of the sample is measured in terms of the elastic storage G' and loss moduli G''. Gels can behave either as viscoelastic liquids or as viscoelastic solids, which is due to the formation of a highly dynamic or a static network structure, respectively. In general, the dynamic moduli depend on the frequency (time scale) of the measurement. The observed frequency dependence gives insight into the relaxation and lifetime of the bonds between the gelator molecules. If the bonds have a permanent character, only a small frequency dependence is expected and $G' \gg G''$ at all frequencies. This behaviour is characteristic of hard gels containing a static network structure, which do not deform under its own weight.[25] Such hard gels can be considered as true gels according to the definition of Jordon Lloyd. Examples of these systems are gel networks formed by derivatives of compound **1**.[70] Another type of gel-like systems are soft-gels, or gellies, which easily deform under their own weight and show gravitational flow. These soft gels are characterised by a typical significant frequency dependency of G' and G'', with $G' > G''$ only at high frequencies while at low frequencies (larger time scales) $G' < G''$, pointing to the slow deformation already at low stress. Typical examples of soft gels are solutions of worm-like micelles.[26,27] Oscillatory rheology experiments are also a reliable method to study the kinetics of gel formation by following G' and G'' as a function of time, after having started gel formation. Here, the gel point or gelling time is taken as the time at which G' starts to exceed G''.

These rheology studies clearly show that viscoelastic properties and fibre and network morphology are closely related, and therefore the study of the fibre and gel morphology form another essential part of the characterisation of supramolecular gels. Because the structural features of fibres and its network are typically in the range between 5–1000 nm, such study requires suitable direct imaging microscopy techniques, or indirect scattering techniques for such long length scales.[19] Over the years a variety of transmission electron microscopy (TEM) and scanning electron microscopy (SEM) techniques have been applied, which mainly vary in sample preparation procedure. In these techniques the gel fibres are not imaged in their native state and artefacts may arise from the sample preparation procedure, *e.g.* freezing of the solvent leading to deformation of the gel or precipitation of dissolved gelator fraction (C_{sol}). Indirect methods used to determine fibre morphology are small-angle X-ray scattering (SAXS) and small-angle neutron scattering (SANS). The particles in the beam are scattered depending on the shape of the fibres and *via* mathematical treatment of the scattering intensity as a function of the scattering angle, the size and morphology of the gel fibres can be determined. The problem with this method is that some prior knowledge of the size and shape of the gel fibres is required to perform the mathematical treatment. Also, if the sample is not homogeneous the mathematical treatment becomes difficult. Although both direct and indirect methods revealed valuable information on fibre and gel-network morphology, they do not add much insight into the supramolecular arrangement within the fibres.

The supramolecular arrangement of the molecules in a gel fibre is one of the most relevant questions in the study of gelling agents. When the supramolecular arrangement in the gel fibre is known it provides direct information on the different intermolecular interactions present in the gel fibre and how these contribute to the gel properties. Usually, NMR spectroscopy and single-crystal X-ray diffraction are applied to determine the organisation in supramolecular structures.[19] However, NMR spectroscopy is of limited use as the gel fibres appear to be in a solid state and therefore give rise to very broad and even unobservable signals in solution-phase NMR, because of the large correlation time of the assemblies with accompanied very short transversal relaxation times T_2.[28] Also single-crystal X-ray diffraction appeared to be unsuitable to determine the structure of gel fibres because of their small size, and presumably also because they contain many defects. As an alternative approach there have been many attempts to grow crystals from gelator molecules for single-crystal X-ray analysis, but crystal growth is in most cases prohibited by the formation of a gel! Only in a few instances has the crystal structure of a gelator been resolved, but even for these cases the question still remains if the crystal structure obtained from the gelator in the crystalline state corresponds to the organisation in the gel fibre. Therefore, powder X-ray diffraction (XRD) is used, which provides information of the organisation of the gel fibre. If the XRD patterns of the crystalline and gel state are comparable, it is safe to conclude that their organisation is similar. Even if a single-crystal X-ray structure is not available, the XRD pattern allows assignment of simple phases such as lamellar and hexagonal phases. However, it has not been possible yet to determine a fibre structure by this technique alone.

If structural information from a single-crystal X-ray study is not available, as is most often the case, then one has to apply other especially spectroscopic techniques to obtain information on the intermolecular interactions within the gel fibre. FTIR and Raman spectroscopy are most widely applied, because they allow the study of hydrogen-bonding interactions, and the investigation of the dominant conformation of alkyl chains, especially by comparison of dissolved gelators with gelators in the gel state. A significant advantage of these techniques is that these can be applied to the native state of the gel.

1.4 Formation of Molecular Gels

Over the years the outcome of many studies on very different kinds of gelator molecules and their gels has led to a fairly coherent view of gel characteristics and gel formation from LMW gelators. Molecular gels from LMW gelators are in most cases not the thermodynamically most stable state. Many gels suffer from long-term instability by slowly developing into heterogeneous systems that contain larger crystals in coexistence with the fibrous network.[29] Hence, the gel state should be regarded as a metastable state, but unfortunately, there is not yet a conclusive study towards the causes of this experimentally observed instability. Nevertheless, two possible causes for this instability have been proposed based on the structure and morphology of the gels. One possible cause is the difference in surface free energy, due to the large interfacial area of the fibrous network compared to crystals from the LMW gelators; another possible cause are differences in lattice energy due to the many defects in the fibrous network or even a different packing within the fibres, compared to crystals.

Because most supramolecular gels are a metastable state, their formation and properties are governed by kinetics and thermodynamics. Several studies have indicated that the gel properties depend on the rate of cooling, and also recent studies in which the rate of gel formation is controlled by catalytic action revealed that the gel properties depend on the rate of formation. For gel formation due to cooling, the gelation process proceeds through the following steps[24] (Figure 1.4): (i) cooling of a solution of gelator to form a supersaturated solution, (ii) formation of nuclei, (iii) growth of nuclei to fibres, (iv) fibre bundling and branching, and (v) fibre entanglement to a 3D fibrous network. The recent work by Meijer *et al.*[84] revealed that fibre formation by self-assembly of small molecular building blocks indeed involves a nucleation step,

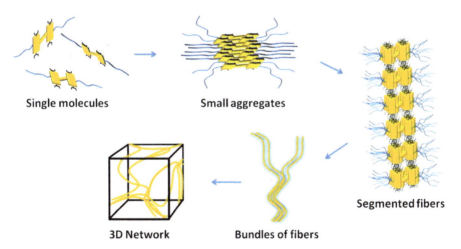

Single molecules **Small aggregates**

Segmented fibers

3D Network **Bundles of fibers**

Figure 1.4 Schematic depiction of the process of supramolecular gel formation. (Reproduced with permission of American Chemical Society from ref. 30).

pointing to a transition from a disordered prenucleus to a stable, more ordered nucleus. Overall, the gelation process has many similarities with the classical process of crystal formation, and therefore gelation is often referred to as a kind of frustrated crystallisation.

The similarities between crystal growth and gelation offers a rational for explaining the effect of gelation kinetics on gel morphology and properties, as the extent of supersaturation is likely to have similar effects on the rate of nucleation, fibre growth and shape, and branching. According to classical crystal growth theory, a larger supersaturation leads to a larger nucleation rate and also to larger deviations from the equilibrium crystal shape. At large supersaturations, fibre growth may approach or even enter the diffusion-limited regime. Diffusion-limited growth causes excessive growth of fibre tip because of its radial diffusion front, thereby amplifying the already elongated shape. Moreover, under diffusion-limited growth conditions a growing interface may become unstable, which leads to tip splitting and hence fibre branching.

There may even be more similarities between crystal growth and gelation.[31] In gels, the fibres have a characteristic elongated shape with a very high aspect ratio, and this shape must arise from a strongly anisotropic growth process. Crystal growth theory states that the growth rate of a crystal plane increases with the attachment energy for that plane, and as a result the planes with the lowest attachment energies are the slowest growing planes and consequently become the largest exposed planes in the resulting crystal shape.[32,33] Interestingly, the attachment energy is directly related by the intermolecular interactions between a growing plane and newly adsorbed molecules. Hence, the observation that fibre growth occurs at their tips suggests that the intermolecular interactions in a direction perpendicular to the fibre tip are much stronger than in other directions. As the attachment energy and interfacial energy are related, this view makes sense because such a process results in crystal or fibre shapes that are dominated by crystal or fibre planes with the lowest interfacial energy.

1.5   How Can We Design Molecular Gels

It is clear that the formation of supramolecular gels by LMW gelators is a complicated phenomenon, controlled by delicate interplay between thermodynamics and kinetics. One possibility to gain further insight into gelation phenomena is to study the subject in even greater detail. However, the characterisation of supramolecular gels also appears to be challenging, and further research along these lines will largely depend on advancements in real-time and *in situ* structural characterisation methods covering a broad range of length scales. Driven by the many potential applications of supramolecular gels, in the meantime scientists have also taken other approaches to develop new LMW gelling systems with improved properties or added functionality like recognition and responsiveness. In the following sections the most prominent approaches with illustrative examples will be discussed. They have in common that they aim for the bottom-up fabrication of new gel materials with tailor-made properties by design of their molecular building blocks.

1.5.1 Property-Based Design

For a long time structural information on LMW gelators was limited and therefore early studies focused on molecular structure–property relationships, for instance by systematic variation of the molecular structure, to identify the structural features that contribute to the gelation behaviour. In what one may call a "property-based design" approach, scientists went one step further to develop new gelators with completely new properties or functions. This property-based design approach is based on the hypothesis that merging of two molecules with different properties may lead to a new molecule with properties inherited from both parents.

An excellent example of such a property-based design in the field of supramolecular gelators started with an elegant and classical structure–property relationship study by Weiss and coworkers, who accidentally found that cholesteryl anthranyl butyrate (CAB) **9** is a potent gelator for a wide range of organic solvents (Figure 1.5).[34,35] After extensive structural modification of CAB into a variety of related anthracene-linker-steroidal (ALS) compounds, it was concluded that the gelation ability only tolerates few structural modifications, and is tightly connected to an overall rod-like molecular shape, a minimum size of the aromatic part, and the presence of the alkyl chain at C17 of the steroid moiety. Further structural investigations indicate that in the fibres the rod-shaped CAB molecules are arranged into helical stacks with partially overlapping anthranyl groups.

The results of these studies inspired Shinkai and coworkers to use the ALS motive as a basis for new gelators with metal ions of photoresponsive functions.[36] Indeed, the gelation ability of **10**, in which the anthranyl group in CAB is replaced by an azobenzene crownether moiety, improves upon binding of metal ions by the crownether moiety, as noted by a marked increase of T_{GS}. Most interestingly, photochemical *trans–cis* isomerisation of the azobenzene moiety in another ALS derivative, compound **11**, resulted in a reversible reduction of T_{GS} by going from the *trans* to *cis* form. This finding allowed the photochemical switching between the liquid and gel state at temperatures in between T_{GS} of the *trans* and T_{GS} of the *cis* state. Later, a more detailed study[37] indicated that this change in gelation ability is caused by a change between a rod-like and a bent molecular shape upon going from the *trans*- to the *cis*-conformation, thereby confirming the finding of Weiss and coworkers on the importance of shape for gelation ability.[35]

Another successful example of property-based design is the development of LMW gelators for liquid CO_2 by Hamilton and coworkers.[38] In their design they combined a strong hydrogen-bonding motive to promote aggregation with CO_2-philic moieties like fluoroalkyl- or fluoroether groups for CO_2 compatibility (Compound **12**, see Figure 1.6). Several of the resulting compounds appeared to be excellent gelators for liquid CO_2 by forming a fibrous network.

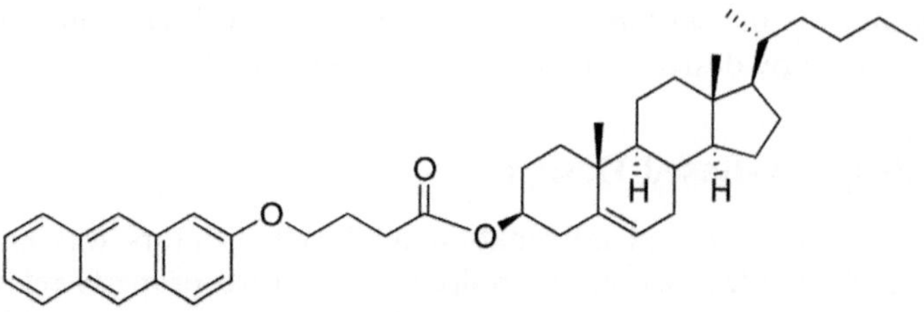

9

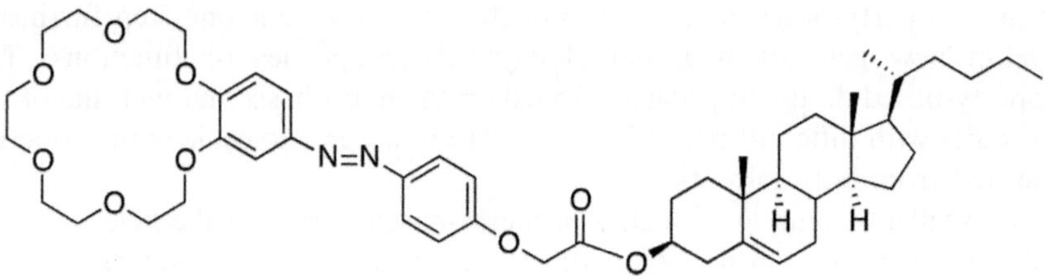

10

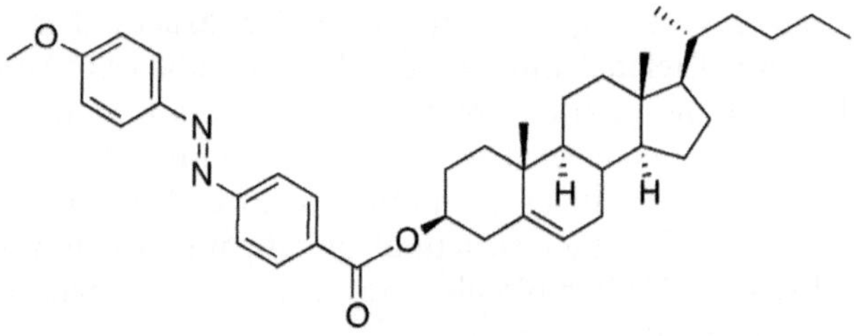

11

Figure 1.5   Structures of some cholesterol-based gelators.

Most interestingly, the liquid–solid gels could be transformed into monolithic aerogels without collapse of the fibrous network by removal of CO_2 above its critical temperature. Later, Hamilton and coworkers were also able to transform the CO_2-gelators into hydrogelators by substitution of the CO_2-philic groups for hydrophilic moieties.[39]

The gelation properties of some long chain alkyl ammonium salts[40] have inspired Weiss and coworkers to develop CO_2 responsive gelators,[41] by exploiting the isothermal and reversible formation of carbamates by reaction of CO_2 with amines. Here, an amine, *e.g.* compound **13**, is used as a "latent" gelator that upon exposure to CO_2 is converted into the corresponding carbamate salt (Figure 1.6). As the ammonium salts have a low solubility, they aggregate and form a gel isothermally. The gel formation is completely reversible as the CO_2 can be removed by flushing with N_2 at elevated temperatures. The CO_2-responsive gelation behaviour of these compounds can for instance be used in carbon dioxide sensors. Another intriguing example are the antibiotic hydrogels from Bing Xu and coworkers.[42,43] They turned

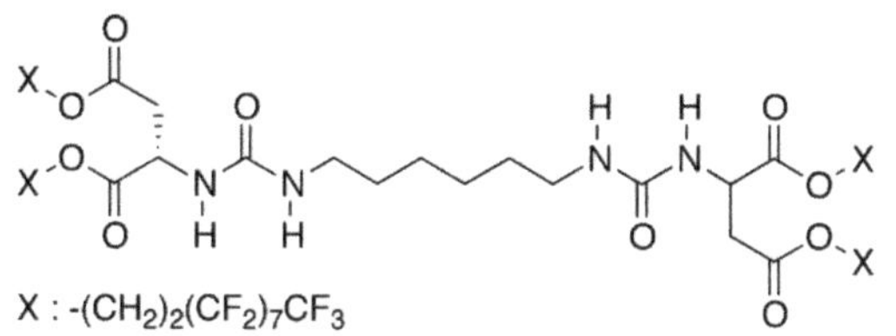

X : -(CH_2)_2(CF_2)_7CF_3

12

$$2\ H_{37}C_{18}-NH_2 \quad \underset{N_2,\,\Delta}{\overset{CO_2}{\rightleftharpoons}} \quad H_{37}C_{18}-\overset{\displaystyle}{\underset{H}{N}}\overset{O}{\overset{\|}{C}}O^- \quad H_3\overset{+}{N}-C_{18}H_{37}$$

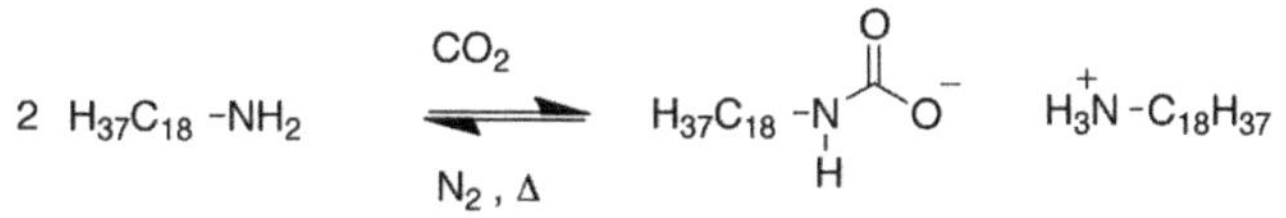

13

Figure 1.6 Structure of a fluorinated gelator capable of gelling liquid CO_2, and an aliphatic amine turned into a gelator by a reversible reaction with CO_2.

vancomycin, a well-known water-soluble antibiotic agent, into a potent hydrogelator by appending it with the hydrophobic pyrene moiety. Interestingly, the vancomycin hydrogels showed largely enhanced antibiotic activity compared to vancomycin itself.

These examples clearly demonstrate that property-based design can be a successful approach to develop new gelators with very different properties, by starting from empirical relationships between gelator structure and macroscopically observable gelation behaviour. Because of the lack of insight on how the macroscopic properties of gel formation are linked to the molecular structure, the property-based design approach depended largely on trial and error: only a few of the synthesised molecules showed the desired properties, whereas the majority of the synthesised derivatives failed as gelator, often without a satisfactory explanation. These failures emphasise once more that gelation ability is very sensitive to the molecular structure. Nevertheless, these negative results can also be very useful, because in combination with the successful newly designed gelators they form a new structure–activity study, which may lead to additional clues on the supramolecular arrangement within these gels.

1.5.2 Library and Selection Approaches for the Discovery of New Gelators

One possibility to accelerate the discovery of new gelator molecules and motives is to screen libraries of potential gelators, which consist of many different compounds sharing a common property or structure. One of the first reports on such an approach is by Shinkai and coworkers, who took the gelation properties of dibenziledene sorbitol as a starting point to prepare a library of potential gelators from a set of different monosacharides by reacting them with benzaldehyde.[44] By screening the sacharide library for gelation properties, they discovered several new carbohydrate-based gelators, and were able to identify the main structural motive responsible for gelation. The library approach appeared especially useful for the discovery of new two-component gelators, because they could be easily prepared *in situ* by mixing of solutions of the components, and the resulting solutions could directly be examined for gel formation, without the need for intermediate purification. Successful examples include libraries of salts from bile acids and amines,[45] and of ureido-type gelators from a reaction of isocyanates with amines.[46] The work by Hamachi and colleagues showed that the library approach is not limited to two-component systems though. They prepared libraries of glycosylated amino acid derivatives by solid-phase synthesis, and examined the compounds for gelation ability after cleavage form the solid support[47,48] (see Figure 1.7). This application of solid-phase synthesis methods extends the library approach to compounds requiring multiple synthetic steps and/or multiple building blocks.

The above library approaches are based on the evaluation of individual library members for gelation ability, but in constitutional dynamic libraries this is no longer the case. In a first example of such a constitutional dynamic gelator library, Lehn and colleagues simultaneously combined multiple building blocks of a two-component gelator system, and examined the resulting mixture for gel formation (Figure 1.8).[49] For the gel-forming mixtures they identified the gelating compound, and thereby also its building blocks. Such a constitutional dynamic library is based on the dynamic formation of all possible gelators from its precursors, together with gel formation as the selection mechanism, which drives the systems towards the most potent gelator. Another, later example of this constitutional dynamic library approach can be found in the work of Smith and colleagues on dynamic salt formation between acidic dendrons and amines.[50]

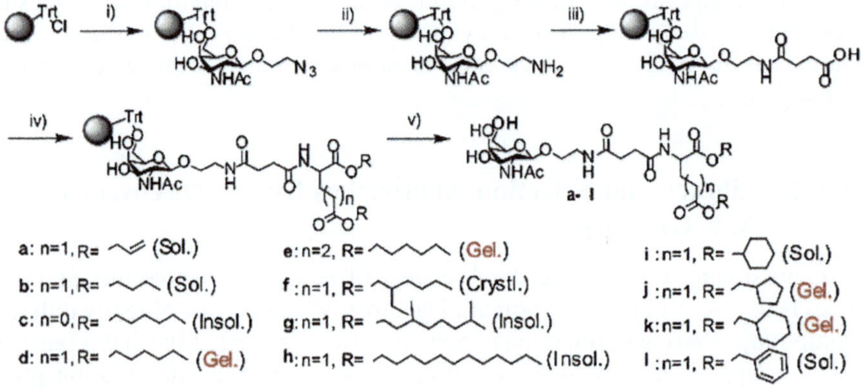

Figure 1.7   Example of a library explored for gelation properties.
(Reproduced with permission of American Chemical Society from ref. 48).

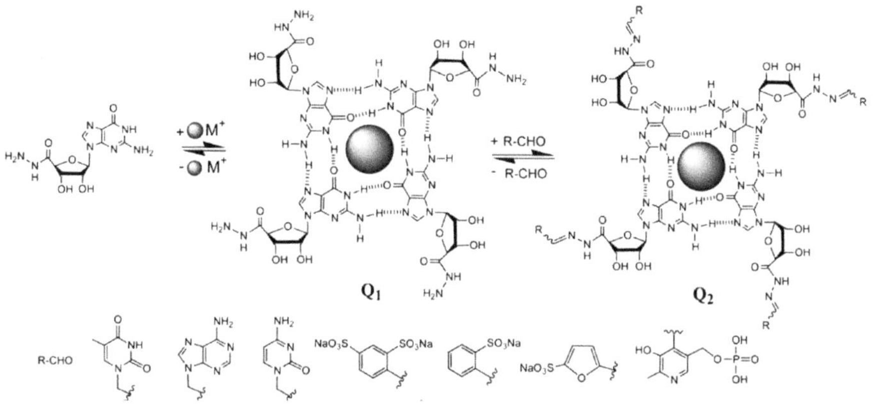

Figure 1.8 Example of a dynamic covalent library.
(Reproduced with permission from ref. 49, Copyright (2005) National Academy of Sciences, U.S.A.).

1.5.3 Structure-Based Design

In their strive to better understand the link between gelation behaviour and molecular structure, scientists took two important steps: they started to formulate design criteria for supramolecular gelation from their knowledge on molecular gelation phenomena, and they connected these criteria to the supramolecular design principles. Together, these steps led to the "structure-based-design" of molecular gelators.

In their 1996 paper[51] Hanabusa and coworkers were among the first to explicitly formulate design criteria for organogelators. According to them a gelator molecule should exhibit the following features: "(i) the existence of intermolecular interactions for building up macromolecule-like aggregates, (ii) the occurrence of intertwining of the aggregates, and (iii) the presence of some factor for preventing the crystallisation of the metastable gel." They came up with a very elegant and simple design of a new organogelator based on the *trans*-1,2-bis(alkyl-amide)cyclohexane core [**14**] with two amides as self-complementary hydrogen bonding groups (Figure 1.9). These compounds are able to gelate a surprisingly broad range of organic solvents at concentrations well below 1 w/v%. Intermolecular hydrogen bonding between the amides provides the driving force for the formation of extended aggregates in nonpolar, organic solvents, whereas the alkyl chains are thought to ensure compatibility with nonpolar solvents to increase solubility and prevent crystallisation.

Around the same time our group exploited the very similar *trans*-1,2-bis(alkyl-ureido)cyclohexane and the related 1,2-bis(alkyl-ureido)benzene core, for the design of functional gelators[31,52] (compounds **15**, **16**). This design was based on two assumptions: (i) the cooperative and self-complementary hydrogen-bonding interactions properties of the urea group would favour its assembly into extended aggregates and (ii) the antiparallel orientation of the hydrogen-bonding urea directs hydrogen bonding along a single molecular

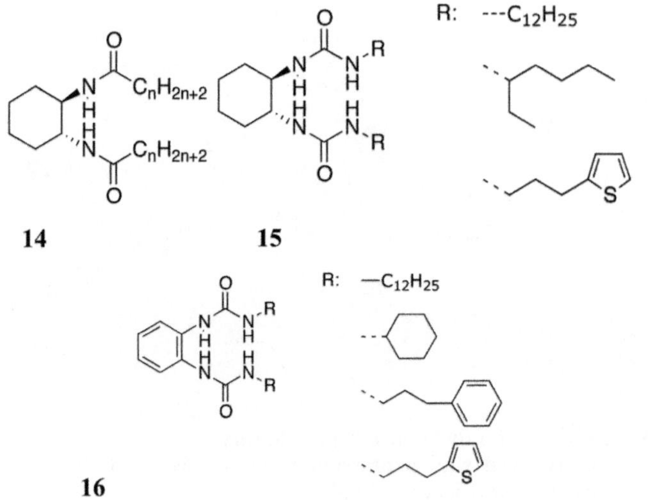

Figure 1.9 Structures of bis-amide and bis-ureido types of gelators.

axis, thereby favouring its 1D self-assembly (Figure 1.10). The design was validated by molecular modeling, which also showed that the alkyl tails could be extended with functional groups without disrupting the 1D self-assembly motive. Indeed, all the synthesised derivatives of these compounds with simple alkyl chains or extended with functional groups appeared to be excellent gelators for organic solvents. These findings confirmed the importance of unidirectional interactions for gel formation.

In the same year Aggeli and colleagues reported on the design of new peptide organogelators.[53] Their design is based on a beta-sheet forming peptide as 1D self-assembly motive, with one face of the sheet occupied by apolar and the other side by polar amino acid residues as solvophobic and solvophilic groups, respectively. These molecular features enforce the formation of bilayer ribbons from two beta-sheets through interactions between the solvophobic sides, while the solvophilic groups prevent further aggregation into thicker bundles. The peptides were potent gelators for moderately polar organic solvents like methanol, and hexanol, and could even be tuned to gel aqueous solutions by introducing more polar amino acids residues at the solvophilic face.

These studies contribute to the formulation of basic design rules for LMW gelators, which can be summarised as follows: 1) the presence of strong self-complementary and unidirectional intermolecular interactions to enforce one-dimensional self-assembly; 2) control of fibre/solvent interfacial energy to control solubility and to prevent crystallisation; and 3) some factor to induce fibre crosslinking for network formation. The value of these rules has become apparent from several successful designs, for instance the tris-amide cyclo-hexane[54,55] and benzene families of gelators, bicomponent gelators made up from porphyrins and fullerenes,[56] and various types of aromatic gelators.[57,58] Noteworthy is the work of Dastidar and colleagues, in which they explored the

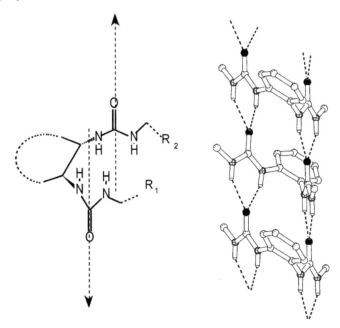

Figure 1.10 (a) Hydrogen-bonding directionality in 1,2-bis-ureido compounds and (b) X-ray structure of 1D hydrogen-bonded stack formed by a 1,2-bisureidobenzene compound. (From ref. 31, reprinted with permission from WILEY-VCH.).

Cambridge Crystallographic Database for 1D self-assembling with hitherto unrecognised gelation ability.[59]

Together, these results show that indeed it is possible to design new LMW gelators by following a few basic design rules, and in particular the first rule, to enforce 1D self-assembly by self-complementary and unidirectional interactions, appeared to be of great value. It also became apparent that the second and third rules are too unspecific for obtaining full control over fibre and network morphology. In particular, the control of the diameter and the prevention of crystallisation, as well as bundling and crosslinking of the fibres remain huge challenges.

An important clue for controlling fibre diameter comes from the seminal work of Israelachvili *et al.* on the self-assembly of amphiphiles,[60] in which they pointed out that the aggregate morphology results from the balance between attractive and repulsive intermolecular forces. In what later became known as the "structure-shape" concept, this balance of intermolecular interactions was ingeniously related to molecular shape *via* the "packing parameter" P, thereby providing a rational for the relation between molecular shape and aggregate morphology (Figure 1.11). For amphiphiles, the packing parameter is defined as $P = V/a_0 \cdot l_c$, in which V is the hydrocarbon core volume, a_0 optimum interfacial area per amphiphile, and l_c the extended length of the hydrocarbon chain of the amphiphile. Here, an increase of repulsive forces leads to an increase of a_0, and hence a decrease of P. Interestingly, the structure-shape concept tells that

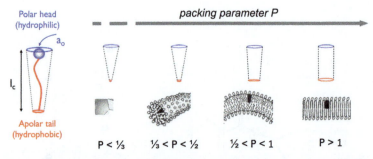

Figure 1.11   The structure-shape concept relates the molecular shape *via* the packing parameter P to the aggregate morphology.

Figure 1.12   (a) Self-assembly of wedge-shaped peptide amphiphiles into fibres with a uniform diameter, and (b) electron micrograph of the peptide fibres. (From ref. 61, reprinted with permission from AAAS).

spherical or cylindrical shaped assemblies are formed from cone- $(P < \frac{1}{3})$ or wedge-shaped molecules $(P < \frac{1}{2})$, respectively. Spherical and cylindrical aggregates have a finite length in three or two of the principle aggregate dimensions, respectively. In terms of intermolecular forces, the repulsive intermolecular forces increase more with aggregate size than the attractive forces do, leading to a minimum of free energy for a specific, discrete aggregate size.

A straightforward implementation of the structure-shape concept was carried out by Stupp and colleagues in their design of new peptide amphiphiles[61] as LMW hydrogelators (Figure 1.12). The peptide amphiphiles form gels with water, which consist of a network of fibres with a uniform width. Self-assembly of the peptide amphiphiles is driven by hydrophobic interactions between the alkyl chains, and hydrogen-bonding interactions between the peptide amide groups. Interestingly, the first three to five amino acid residues after the alkyl chain are involved in a beta-sheet hydrogen-bonding network along the long fibre axis, thereby providing the necessary monodirectional interactions for fibre formation.[62] The uniform diameter of the fibres can be related to the wedge shape of the peptide amphiphile, which originates from the larger cross section of the peptide fragment compared to the alkyl chain, and repulsive electrostatic interactions between the charged amino acid residues.

Another implementation of the structure shape concept is found with gelators that are based on dendrons.[63–65] In these molecules, dendritic branches are used as solvophilic headgroups with a large cross-sectional area, which are connected to a solvophobic core with a smaller cross-sectional area. This design results in either cone-shaped or dumbbell-shaped molecules, which were found to form gels with selected solvents. In most cases the wedge or dumbbell shape of the dendron gelators indeed resulted in the formation of fibres with a uniform diameter, but aggregation of the fibres to bundles with a polydisperse diameter also occurred.

In the above examples, the observed finite fibre diameter results from a balance between attractive (van der Waals, hydrogen bonding, hydrophobic) and repulsive (steric, electrostatic) intermolecular interactions, acting at the molecular length scale. There are also some interesting examples of self-assembled fibres, in which the finite diameter is the result of a balance of forces acting at the length scale of the aggregate. In one example, Oda and colleagues[66] reported on gel formation from gemini quarternary ammonium surfactants in the presence of tartrate as a chiral, divalent counterion. These gels were made up from chiral twisted ribbons of the gemini surfactants, and they showed that the twisting can be tuned by the enantiomeric excess of the tartrate counterions (Figure 1.13).[67] Interestingly, at high enantiomeric excess the twisted ribbons have a defined, finite width. From a theoretical model they were able to show that this finite ribbon width results from a balance of line forces (the one-dimensional equivalent of interfacial forces, here associated with the edges of the ribbons) and chiral twisting forces caused by chiral deformation of the ribbons.

Building on their earlier work on peptide gels consisting of beta-sheet ribbons (*vide supra*), Aggeli and colleagues found that related peptides can form a hierarchy of supramolecular structures,[69] starting with the formation of

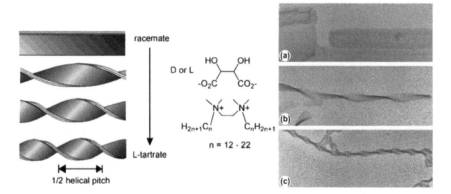

Figure 1.13 The presence of chiral counterions causes twisting of ribbons formed by gemini surfactants. The panel on the right shows electron micrographs of ribbons formed from gemini surfactants with (a) a racemic mixture (b) 50% enantiomeric excess and (c) 100% enantiomeric excess of counterions. (From ref. 67, reprinted with permission from Nature, and reprinted with permission from ref. 68. Copyright (2002) American Chemical Society).

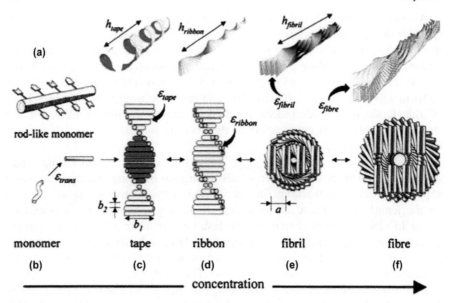

rod-like monomer

monomer   tape   ribbon   fibril   fibre

(b)   (c)   (d)   (e)   (f)

——————————— concentration ———————————▶

Figure 1.14   Hierarchical self-assembly of peptides: monomers (a,b) self-assemble to tapes (c), that dimerise to twisted ribbons (d). Subsequently, these ribbons may stack into helically twisted fibrils (e), which ultimately entwine to helical twisted fibres (f).
(Reproduced with permission from ref. 69, Copyright (2001) National Academy of Sciences, U.S.A.).

helically twisted beta-sheet tapes, that dimerise to twisted ribbons (Figure 1.14). Subsequently, these ribbons may stack into helically twisted fibrils, which ultimately entwine to helical twisted fibres. Like with the chiral gemini surfactant tapes, aggregation of twisted ribbons into a fibril is accompanied by a bending distortion of the most favourable ribbon geometry, which becomes larger with increasing fibril width. Conversely, the contribution from the attraction energy between ribbons increases with increasing fibril width. Hence, the balance between the associated elastic deformation energy penalty and ribbon attraction energy leads to formation of fibrils of discrete width, formed by a defined number of ribbons.

In the above systems, the defined fibre width results from opposing interactions. In fact, the balance between these opposing interactions leads to a minimum of the free energy for a given aggregation number, and consequently these fibres may represent a thermodynamically stable gel,[29] which would not convert over time to crystals. A comparable mechanism may be active in other supramolecular gel systems consisting of fibres with discrete width as well, and would place these systems apart from the much more common kinetically trapped supramolecular gels.

In supramolecular gels, the morphology and strength of gels of fibrous networks does not only depend on the fibre diameter. In fact, the mesh size of the network is one of the primary characteristics of network morphology and is largely determined by fibre density and the number or density of the crosslinks between fibres.[7] The density of the crosslinks together with their strength also has a large effect on the viscoelastic properties of the gels,[70] and therefore, the control of crosslink formation is one of the primary design parameters of fibrillar gel networks.

A closer inspection of various fibrillar gel systems reveals that for most gel systems the crosslinks between fibres consist of branched fibres or bundled fibres, and intertwined fibres. Together, these different types of crosslinks are also known as junction zones, and the branched and bundled types together are grouped as microcrystalline domains. In the microcrystalline domains the fibres are held together by similar intermolecular interactions as between fibre filaments, whereas with intertwined fibres also mechanical interlocking contributes to the stability of the crosslinks. For both cases, the formation of the crosslinks takes place during formation of the gels,[71] and is therefore governed by the same complicated interplay of thermodynamics and kinetics.

There are few examples of studies in which fibre branching has been controlled at the kinetic level. Liu and coworkers[72] have been able to control fibre branching of a steroid-based organogel system by adding a tiny amount of a polymer (Figure 1.15). They have been able to show that during fibre growth, the polymer adsorbs at the growing tip and causes a lattice mismatch, which leads to branching. More recently, our group found that fibre morphology and in particular branching of dynamic covalent gels can be tuned by controlling the gel growth kinetics by catalytic action.[73] It should be noted that these approaches are independent of the molecular design of the gelator, but are aimed at influencing the gel formation mechanism.

A very different approach to generate and control crosslink formation between fibres is the introduction of specific noncovalent or covalent bonds between fibres. In an early example our group reported the covalent crosslinking of fibrous networks of cyclohexyl bis-urea type of gelators (**17**), by polymerisation of appended methacrylate moieties (Figure 1.16).[52] This approach preserved the original network morphology and thereby did not offer control over crosslinking density. Moreover, the gels lost their thermoreversibility, indicating

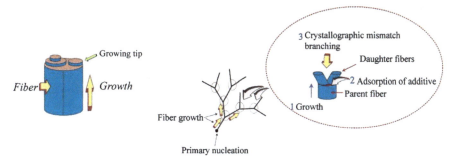

Figure 1.15 The adsorption of additives causes a crystallographic mismatch at a growing fibre tip, leading to branching.
(Reprinted with permission from ref. 74. Copyright (2002) American Chemical Society).

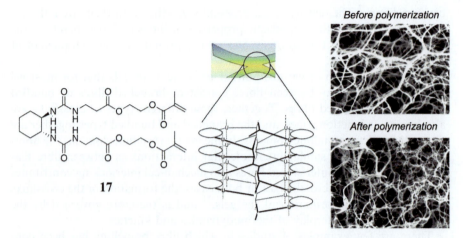

Figure 1.16   Photo-polymerisation of a gel of cyclohexyl bis-urea **17**, greatly enhances its mechanical and thermal stability, while preserving the network morphology.

that the crosslinking reaction has transformed the supramolecular network into a covalent network. Later, Finn and coworkers utilised the azide-alkyne click reaction for covalent crosslinking of gelator molecules during gelation, leading to the formation of strong but brittle gels.[75] Interestingly, despite the irreversibility of the click reaction the gel formation was still thermoreversible, presumably because of the low ratio of crosslinking agent to gelator molecules. Nevertheless, also in this example there was no control over the crosslinking density between fibres, because the presence of reactive groups at all gelator molecules enabled both intra- and interfibre bond-forming reactions.

In different approaches, Shinkai and coworkers exploited the formation of reversible connections to reinforce gel networks. In one example, they used the formation of metal coordination complexes to improve the thermal stability of gels,[76] but in this example the formation of the coordination complexes also led to increase of the fibre diameter, indicating that coordination bond formation was not limited to crosslinking of fibres. In another example they explored dynamic covalent bonds to reinforce gels. They used a phenyl boronic acid appended polymer with carbohydrate-based gelators to crosslink the gelator supramolecular assemblies, leading to a gel network (Figure 1.17).[77] The carbohydrate gelator by itself does not form fibres or gels but instead assembles into 100–500 nm large vesicles, which are then crosslinked by reaction with the boronic-acid appended polymer to a vesicular network or vesicle gel.[78] Formation of these vesicle gels remained thermoreversibility, most likely due to the dynamic nature of boronic acid esters. Interestingly, the gels showed a maximal thermal stability for a specific gelator/polymer ratio. These preliminary results clearly demonstrate that selective crosslinking of supramolecular gels can be achieved by using polymers as crosslinking agents, and that the crosslinking density can be tuned by the gelator-polymer ratio.

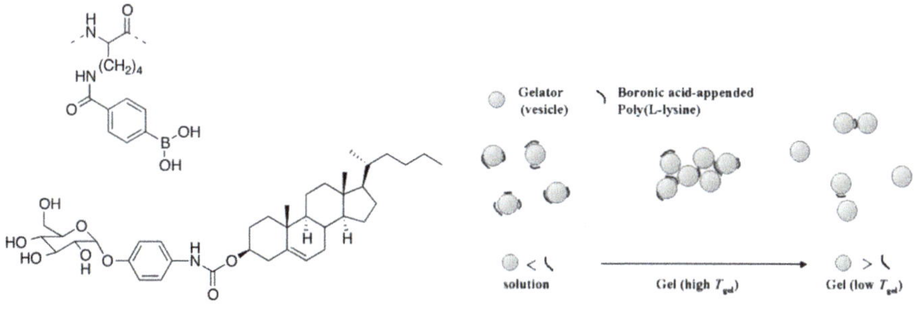

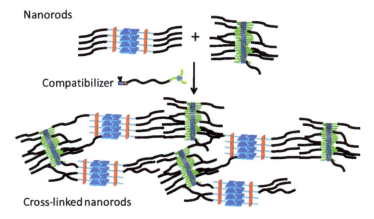

18 + 19

Figure 1.17   The formation of dynamic covalent bonds between boronic acid functionalised polymer **18** and vesicles from glucose-cholesterol compound
19 leads to the reversible formation of a vesicle gel.
(Reproduced from Ref. 77 with permission from The Royal Society of
Chemistry).

Figure 1.18   Addition of a ditopic crosslinking agent to a orthogonal self-assembling
systems results in the specific formation of crosslinks between fibres.
(Reprinted with permission from ref. 79. Copyright (2011) American
Chemical Society).

Recently, Palmans and Meijer and colleagues took a next step towards full
control over crosslinking in supramolecular gels.[79] They used ditopic crosslinking agents for an orthogonal self-assembling system[80] of two supramolecular polymers (Figure 1.18). Monomers of two different supramolecular
motives each self-assemble into short supramolecular polymeric fibrils leading
to a viscous solution. The ditopic crosslinking agent consists of a telechelic
polymer capped with the two supramolecular monomers, one of each motive, at
its ends. Addition of the ditopic crosslinker transformed the viscous solution
into a stiff supramolecular gel, due to the binding of the monomers at each
polymer end to the corresponding supramolecular fibres. Interestingly, the use
of an orthogonal self-assembling system for crosslinking prevents back-folding

and binding of the other polymer end at the same fibre, thus affording greater control over the crosslinking density *via* added crosslinking agent.

The above examples demonstrate the successful application of design principles at different levels of supramolecular gel formation, starting from 1D self-assembly, *via* control of fibre diameter, up to the control of crosslink formation in fibrous networks. Still, we are far away from full control over gel morphology and properties by design, and many challenges lay ahead, for example how to master fibre stiffness, crosslink distance and geometry, and how to design of gels with nonfibrous morphologies.

1.5.4 Gelator Scaffolds as Starting Point for Functional Gelators

Although controlling gel structure and properties by design remains a challenge, there has been a remarkable progress in the ability to develop new gelling systems with tailor-made properties during the past decades. This progress has been possible because many researchers took a pragmatic approach for the design of functional gelators, by using known gelator systems as a scaffold for the incorporation of functional moieties at well defined positions within the gel fibres. In many cases, the success of this so called "gelator scaffold approach" is owed to, first, the well conserved supramolecular aggregation mode of several well-known gelators, and second, by considering possible solubility issues and steric constraints in the design of the functional gelator. In an early example, our group exploited the linear aggregation motive of bis-urea gelators to arrange thiophene moieties in linear π-stacked assemblies.[81] The resulting compounds **20** indeed form elongated fibres which gel certain organic solvents, and more detailed structural characterisation clearly indicate that the 1D aggregation mode of the urea groups has been preserved (Figure 1.19). Most

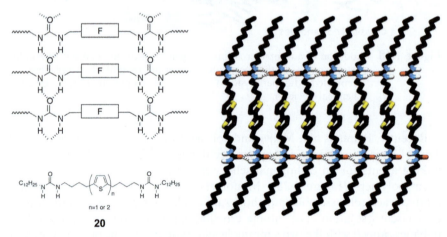

20

Figure 1.19   The linear aggregation mode of urea groups can be used to organise other functional moieties (F) in linear arrays. The image on the right shows such an array of bis-urea compound **20** with stacked thiophene groups, which showed strongly enhanced charge mobility in the fibres. (From ref. 81, reprinted with permission from WILEY-VCH).

(a)

(b)

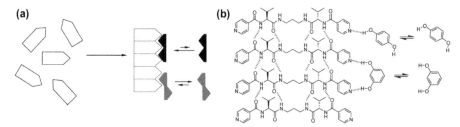

Figure 1.20 The organisation of multiple interacting groups by a gelator scaffold enables multivalent interactions with guest molecules having complementary interacting groups in the correct orientation.
(Reproduced from Ref. 82 with permission from The Royal Society of Chemistry).

remarkably, the isolated fibres showed much larger charge mobility than other mono- and bisthiophene molecules, which was most likely due to close π-interactions between the thiophene moieties within the 1D stacks.

A very different application of the gelator scaffold approach is found in the work of Escuder, in which they used a peptide bolamphiphile type of gelator as a scaffold for the organisation of receptor groups at the fibre interface (Figure 1.20).[82] The close proximity of multiple receptor groups enables multivalent binding and together with the well-defined spatial arrangement of the receptor groups, this leads to strongly enhanced affinity of the fibres for guest molecules with complementary interaction groups, provided that the interacting groups at the guest molecules have the correct orientation to allow simultaneous binding.

Recognition at the fibre-liquid interface has many applications, like in separation technology and catalysis. In recent years there has been a strong interest in the use of hydrogels as artificial extracellular matrix for tissue engineering. In their pioneering work, Stupp and coworkers exploited gel from the peptide amphiphile gelators (see Figure 1.12) as artificial extracellular matrix to for the differentiation and proliferation of stem cells.[83] By equipping the peptide amphiphile with a neurite promoting pentapeptide (IKVAV), they have been able to promote the differentiation of neural progenitor cells to neurons if grown on gels of the peptide amphiphile.

These few examples already demonstrate the value of the gelator scaffold approach for the spatial organisation of functional moieties. Further examples are discussed in more detail in following chapters. It should be noted that the gelator scaffold approach resembles the property-based design approach, but the prior knowledge on the aggregation mode of potential gelator scaffolds will greatly facilitate its use as organising group.

1.6   Conclusion

It is clear the formation of molecular gels by LMW gelators is a complicated phenomena, controlled by a delicate balance of intermolecular interactions between gelator molecules and between gelator and solvent. Moreover, detailed structural knowledge of actual supramolecular arrangement within the fibres is still very limited because of lack of suitable methods for the structure elucidation. This gap in knowledge, together with the complicated interplay between thermodynamics and kinetics, has made the design of novel molecular gelators a true challenge. Over the years, scientists have conquered this challenge, and the success of the different design approaches have increased the understanding of the phenomena of supramolecular gelation, and has even allowed the identification of several critical design parameters. Still, many challenges in the area of molecular gels lay ahead, some of them will be within the realm of supramolecular chemistry, and others will be found at interfaces with other disciplines, like material science, physics and biology. Science is about asking the right questions, and answering these question will sprout new questions. Clearly, molecular gels are a perfect example of a vivid scientific field.

References

1. These numbers are based on a Web-of-Science search in the TOPIC field, using the search terms (supramolecular gel*) OR (gelator*) OR (organogel*) OR (supramolecular hydrogel*).
2. N. Pilpel, *Chem. Rev.*, 1963, **63**, 221–234.
3. P. J. Flory, *Far. Discus. Chem. Soc.*, 1974, **57**, 7–18.
4. T. Graham, *Philos. Trans. Royal Soc. London*, 1861, **151**, 183–224.
5. D. J. Lloyd and J. Alexander in *Colloid Chemistry*, Chemical Catalogue Company, New York, 1926, p. 767.
6. M. Rubinstein and R. H. Colby, *Polymer Physics*, Oxford University Press, New York, 2003.
7. P. Terech and R. Weiss, *Chem. Rev.*, 1997, **97**, 3133–3159.
8. J. M. Lehn, *Angew. Chem. Int. Ed. Eng.*, 1990, **29**, 1304–1319.
9. P. Terech, V. Rodriguez, J. Barnes and G. McKenna, *Langmuir*, 1994, **10**, 3406–3418.
10. L. Lu and R. Weiss, *Chem. Commun.*,1996, 2029–2030.
11. R. Twieg, T. Russell, R. Siemens and J. Rabolt, *Macromol.*, 1985, **18**, 1361–1362.
12. T. Brotin, R. Utermohlen, F. Fages, H. Bouas-Laurent and J. Desvergne, *J. Chem. Soc. Chem. Commun.*, 1991, 416–418.
13. Y. C. Lin and R. G. Weiss, *Macromol*, 1987, **20**, 414–417.
14. H. Willemen, T. Vermonden, A. Marcelis and E. Sudholter, *Eur. J. Org. Chem.*, 2001, 2329–2335.
15. J. Wolfe, R. Hann and C. Hudson, *J. Am. Chem. Soc.*, 1942, **64**, 1493–1496.
16. M. Watase, Y. Nakatani and H. Itagaki, *J. Phys. Chem. B*, 1999, **103**, 2366–2373.
17. K. Hanabusa, J. Tange, Y. Taguchi, T. Koyama and H. Shirai, *J. Chem. Soc. Chem. Commun.*, 1993, 390–392.
18. F. Menger, Y. Yamasaki, K. Catlin and T. Nishimi, *Angew. Chem. Int. Ed. Eng*, 1995, **34**, 585–586.

19. G. Yu, X. Yan, C. Han and F. Huang, *Chem. Soc. Rev.*, 2013, doi:10.1039/c3cs60080g.
20. P. Terech, C. Rossat and F. Volino, *J. Colloid Interface Sci.*, 2000, **227**, 363–370.
21. A. R. Hirst, I. A. Coates, T. R. Boucheteau, J. F. Miravet, B. Escuder, V. Castelletto, I. W. Hamley and D. K. Smith, *J. Am. Chem. Soc.*, 2008, **130**, 9113–9121.
22. K. Murata, M. Aoki, T. Suzuki, T. Harada, H. Kawabata, T. Komori, F. Ohseto, K. Ueda and S. Shinkai, *J. Am. Chem. Soc.*, 1994, **116**, 6664–6676.
23. S. R. Raghavan and B. H. Cipriano, *Chapter 8 in Molecular Gels: Materials with Self-Assembled Fibrillar Networks*, ed. R. G. Weiss and P. Terech, Springer, Dordrecht, 2006.
24. R. G. Weiss and P. Terech, *Molecular Gels: Materials with Self-Assembled Fibrillar Networks*, Springer, Dordrecht, 2006.
25. J. Brinksma, B. L. Feringa, R. M. Kellogg, R. Vreeker and J. van Esch, *Langmuir*, 2000, **16**, 9249–9255.
26. P. Terech, V. Schaffhauser, P. Maldivi and J. M. Guenet, *Europhys. Let.*, 1992, **17**, 515.
27. P. Terech, V. Schaffhauser, P. Maldivi and J. M. Guenet, *Langmuir*, 1992, **8**, 2104–2106.
28. B. Escuder, M. Llusar and J. Miravet, *J. Org. Chem.*, 2006, **71**, 7747–7752.
29. P. Terech, *Langmuir*, 2009, **25**, 8370–8372.
30. E. Krieg, E. Shirman, H. Weissman, E. Shimoni, S. G. Wolf, I. Pinkas and B. Rybtchinski, *J. Am. Chem. Soc.*, 2009, **131**, 14365–14373.
31. J. van Esch, F. Schoonbeek, M. de Loos, H. Kooijman, A. Spek, R. Kellogg and B. Feringa, *Chem. Eur. J.*, 1999, **5**, 937–950.
32. P. Hartman and P. Bennema, *J. Cryst. Growth*, 1980, **49**, 145–156.
33. I. Weissbuch, R. Popovitz-Biro, M. Lahav and L. Leiserowitz, *Acta Crystallogr. B: Struct. Sci.*, 1995, **51**, 115–148.
34. Y. C. Lin and R. G. Weiss, *Macromol.*, 1987, **20**, 414–417.
35. Y. Lin, B. Kachar and R. G. Weiss, *J. Am. Chem. Soc.*, 1989, **111**, 5542–5551.
36. K. Murata, M. Aoki, T. Nishi, A. Ikeda and S. Shinkai, *J. Chem. Soc. Chem. Commun.*, 1991, 1715–1718.
37. K. Murata, M. Aoki, T. Suzuki, T. Harada, H. Kawabata, T. Komori, F. Ohseto, K. Ueda and S. Shinkai, *J. Am. Chem. Soc.*, 1994, **116**, 6664–6676.
38. C. Shi, Z. Huang, S. Kilic, J. Xu, R. Enick, E. Beckman, A. Carr, R. Melendez and A. Hamilton, *Science*, 1999, **286**, 1540–1543.
39. L. Estroff and A. Hamilton, *Angew. Chem. Int. Ed.*, 2000, **39**, 3447–3450.
40. D. Abdallah and R. G. Weiss, *Chem. Mater.*, 2000, **12**, 406–413.
41. M. George and R. G. Weiss, *J. Am. Chem. Soc.*, 2001, **123**, 10393–10394.
42. B. Xing, C. Yu, K. Chow, P. Ho, D. Fu and B. Xu, *J. Am. Chem. Soc.*, 2002, **124**, 14846–14847.
43. F. Zhao, M. L. Ma and B. Xu, *Chem. Soc. Rev.*, 2009, **38**, 883–91.

44. K. Yoza, N. Amanokura, Y. Ono, T. Akao, H. Shinmori, M. Takeuchi, S. Shinkai and D. Reinhoudt, *Chem. Eur. J.*, 1999, **5**, 2722–2729.
45. K. Nakano, Y. Hishikawa, K. Sada, M. Miyata and K. Hanabusa, *Chem. Let.*, 2000, 1170–1171.
46. M. Suzuki, Y. Nakajima, M. Yumoto, M. Kimura, H. Shirai and K. Hanabusa, *Org. Biomol. Chem.*, 2004, **2**, 1155–9.
47. S. Kiyonaka, S. Shinkai and H. Hamachi, *Chem. Eur. J.*, 2003, **9**, 976–983.
48. S. Kiyonaka, K. Sugiyasu, S. Shinkai and I. Hamachi, *J. Am. Chem. Soc.*, 2002, **124**, 10954–10955.
49. N. Sreenivasachary and J. M. Lehn, *Proc. Natl. Acad. Sci. USA*, 2005, **102**, 5938–5943.
50. A. Hirst, J. Miravet, B. Escuder, L. Noirez, V. Castelletto, I. Hamley and D. Smith, *Chem. Eur. J.*, 2009, **15**, 372–379.
51. K. Hanabusa, M. Yamada, M. Kimura and H. Shirai, *Angew. Chem. Int. Ed. Eng.*, 1996, **35**, 1949–1951.
52. M. de Loos, J. van Esch, I. Stokroos, R. Kellogg and B. Feringa, *J. Am. Chem. Soc.*, 1997, **119**, 12675–12676.
53. A. Aggeli, M. Bell, N. Boden, J. Keen, P. Knowles, T. McLeish, M. Pitkeathly and S. Radford, *Nature*, 1997, **386**, 259–262.
54. K. Hanabusa, A. Kawakami, M. Kimura and H. Shirai, *Chem. Lett.*, 1997, 191–192.
55. K. J. C. van Bommel, C. van der Pol, I. Muizebelt, A. Friggeri, A. Heeres, A. Meetsma, B. L. Feringa and J. van Esch, *Angew. Chem. Int. Ed.*, 2004, **43**, 1663–1667.
56. M. Shirakawa, N. Fujita and S. Shinkai, *J. Am. Chem. Soc.*, 2003, **125**, 9902–9903.
57. A. Ajayaghosh, S. George and V. Praveen, *Angew. Chem. Int. Ed.*, 2003, **42**, 332–335.
58. J. P. Hill, W. Jin, A. Kosaka, T. Fukushima, H. Ichihara, T. Shimomura, K. Ito, T. Hashizume, N. Ishii and T. Aida, *Science*, 2004, **304**, 1481–1483.
59. P. Dastidar, *Chem. Soc. Rev.*, 2008, **37**, 2699–2715.
60. J. N. Israelachvili, D. J. Mitchell and B. W. Ninham, *J. Chem. Soc. Faraday Trans. 2*, 1976, **72**, 1525–1568.
61. J. D. Hartgerink, E. Beniash and S. I. Stupp, *Science*, 2001, **294**, 1684–1688.
62. S. E. Paramonov, H. W. Jun and J. D. Hartgerink, *J. Am. Chem. Soc.*, 2006, **128**, 7291–7298.
63. G. Newkome, C. Moorefield, G. Baker, R. Behera, G. Escamillia and M. Saunders, *Angew. Chem. Int. Ed. Eng.*, 1992, **31**, 917–919.
64. E. Zubarev, M. Pralle, E. Sone and S. Stupp, *J. Am. Chem. Soc.*, 2001, **123**, 4105–4106.
65. C. S. Love, A. R. Hirst, V. Chechik, D. K. Smith, I. Ashworth and C. Brennan, *Langmuir*, 2004, **20**, 6580–6585.
66. R. Oda, I. Huc and S. Candau, *Angew. Chem. Int. Ed.*, 1998, **37**, 2689–2691.

67. R. Oda, I. Huc, M. Schmutz, S. Candau and F. MacKintosh, *Nature*, 1999, **399**, 566–569.
68. D. Berthier, T. Buffeteau, J. Leger, R. Oda and I. Huc, *J. Am. Chem. Soc.*, 2002, **124**, 13486–13494.
69. A. Aggeli, I. Nyrkove, M. Bell, R. Harding, L. Carrick, T. McLeish, A. Semenov and N. Boden, *Proc. Natl. Acad. Sci. USA*, 2001, **98**, 11857–11862.
70. P. Terech, D. Pasquier, V. Bordas and C. Rossat, *Langmuir*, 2000, **16**, 4485–4494.
71. see a.o. X. Liu and P. Sawant, *Appl. Phys. Let.*, 2001, 79, 3518–3520 and references therein.
72. X. Liu, P. Sawant, W. Tan, I. Noor, C. Pramesti and B. Chen, *J. Am. Chem. Soc.*, 2002, **124**, 15055–15063.
73. J. Boekhoven, J. M. Poolman, C. Maity, F. Li, L. van der Mee, C. B. Minkenberg, E. Mendes, J. H. van Esch and R. Eelkema, *Nature Chem.*, 2013, **5**, 433–437.
74. X. Liu, P. Sawant, W. Tan, I. Noor, C. Pramesti and B. Chen, *J. Am. Chem. Soc.*, 2002, **124**, 15055–15063.
75. D. D. Díaz, K. Rajagopal, E. Strable, J. Schneider and M. G. Finn, *J. Am. Chem. Soc.*, 2006, **128**, 6056–6057.
76. S. Kawano, N. Fujita, K. V. Bommel and S. Shinkai, *Chem. Let.*, 2003, **32**, 12–13.
77. H. Kobayashi, M. Amaike, J. Jung, A. Friggeri, S. Shinkai and D. Reinhoudt, *Chem. Commun.*, 2001, 1038–1039.
78. M. A. Gradzielski, *J. Phys. Condens. Matter*, 2003, **15**, R655.
79. T. Mes, M. M. Koenigs, V. F. Scalfani, T. S. Bailey, E. W. Meijer and A. R. Palmans, *ACS Macro Letters*, 2011, **1**, 105–109.
80. A. Brizard, M. Stuart, K. van Bommel, A. Friggeri, M. de Jong and J. van Esch, *Angew. Chem. Int. Ed.*, 2008, **47**, 2063–2066.
81. F. Schoonbeek, J. van Esch, B. Wegewijs, D. Rep, M. de Haas, T. Klapwijk, R. Kellogg and L. Feringa, *Angew. Chem. Int. Ed.*, 1999, **38**, 1393–1397.
82. B. Escuder, J. F. Miravet and J. A. Sáez, *Org. Biomol. Chem.*, 2008, **6**, 4378–4383.
83. G. A. Silva, C. Czeisler, K. L. Niece, E. Beniash, D. A. Harrington, J. A. Kessler and S. I. Stupp, *Science*, 2004, **303**, 1352–1355.
84. P. Jonkheijm, P. van der Schoot, A. P. H. J. Schenning and E. W. Meijer, *Science*, 2006, **31**, 80–83.

CHAPTER 2

Techniques for the Characterisation of Molecular Gels

VICENT J. NEBOT AND DAVID K. SMITH*

Department of Chemistry, University of York, Heslington, York,
YO10 5DD, UK
*Email: david.smith@york.ac.uk

2.1 Molecular Gels – Hierarchical Materials

As explained in Chapter 1, molecular gels are fascinating hierarchical self-assembled materials, in which molecular-scale building blocks self-assemble, usually into fibrillar structures as a consequence of controlled noncovalent interactions (Figure 2.1).[1,2] These fibrils, which have molecular-scale diameters, then bundle together to yield nanoscale fibres. Interaction between these nanoscale fibres, usually on the microscale level, then gives rise to an entangled sample-spanning network that prevents flow of the bulk solvent, as such leading to the gel-type macroscopic behaviour. As a result of the hierarchical nature of gel assembly, full characterisation of gel-phase materials is a fascinating challenge, which relies on using a variety of techniques that can span all of the different length-scales involved.[3] Indeed, the need to characterise self-assembled gels in a hierarchical manner is one of the most interesting, challenging (both practically and intellectually) and fun aspects of research in the field and provides outstanding training for young researchers. This chapter aims to provide a primer into how different techniques can be used to provide

RSC Soft Matter No. 1
Functional Molecular Gels
Edited by Beatriu Escuder and Juan F. Miravet
© The Royal Society of Chemistry 2014
Published by the Royal Society of Chemistry, www.rsc.org

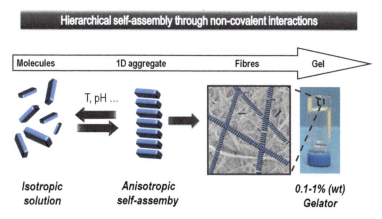

Figure 2.1   Schematic of molecular gel self-assembly.

insight at different levels of the assembly process using selected examples from the gel literature. As such, we will consider the characterisation of a gel from the "top down". This is often what happens in the laboratory, as many researchers first discover gelators (often serendipitously) by observations of the materials behaviour on the macroscopic level, and then begin to characterise what is happening on the nanoscale, and ultimately understand the materials on the molecular level.[4]

2.2   Gel Preparation and Storage

The first important piece of characterisation data about a gel is the way it has been made and for how long, and under what conditions, it has been stored. Such information is often absent in literature reports, but can make a significant difference in all aspects of gel behaviour. Molecular gels are most often formed by heating the solid low molecular mass gelator in an appropriate solvent to achieve complete solubilisation, and then cooling to allow gelation to occur. Cooling can be performed in a number of different ways, which can directly impact on the gel morphology and properties – for example rapid cooling in ice and slow controlled cooling (*e.g.* 1 °C/min) have been reported in some cases to give rise to completely different materials.[5,6] This can be rationalised based on the amount of energy present during gel formation, and it has been suggested that this is analogous to kinetic (fast cooling, low energy) and thermodynamic (slow cooling, high energy) control. In a number of cases, sonication is applied to achieve gelation – either with or without heating, and in some cases is essential, with other methods being ineffective in encouraging gelation.[7–9] Typically, sonication allows efficient dispersion and solubilisation of the low-molecular-weight gelator. The energy provided by input of ultrasound can also encourage self-assembly processes – and has also been reported to trigger transitions from one form of a gel to another.[10] In some cases, a gel can form in the absence of heating/sonication, simply on dispersing the gelator in the solvent – for example, *in situ* gels, in which the chemical reaction between two components (isocyanates and amines) can give rise to ureas that instantly form gels in organic solvents.[11,12] Instant gelation indicates that the energy barrier to solubilisation and self-assembly is very low, and the gel state can therefore be accessed even at room temperature.

Once the gel has been formed it is important to realise that these are dynamic materials that are capable of changing over time. Obviously, on a simplistic level, solvent can evaporate and change the composition. However, the nanoscale structuring of these systems can also significantly change on standing, leading to major changes in properties. Gels are metastable, kinetically trapped materials, and their stability therefore depends on the energy barriers to further evolution. There have been a number of reports in which gels have been shown to be unstable with respect to aggregation, which can lead to microcrystallisation and destabilisation of the gel on standing – in some cases, this process occurs due to lateral aggregation of the gel nanofibres into large microcrystalline objects.[13,14] However, gels can also improve on ageing.[15,16] This can be a consequence of molecular reorganisation of the molecular-scale building blocks within the fibres after gelation has taken place in order to form a thermodynamically more stable assembly, or as a result of the overall re-organisation/equilibration of different fibre structures within the gel network on the nanoscale over extended periods of time as a gel ages.

In the light of the above considerations, it is absolutely essential that the method of gel preparation is accurately reported, as gel properties can vary dependent on the precise details of preparation and solvent chosen. Once the gel-forming conditions have been chosen, it is important to carefully choose appropriate techniques to characterise the gel across all hierarchical levels, and the ways of doing this are described in the following sections.

2.3 Analysis of Macroscopic Behaviour

2.3.1 "Table-Top" Rheology

Quick and simple ways of visually assessing the physical (*i.e.* macroscopic) behaviour of the gels are referred to as "table-top" rheology,[17] and can be readily carried out in any laboratory without sophisticated equipment. These techniques are particularly useful for exploring the gel–sol phase boundary, as the conversion of the material from a gel to a sol can be simply and visually assessed. There are two simple parameters that are widely used to define the macroscopic properties of molecular gels:

- *Minimum gelation concentration* (MGC) – the minimum concentration of gelator required to form a sample-spanning, self-supporting gel at a given temperature (usually 25 °C),
- *Gel–sol transition temperature* (T_{gel}) – the temperature at which the gel is converted into a sol on slow heating.

There are several ways in which these parameters can be assessed, and these are described in more detail below.

Tube Inversion Methodology. The simplest method of monitoring the gel–sol transition involves inverting a vial of the gel and watching to see whether any flow occurs. By using a thermoregulated oil bath with controlled heating and cooling, it is possible to accurately determine the T_{gel} value. What is actually being measured here is the point at which the yield stress exerted by the gel under the force of gravity prevents the material from becoming self-supporting. It is essential to use the same sample mass and vial type when performing comparative studies, to ensure the yield stress remains constant. For this reason, T_{gel} values can, for example, appear different if measured in NMR tubes, rather than sample vials. This limitation means it can sometimes be difficult to compare results recorded in different laboratories in a quantitative way.

Dropping Ball Method. A small metal ball is placed onto the gel, and the dropping of the ball through the gel is observed.[18] In the ideal scenario, the ball should be immobile in the gel, but drop rapidly in the sol. By varying the temperature, the T_{gel} value at any given concentration of gelator can be determined. The yield stress is dependent on the density of the ball and its radius, and therefore, the ball must be kept constant across experiments. The sample tube should also be significantly larger than the ball, otherwise the presence of the walls can affect the motion of the ball.

The thermal stability of the gel is a concentration-dependent parameter, and determining T_{gel} values at different concentrations allows the construction of a "phase diagram" (Figure 2.2).[19] It is possible to analyse this in order to

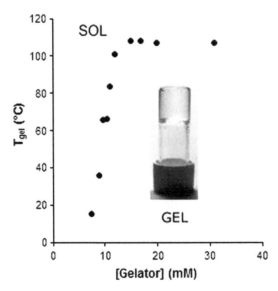

Figure 2.2 Typical phase diagram for concentration dependent T_{gel} values. Adapted from ref. 19 with permission of the American Chemical Society.

estimate thermodynamic parameters associated with gelation. Schrader's relationship[20] can be used to generate plots of ln[gelator] against $1/T_{gel}$, in which the gradient is $-\Delta H/R$, where ΔH can be approximated to the enthalpy associated with the gel–sol transition. This value can provide some insight into the thermodynamics of the interactions between molecular-scale building blocks, although the assumptions inherent in this treatment mean the data are best used to compare the behaviour of related gelators. It should be noted that many gels apparently exhibit a plateau region in T_{gel} above a certain concentration – corresponding to a point at which the presence of additional gelator does not appear to enhance the thermal macroscopic properties of the gel. This behaviour can be understood based on gelation being related to solubility phenomena, with the gelator molecule achieving a saturated solution and then giving rise to self-assembled nanostructures.[21]

2.3.2 Differential Scanning Calorimetry

Differential scanning calorimetry (DSC) provides an instrumental way of directly measuring the temperature at which the gel–sol transition takes place (both on heating and cooling) and the phase-change enthalpy ($\Delta H_{gel-sol}$), and as such, can provide an insight into the thermodynamics of the gelator–gelator interactions. In DSC, the difference in the amount of heat energy required to increase the temperature of a sample and reference is measured as a function of temperature.[22] When the sample of interest undergoes a physical transformation, such as a gel–sol phase transition, more heat will need to flow to it than the reference in order to maintain both at the same temperature (endothermic). On cooling, as the material undergoes an exothermic sol-gel transition, less heat will need to flow to it than the reference. Observing the difference in heat flow associated with a phase transition enables the measurement of the amount of heat associated with it by integration of the DSC trace. In general, DSC is performed in both heating and cooling modes in order to assess both endothermic and exothermic transitions and assess the thermoreversibility of the gel–sol phase change. Furthermore, multiple cycles of heating and cooling can be performed, and can be very informative. It should be noted that the second cycle of the sample will correspond to a gel that has formed under the cooling conditions of the DSC cooling ramp. As such, this material may be different to the gel initially put into the calorimeter. In general terms, DSC gives more effective data when using gels at relatively high concentrations.

DSC has been used to correlate the performance of gels in terms of T_{gel} and enthalpy with the molecular structure of the gelators, and their ability to form hydrogen-bond interactions.[23] DSC can also be used to probe other aspects of self-assembly in addition to the simple gel–sol transition, such as the formation of liquid-crystalline mesophases.[24] A nice example of the application of DSC methods to provide deeper insight into gelation demonstrated that this approach could be used to characterise hybrid materials consisting of two different thermally responsive networks.[16] Two different gelators with significantly different T_{gel} values were each characterised by DSC individually. A gel

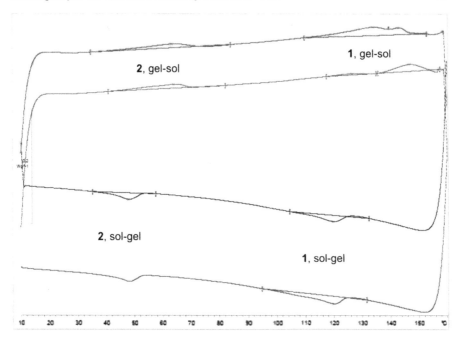

Figure 2.3 DSC traces of a hybrid gel that contains two independent thermally addressable nanoscale gel networks each based on a different self-sorting gelator. Two heat–cool cycles were applied to the sample.
Figure adapted from ref. 16 with kind permission of the Royal Society of Chemistry.

was then assembled from the mixture of components, and it was shown to possess a combination of the thermal signatures of each of the individual component gelators (Figure 2.3). As the sample is heated, the endotherms can be observed first for gelator "2" being converted from gel to sol, followed by gelator "1". On cooling, each gelator then forms its own independent network at a distinctive temperature with the two exotherms representing the sol-gel transitions. As such, it was clear that these gelators were orthogonal to one another and could assemble their own self-sorted nanoscale networks, each of which was individually thermally addressable.

2.3.3 Rheology

As soft materials, gels can ideally be explored using rheological methods that more fully explore the response of materials to applied stress.[25] Although this requires the use of specialist equipment it is the best way of gaining subtle and detailed insight into the macroscopic performance of gel-phase materials. Deformation is the relative displacement of points of a body. It can be divided into two types: flow and elasticity.[26] Flow is irreversible deformation; when the stress is removed, the material does not return to its original form.

Flow properties are defined by resistance to flow (*i.e.* viscosity) and can be measured by determining the resistance to flow when a fluid is sheared between two surfaces. Elasticity, on the other hand, is reversible deformation; *i.e.* the deformed body recovers its original shape. The mechanical properties of an elastic solid may be studied by applying a stress and measuring the deformation of strain. Fluids that exhibit both flow and elastic behaviour are referred to as viscoelastic.

The response of materials to stress–strain and stress–relaxation measurements helps define materials properties. By subjecting a material to oscillatory stress and determining the response, both the elastic and viscous characteristics can be obtained. Elastic materials store energy, whereas liquids dissipate it as heat. A sinusoidal stress applied to an ideal elastic material produces a sinusoidal strain proportional to the stress and in phase with it (0°). For ideal viscous materials the stress and strain are out of phase by 90°. In real viscoelastic materials, the strain is proportional to the stress, but lags behind it by some angle (δ) between 0° and 90°. In brief, the complex dynamic modulus, G^*, can be resolved into two components: $G' = G^* \cos\delta$ and $G'' = G^* \sin\delta$. The parameter G' is called the storage modulus and is a measure of the elasticity associated with the energy stored in elastic deformation. Conversely, G'' is called the loss modulus and represents the ability of the system to flow under stress (*i.e.* is associated with energy dissipation). As such, the ratio G'/G'' is the ratio of energy stored to energy dissipated. For a viscoelastic material $G' > G''$ (*i.e.* more energy is stored than dissipated) and the point at which $G' = G''$ is the point at which the material loses its viscoelastic properties and simply behaves as a viscous material (*i.e.* the gel is converted into a sol). Usually, measurements are carried out in a rheometer with a parallel (or conical) two-plate geometry in which a back-and-forth oscillatory stress is applied (Figure 2.4), and the data are directly converted into G' and G'' values. Plots of G' and G'' versus applied oscillatory stress, time and temperature allow the determination of different parameters useful for gel characterisation, such as σ_y (yield stress, Pa), T_{gel} (thermal stability) or the recovery time after destructive stress (thixotropy).

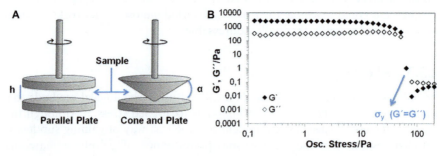

Figure 2.4   Geometry of typical rheometers (A) and sample data illustrating how G' and G'' vary with applied stress (B), with the point at which the gel breaks down being clearly marked (yield stress point, σ_y).

Rheological measurements are relatively straightforward to perform, however, a careful optimisation of the experimental parameters is crucial. First, the above considerations are only true within the linear viscoelastic regime (LVR), in which both moduli must remain independent of the frequency and stress range applied, and therefore the determination of the LVR is a key point. Furthermore, during sample preparation gel materials are often partially destroyed and therefore an equilibration time is often required for the gel to completely recover before performing the measurements. The recovery time can be easily determined by simply running a time sweep at low and constant oscillatory stress and frequency (within the LVR). Consequently, the optimisation of the experimental parameters is often a tedious iterative task.

To exemplify the use of rheology, Menger and Caran studied the rheological properties of a family of gelators derived from cystine.[3] The rheological study allowed the authors to establish a relationship between gelator molecular structure, gelation behaviour and gel properties. Using this approach they were able to determine the linear viscoelastic regime as a function of the frequency for a small stress, the yield stress (σ_y) by performing a stress sweep at constant strain and frequency, and the T_{gel} value, by a temperature sweep at low and constant oscillatory stress and frequency.

Perhaps some of the most effective and detailed rheological work on gel-phase materials has been carried out by Terech and coworkers. In a key paper, the use of different approaches to measure phase transitions was probed for a single gelator.[27] It was suggested that a full rheological study was the more accurate and convenient technique. However, there was good qualitative agreement between rheology and dropping ball methods in terms of the concentration dependence and the extracted thermodynamic parameters, although there were differences in the precise measured T_{gel} values of 5–10 °C. NMR methods (see below) were in good agreement with rheology, with the added advantage of being easily used to extract kinetic information associated with self-assembly and gelation (see Section 2.8).

In addition to providing information about macroscopic phase transitions, rheology is a useful way of probing subtle changes within a gel, where the material strength may change, but the overall gel remains intact.[28] In this way, rheology provides the most detailed insight into the overall macroscopic performance of the gel, and the way in which it responds to molecular composition. For example, by using rheology, the responsiveness of gels to different stimuli by adapting their internal nanostructures can be read out by changes to the macroscopic performance.[29]

As an increasing number of gels have been studied by detailed rheology, it is clear that a variety of theoretical models can be employed.[30,31] Evidently, there is the need for continued application of rheological methods to supramolecular gels in order to provide a better insight into the way in which modifying gelator structures on the molecular scale and the processing of the macroscopic materials can control the rheological properties of the self-assembled gel.

2.4 Analysis of Nanoscale Structure

2.4.1 Electron Microscopy

Perhaps more than any other method, electron microscopy is the technique that opened up the nanoworld for closer inspection, and as such equipment has become increasingly standard in many laboratories, the field of "nanotechnology" has become dominant.[32–34] In the simplest use of electron microscopy to image the nanostructure of gel-phase materials, a thin layer of gel is first allowed to dry on a substrate (either under ambient conditions or *in vacuo*). The sample is coated under vacuum with a (ca. 2 nm) metallic layer and then imaged by scanning electron microscopy (SEM). The SEM image obtained in this way therefore represents a dried and treated sample. Usually, the network structure of the gel collapses onto itself during drying to yield a xerogel (if collapse does not occur, the structure is referred to as an aerogel). It should be noted that structural changes other than collapse may also occur during drying, however, it is often assumed that such effects are minor – at least when comparing related families of gelators. Transmission electron microscopy (TEM) can also be applied to gel imaging, although it is often necessary to apply a heavy-metal staining agent to enhance image contrast.

A range of different gel morphologies can be observed using SEM.[35] In general terms, transparent gels often exhibit nanoscale structuring, whilst opaque gels, which scatter light, have larger microscale features. Typically, assembled supramolecular polymer nanofibres are observed. Other types of "one-dimensional" objects such as tapes/ribbons have also been reported (Figure 2.5). However, it should be noted that generally, the observed nanostructure diameters are much larger than the molecular diameter. This is a consequence of the fact that molecular-scale fibrils often bundle together in some way to generate larger nanofibres. However, the mechanism by which molecular-scale fibrillar objects hierarchically assemble into the nanoscale fibres observed by electron microscopy is often somewhat unclear – although in some cases, detailed models to help understand this process can be developed (see Section 2.5.4).[36]

In some cases, the chirality inherent at the molecular level is transcribed into the nanoscale assembled objects and can be directly observed by electron microscopy – usually in the form of helicity.[37] Indeed, in early work on gels, the two enantiomers of lithium 12-hydroxystearate were shown by transmission electron microscopy to assemble into left- or right-handed helical fibres, depending on the molecular-scale chiral information (Figure 2.6). Remarkably, this work was reported as early as the 1960s, and as such predates most of supramolecular chemistry and the genesis of "nanochemistry".[38] It is worth noting in passing that there is lots of precedent for studying gels within the grease/lubrication literature, in which such materials have been applied for over 100 years,[39] and some of this has often been overlooked by academic chemists.

Cryo-electron microscopy techniques are used to try and minimise disruption to the self-assembled network on drying.[40] A rapid freezing step is used in an attempt to limit thermal motion of the self-assembled gelator network. Freezing

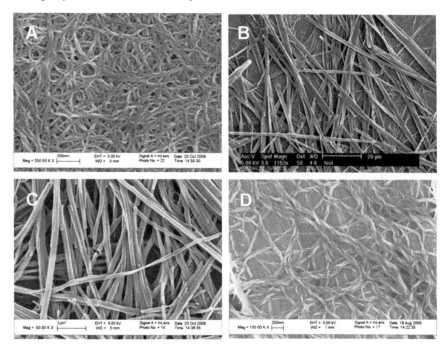

Figure 2.5 Variety of morphologies observed for supramolecular gels by using SEM on the "collapsed" xerogel state.
Figures adapted from ref. 35 with kind permission of Elsevier.

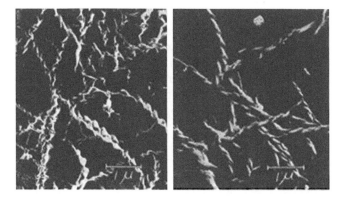

Figure 2.6 Chiral nanostructures as imaged by transmission electron microscopy in the 1960s.
Figure adapted from ref. 38 with kind permission of the American Chemical Society.

can be achieved by sample cooling at liquid nitrogen temperatures. In some cases, solvent is then sublimed from the sample by freeze drying (of course this may modify the gelator structure as desolvation occurs, but the low temperature provides less thermal energy for network reorganisation). Usually, this is done with the electron microscope – but freeze drying of samples can also be performed prior to standard SEM in an attempt to better preserve the nano-morphology.[41,42] Overall, cryo-electron microscopy in which a freezing step is employed leads to significantly less disruption of the nanostructure, and therefore a typical cryo-EM image shows a more expanded and "solvated" network, in which the solvent pockets can be readily visualised as cavities within the network (Figure 2.7).[43,44]

In some rare examples, nonfibrillar gel morphologies have been observed by electron microscopy. For example, platelet-type morphologies can constitute the "solid-like" phase (Figure 2.8).[45–49] These platelets form honeycomb-like networks leading to solvent viscosification. It has been argued that in these systems, growth of an ordered structure occurs in two dimensions (and is prevented in the third). This would therefore appear to be related to the for-mation of gel fibres, in which growth is only allowed in one direction, and prevented in the other two. As such, it can be argued that gelation is a type of crystallisation process that is frustrated in at least one-dimension of growth, with the nanostructures that form generating a three-dimensional structure as a consequence of weak interactions and entanglements between one-dimensional or two-dimensional nanocrystalline morphologies. In other cases, electron microscopy has shed light on weak gel structures that form from vesicle-vesicle aggregation – with the vesicles being the self-assembled units that lead to the formation of networks and hence inducing the same behaviour as in supra-molecular gels (Figure 2.8).[50–53]

Obviously there is great interest in visualising solvated gel networks and increasingly advances in electron microscopy methods can make this possible. Environmental scanning electron microscopy (ESEM)[54] has been used to image gels in the native state under atmospheric pressure without having to dry them first. The humidity within the ESEM experiment can be precisely controlled in order to stabilise the sample. It is also possible in this technique to add free-radical scavengers to the sample in order to try and prevent decomposition within the electron beam. In an elegant example of this approach, ESEM was used to monitor a metastable hydrogel that evolved from a nanofibrillar

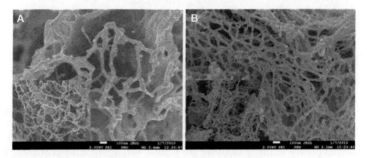

Figure 2.7   Electron microscopy on freeze-dried samples can provide a visualisation of gel samples in a more solvated and expanded state.
Figure comparing linear (A) and dendritic (B) peptide-isomer gelator morphologies adapted from ref. 42 with kind permission of the Royal Society of Chemistry.

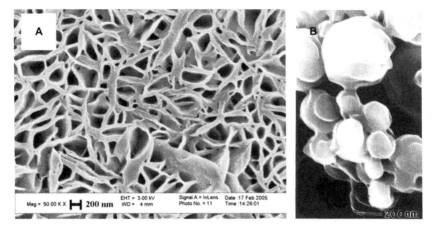

Figure 2.8 Some example nonfibrillar self-assembled gel morphologies observed by electron microscopy.
Figure adapted from refs. 48 and 50, with kind permission of Wiley-VCH and the American Chemical Society, respectively.

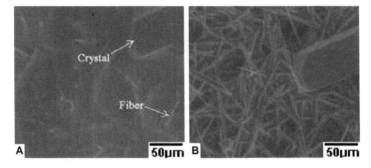

Figure 2.9 ESEM image (A) of a gel that slowly evolves into a microcrystalline state compared with the dried SEM image (B).
Figure adapted from ref. 13 with kind permission of the American Chemical Society.

morphology into a microcrystalline one (Figure 2.9).[13] As can be seen, the images are much hazier than the equivalent sample imaged with SEM, as the image is measured within a fluid-like environment and the self-assemblies are hence considerably more dynamic. However, nanofibre dimensions and morphologies can be estimated from this method, which can, as a consequence, provide useful visual corroboration of precisely what is happening in the solvated state. It therefore seems likely that ESEM will increasingly be applied in future studies of soft gel-phase materials.

2.4.2 Atomic Force Microscopy

Atomic force microscopy provides an effective way of visualising nanoscale morphologies, with the added benefit that it is able to provide dimensional information such as fibre width and height.[55] In order to achieve effective imaging, however, individual fibres should be present on the surface – not mats of fibres. As such, it is usually necessary to use dilute solutions so these materials can be imaged, and samples are often dried at concentrations below the critical gelation concentration. It is not always clear whether the fibres even still exist at these low concentrations, with gelation not being observed because they cannot form an interpenetrated network, or whether fibres are not initially present, but instead, self-assemble as the sample dries and the concentration increases. There is often considerable optimisation required to find ideal conditions for AFM imaging, and it should be remembered that the drying conditions can have an impact on the morphologies that form. In addition, interactions between the solid support and the nanostructures being visualised can be quite significant. It is well known that objects have a tendency to appear flattened by AFM owing to interactions with the surface combined with the effect of the tip exerting pressure on the sample – this has the effect of increasing the apparent fibre width and decreasing the measured fibre height. This is particularly marked for softer nanostructures, which have more compressible structures. In spite of these limitations, however, it is possible to achieve useful information about morphology from AFM,[19] including the ability to observe helical fibres, with the chirality being translated from the molecular scale to the nanoscale (Figure 2.10).[56–58]

In a key paper, AFM was used not only to determine that gel fibres had formed from a TTF-derivative, but also to demonstrate they were conductive "wire-like" objects after doping by exposure to iodine vapour.[59,60] This was achieved by using current-sensing (conductive) AFM on a graphite surface. In this experiment, topological and electrical signals are recorded as a metal-coated AFM tip is passed over the material. Using this approach, fibre-like morphologies were observed – including regions of parallel nanofibres. In some areas of the AFM image, $I–V$ responses from the AFM tip indicating nanostructures with an apparently metallic character were observed (Figure 2.11).

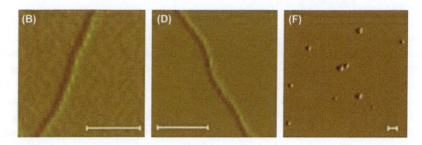

Figure 2.10   AFM images of chiral fibres with helicity (B and D) reflecting the molecular-scale chirality. Image (F) represents the nonfibrillar morphology obtained from a racemic mixture of gelators.
Figure adapted from ref. 57 with kind permission of the American Chemical Society.

Figure 2.11 Current-sensing AFM used to detect the conductivity of doped TTF-
derived nanowires.
Image adapted from ref. 59 with permission from Wiley-VCH.

2.4.3 X-Ray/Neutron Scattering Methods

The irregular packing of fibrils and fibres within solvated gels means that rather
than giving well-defined diffraction patterns, these materials generally broadly
scatter X-rays and/or neutrons (Figure 2.12).[61–63] However, on some occasions
at low angles of scattering, the data for solvated gels can sometimes be fitted to
an appropriate model. For example, the scattering data obtained from a gel can
be modelled as a collection of cylinders or tapes and the output from this fitting
can provide useful information about nanoscale dimensions, such as the cy-
linder diameter. This can thus provide a direct insight into the smallest repeat
unit, *i.e.* the dimensions of molecular fibrils. This is in contrast to electron
microscopy methods, which focus instead on the bundled/aggregated larger
nanoscale fibres. As such, small-angle X-ray or neutron scattering methods
can provide unique information about the morphologies responsible for gelation.[64–66] It is also possible to estimate the volume fraction of connection
points (nodes) within the gel. In general, for useful data to be collected on
solvated samples, given their low levels of order, it is necessary to use a
synchrotron source.

X-ray methods can also be used on dried gels (xerogels), which have significantly more order than the solvated systems and for which standard small-
angle X-ray diffractometers can be employed. For xerogels, diffraction peaks
are often observed, which can directly give parameters such as the long *d*
spacing, corresponding to the longest repeat distance in the gel – often the fibre
diameter. For example, analysis of *d* spacing from crystallographic peaks
allowed the proposal of an inter-digitated bilayer model of self-assembly for a

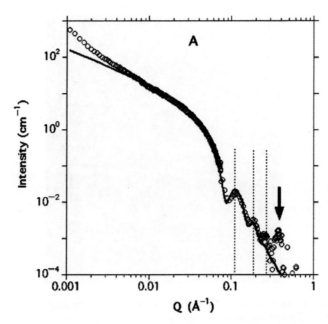

Figure 2.12   Example neutron-scattering data from a solvated gel showing the agreement in fit between experimental data (points) and theoretical form-factor model for rigid homogeneous cylindrical fibres (line). The arrow represents the Bragg peak. The dotted vertical lines represent three form-factor oscillations. This scattering plot contains a relatively large amount of information for a solvated gel.
Figure adapted from ref. 64 with kind permission of the American Chemical Society.

gelator, with characteristic peaks being observed at 1, 1/2 and 1/3 of the extended molecular length of two gelator molecules – the first example of a well-ordered bilayer-based aqueous gel (Figure 2.13).[67] Although this can lead to a more precise understanding of the molecular packing within fibrils, results must be treated with some care, as drying can lead to morphological changes.[68]

It is also possible to use detailed computer-aided modelling of X-ray data to achieve a degree of molecular structure prediction within the fibre – a powerful approach for trying to identify in more detail the molecular-scale interactions that underpin self-assembly (see Section 2.5). The interactions between molecules can be modelled (for more details see Section 2.5.4) and various putative crystal structures can be tested for their degree of fit to the observed SAXS data. In elegant work, Steed and coworkers have demonstrated that this approach can give powerful insights into the way in which molecular building blocks support gelation.[69]

2.4.4   Light-Scattering Methods

Light-scattering methods can also provide insight into supramolecular assembly processes.[70] In static light scattering the absolute value of the intensity

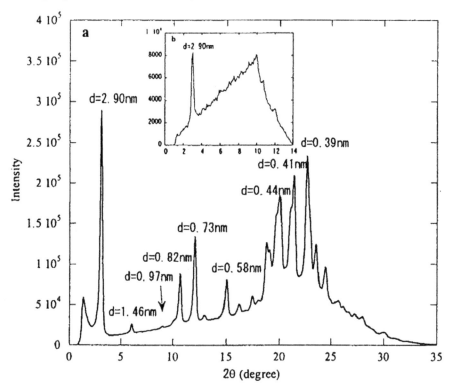

Figure 2.13 Powder X-ray scattering data on a dried xerogel sample used to define the *d* spacing of molecular-scale repeat units in a self-assembled gelator. Figure adapted from ref. 67 with kind permission of the American Chemical Society.

of scattered light is measured. More useful in studying gels, however, is dynamic light scattering (DLS), where instantaneous variations in intensity are recorded. In summary, DLS works because photons interact with the electron clouds of the analyte resulting in energy transfer to the electrons. This results in the electrons re-emitting photons. The analysis of this "scattered light" provides information about the structure and shape of the scatterer. The intensity of observed scattered light depends on both concentration and the angle of detection. When carrying out DLS, it is essential that the solution should be dust-free and must be studied at several different concentrations and angles. It is also true that meaningful data can only be obtained for transparent gels – if the gel is too opaque, the intensity of scattered light will drop too low. In general terms, the simplest way of handling DLS data is to assume the particle is spherical – clearly this is not a good model for a supramolecular gel, but the DLS data will give the equivalent hydrodynamic radius. By careful multiangle dynamic light scattering, it is possible, in principle, to gain more detailed information about nonspherical morphologies – this has rarely been done for gels. It should be noted therefore, that in general, DLS studies on gels tend to be interpreted qualitatively. However, the key advantage of dynamic light scattering is that it is a true solution-phase technique, and reports directly on what is happening in the solvated gel-phase – as such it can provide useful insights into processes within gels.

Dynamic light scattering has been used to follow the gelation process in real time for a variety of physical and chemical gels[71,72] – although has been less often exploited to study low molecular weight gelators.[73] Monitoring intensity against time (or concentration of gelator) provides useful insight into the build-up of the gel network. In general, these studies have determined that there is a "gelation threshold" at which the gel rapidly forms a gel network. This is in agreement with molecular-scale techniques (see Sections 2.5 and 2.6) that indicate that in many cases, gelation is a cooperative process that mimics nucleation and crystal growth. Time-resolved gel studies can also be performed with a temperature gradient (*i.e.* using controlled cooling to induce slow gelation). In elegant work,[74–76] Shibayama and coworkers have combined DLS and SANS studies. In one case,[74] they determined that for their two-component gelator system, a three-step gelation process took place: (i) induction stage (no gelation), (ii) association stage (pre-gel), (iii) freezing stage (gel). It was noted that the association stage (ii) consisted of medium-sized flexible fibrous clusters loosely associated with one another – difficult to observe using other techniques.

2.5   Analysis of Molecular-Scale Assembly

2.5.1   NMR Methods

NMR spectroscopy is a useful approach to try and understand how, on a molecular level, gelators assemble into fibrillar architectures, and to probe the intermolecular interactions between the molecular building blocks that lead to self-assembly.[77,78] In general, in assembled gels the free monomers and small oligomers have suitable relaxation times for study by NMR, and are observable. Conversely, aggregates have reduced transversal relaxation times (T_2) as a consequence of their molecular size and slow diffusion, resulting in long correlation times. As such, the NMR signals of aggregated material are broadened and are often hidden under the baseline (Figure 2.14). This offers powerful possibilities for quantifying the behaviour of gels, as integration of the signals, combined with the use of a mobile, small-molecule, internal standard, allows for quantification of precisely what is immobilised within the "solid-like" gel network, and what is mobile within the solvent voids. This can be particularly useful for probing the self-assembly and gelation of complex mixtures of components. For example, in a two-component gel, it was demonstrated using NMR that certain components were taken preferentially into the gel-phase network and "immobilised" – these were the building blocks that formed more effective gels.[79]

At concentrations below the minimum gelation concentration (MGC), NMR methods can often provide some insight into the monomer–oligomer equilibria that occur within the "mobile" solution phase. Often, as concentration

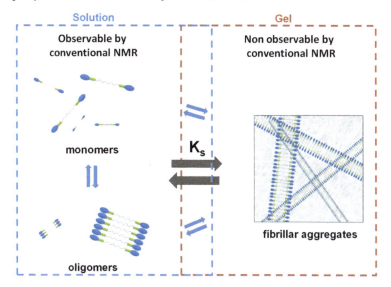

Figure 2.14 The use of NMR methods to define mobile and immobile molecular building blocks within a dynamic gel-phase material.

increases, the NMR resonances broaden and exhibit changes in chemical shift, indicative of the noncovalent interactions forming between them – for example, hydrogen-bond interactions typically induce downfield shifts as concentration increases. It is possible to use the data from this type of experiment to gain insight into the thermodynamics and mechanism of gelation (*e.g.* isodesmic or cooperative fibre growth[21] – this kind of data analysis is discussed in more detail in Section 2.6.2). A similar effect on chemical shift is observed when changing the temperature of a system (below its MGC), because the position of the monomer–oligomer equilibrium can be changed thermally.

Once the full gel begins to form, it is possible using NMR to estimate at each concentration how much of the gelator has been immobilised within the "solid-like" network. This experiment has been performed on several different systems, and in each case, as the gelator is added to the solvent, the gelator is initially all visible by NMR – indicating no aggregation. This continues until a critical concentration is reached, and then from that point on, all of the gelator becomes incorporated into a gel-phase aggregate (Figure 2.15). This indicates that gelation is somewhat like crystallisation – *i.e.* there is a maximum solubility at which the solution becomes saturated (K_s), and at this saturation concentration, aggregation into gel nanostructures begins. Furthermore, by comparing the visual results of gelation with the NMR results it is possible to work out precisely how much immobilised gelator is required to underpin a sample-spanning macroscopic gel – a new parameter only accessible using this approach, and defined as the critical network concentration (CNC). This provides direct insight into how good a particular gelator is at immobilising a specific solvent and can be a powerful way of comparing and quantifying the performance of related families of gelators.[21]

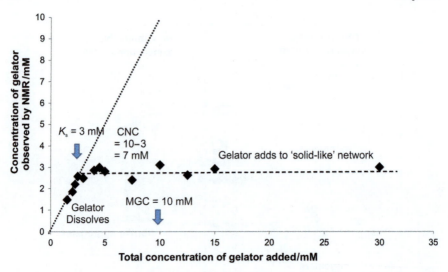

Figure 2.15 Use of NMR and knowledge of the experimentally observed minimum gelation concentration (MGC) to determine the maximum gelator solubility (K_s) and the critical network concentration (CNC) for a gelator.

By monitoring the influence of temperature on the saturation solubility of the gelator, K_s, it is possible to estimate the thermodynamic parameters responsible for aggregation by using the van't Hoff equation (2.1)

$$\ln K_s = -(\Delta H / RT) + (\Delta S / R) \tag{2.1}$$

where ΔH and ΔS represent the molar enthalpy and entropy for the dissolution process (*i.e.* in this case, gel to sol transition). By plotting $\ln K_s$ *vs.* $1/T$ it is possible to extract the thermodynamic parameters which provide insight into the driving forces responsible for aggregation and nanoscale assembly.[21] Escuder and coworkers made use of this approach to provide insight into the thermodynamics of hydrogelation and reported that by careful control of molecular design, and in particular the hydrophobic component of the structure, it is possible to develop gels that are either entropy controlled or enthalpy controlled, with each of these gels presenting quite different temperature-induced responses in properties such as the minimum gelator concentration (MGC) or the rheological moduli.[80]

Fibrillar gel networks can be considered to have analogy with many important biological molecules and structures, which also have large sizes and as a consequence, exhibit reduced relaxation times and significantly broadened peaks. However, this means that a number of techniques that can be useful for studying biosystems[81] can also be directly applied in the field of self-assembled gels. For example, nuclear Overhauser effects can also usefully shed light on gelation behaviour.[77,78,82] The reduced transverse relaxation times (T_2) of gel fibres can give rise to transfer NOE phenomena. The transfer NOE phenomenon is often observed in the study of interactions between small molecules and macromolecules (*e.g.* proteins) where fast exchange on the NMR timescale of the small molecules between free and bound states reduces T_2 and gives negative NOE correlations. These negative NOE signals are therefore characteristic of the small molecule bound to the macromolecule and provide valuable structural information about the complex. This approach can be used to demonstrate that certain molecules are able to interact with gel fibres, whereas others remain more mobile within the solvent phase of the gel – for example, in one example, resorcinol showed a significant decrease in its T_2 value in the presence of gelator, whereas other phenols did not, indicative of selective interactions between resorcinol and the gel fibres.[83,84] As such, this is a useful approach for probing subtle mobility effects within gel-phase networks. It is important in this technique to use short mixing times in order to avoid problems related to spin diffusion phenomena, in which the magnetisation transfer can spread throughout the molecule by crossrelaxation until every spin system has been affected – as such indirect NOE signals can be observed at longer mixing times. A curve of the NOE intensities versus mixing time is necessary to discard spin-relaxation processes. At short mixing times the NOE intensities exhibit no dependence on mixing time, and it is certain that no spin diffusion takes place. However, in some cases, spin diffusion still takes place at very low mixing times, and this NOE approach is not useful.

High-resolution magic-angle spinning (HRMAS) NMR[85–87] has also been applied to the study of supramolecular gels. This technique is widely employed to provide information about interface systems, more particularly for detecting NMR resonances from a conformationally mobile chemical moiety, grafted to or interacting with an immobile phase. If the mobile moiety displays sufficient isotropic rotational mobility (correlation time of *ca.* 10^{-11} s or less) because it partially experiences a liquid-like phase, the dipolar interactions for its particular NMR signals are naturally eliminated, leading to the loss of dipolar broadening, resulting in reasonably sharp resonances. By using a diffusion filter in pulsed field gradient spin echo (PFGSE) experiments, it is possible to eliminate signals from completely freely moving solution-phase molecules and hence focus attention on the parts of the fibrillar network that have any degree of mobility. Miravet and coworkers applied this technique to some peptide-derived supramolecular gels and explored their conformational behaviour as a function of solvent polarity.[88] Remarkably, only some parts of the gelator could be detected using this technique, behaviour that could be related to the differentiated mobility of different gelator moieties – *i.e.* only the flexible parts of the molecules within the fibrillar network are observable. As such, this technique provides a unique insight into the dynamics of gels.

Saturation-transfer-difference (STD) NMR experiments can also, in principle, be used to detect interactions between a gel network and individual gelators and/or other components within the overall gel.[89] The selective saturation of a single proton-resonance within a macromolecule (or in this case, within a gel network) results in the rapid spread of the magnetisation through the entire system due to spin diffusion. However, by subtracting the spectrum without saturation (off-resonance) with a spectrum with selective saturation (on-resonance), it is possible to visualise whatever small molecules are bound to the gel network. Furthermore, any nuclei in direct contact with the gel network will have larger intensity signals in the STD spectrum. This approach has been used to characterise a gelation system by irradiating at 100 ppm ("off-resonance") and –1 ppm ("on-resonance") in the ^{1}H NMR spectrum.[90] However, STD signals were seen for all gelator resonances, and also the solvent molecules (in this case, acetonitrile and water), suggesting that the whole structure was involved in gelation, as was the solvent, playing an explicit role in mediating assembly. This approach may, in the future, prove useful for probing molecular recognition processes within complex gel mixtures in more detail.

2.5.2 Infrared Spectroscopy

Infrared (IR) spectroscopy is very useful for probing hydrogen bond interactions between the molecular building blocks. In particular, O–H, N–H and C=O stretches all show distinctive responses to hydrogen bonding – with the bonds changing in strength as hydrogen bonds form resulting in a shift in wave number. Van der Waals interactions can also sometimes be detected by looking for changes in C–H stretching interactions. Measuring the IR spectra of gels can be complicated by the fact that they are solvated materials, with the solvent potentially obscuring key parts of the IR spectrum. Typically, it is necessary to compare the IR spectra of the gelator in both the sol and the gel in order to determine the key noncovalent interactions responsible for assembly. Variable-temperature IR spectra can provide a very useful way of probing the change in these interactions as gelation occurs in response to temperature changes.[91] Variable-concentration IR spectroscopy can be used to follow the build-up of the gel network and provide insights into error-checking processes during self-assembly.[92] In some cases, comparing the IR spectrum of the solid gelator with that of the dried nanostructured xerogel can demonstrate how noncovalent interactions have changed when gelation has occurred and organised the molecules into nanoscale morphologies prior to sample drying.[93] IR spectroscopy has also been used to demonstrate that a gel was broken down by Ag$^+$ cations because they complexed to alkene groups on the gelator periphery, with the distinctive change in alkene IR bends reporting on the binding mode.[94]

2.5.3 Optical Spectroscopy and Fluorescence

Optical spectroscopy can be used to report on gelation processes – once again variable concentration and temperature studies can provide valuable information about the assembly process. As a simple example, donor–acceptor systems often form a red colour on complex formation – this visible spectroscopic output acts as a direct signature that donor–acceptor interactions are occurring.[95–97] It has been reported that slowly cooling a two-component gelation system consisting of donor and acceptor components initially gave rise to a red colour, indicative of the formation of a donor–acceptor complex, but no gelation.[98] Only on further cooling did gelation take place. Combination of visible spectroscopic observations with NMR studies, allowed the conclusion that the donor–acceptor complexes were helping align the molecules so that hydrogen-bond networks could then drive the main gelation event. Spectroscopic studies (UV-Vis near-infrared) have also provided evidence for the formation of columnar TTF stacked cores within gel-phase materials, which existed in mixed-valence states and provided an efficient pathway for electron conduction, materials that exhibited moderate conductivity levels on oxidation with iodine.[99]

Gelators that include fluorophores can exhibit useful changes in their spectra on aggregation. Also, fluorophores can often enhance gelation as a consequence of their large hydrophobic surfaces, which are also capable of forming π–π stacking interactions. Some fluorophores emit as excimers when in close proximity (such as in a gel fibre) but emit as monomers when present in dilute solution. For example, pyrene-modified vancomycin gels were proposed to assemble in part due to π–π stacking as a consequence of the fluorescence spectrum in the gel state, which resembled that expected for pyrene excimers.[100] However, it is not always the case that gelation enhances excimer formation – in one example, excimers were present in the sol, but not in the gel, with gelation limiting the ability of the fluorophores to come into close proximity.[101] Fluorescence can also detect smaller changes associated with differences in polarity between the gelator in aggregated and nonaggregated states. Once again, VT fluorescence can be a useful technique in order to determine which spectral features are responsive to the aggregation process.[102] However, it should be noted that many gels only form at concentrations well above the optimum level for optical spectroscopy, and this can potentially limit the usefulness of these techniques.

An alternative way of using fluorescence spectroscopy to probe gels is to employ a small amount of fluorescent additive and monitor changes in its intensity/wavelength to report on its local environment. For example, 8-anilino-1-naphthalenesulfonic acid (ANS) has been used to report on hydrophobic domains within hydrogels.[103] In one example, this has been used to follow the evolution of the gel over time, and it was demonstrated using the ANS reporter that gelation was, in that particular case, not just a simple one-step process.[104]

2.5.4 Computational Modelling of Self-Assembly

Computational modelling[105] can help provide insight into self-assembly process at various hierarchical levels. It is usual to compare the results of modelling with experimental data in order to validate whether the model is an appropriate one to help understand self-assembly.

Molecular dynamics (MD) and, at a higher level of theory, density functional theory (DFT) can be used to probe the gelator–gelator interactions that might be responsible for underpinning self-assembly.[69,106–109] MD models the molecules at an atomistic level, with each atom in the structure being parameterised and minimised for its covalent and noncovalent interaction energies, while DFT is an *ab initio* approach that simulates atomic-level information directly from the first principles of quantum mechanics. These approaches can yield information about the molecular recognition pathways within gels – this can be used, for example, with powder X-ray diffraction data in order to generate structural packing diagrams for the unit cell within the xerogel. However, caution must be applied, as such computational studies are frequently performed *in vacuo*, yet in gels, the solvent plays a vital, and often explicit role (see Section 2.7). As such, for higher-quality modeling, refinement should be carried out within a solvent box. This may allow solvation to be accounted for – however, many solvation effects do not become dominant until larger nanoscale surfaces are formed. Because of the high levels of computational power required for MD/DFT methods, it is not readily possible to model larger fibrillar aggregates using this approach, and therefore longer-range interactions, solvation effects, or fibril–fibril packing effects may be missed.

In order to model the nano- and the mesoscale behaviour of gels, it is possible to use mesoscale modelling techniques, in which a degree of coarse graining is employed.[110] In eye-catching early work this approach was used to provide insight into the self-assembly of a peptide gelator, which were treated in a simplified, coarse-grained approach as chiral rods.[36] A series of assumptions were made about the relative interactions between different amino acids both with solvent and with other gelator molecules, and a statistical-mechanical model was constructed for each hierarchical step in the assembly process (Figure 2.16). This led to the proposal that this gelator sequentially assembled into tapes, ribbons, fibrils and fibres as the gelator concentration was increased. The insight provided by this kind of modelling had a significant impact on the general understanding of fibre assembly and gelation.

In other mesoscale approaches, precise atomic structural detail, which would be employed in the MD/DFT level, is converted into approximate potentials, for example using the united atom model[111] (or other coarse-graining approaches). These potentials allow for electrostatic forces, solvophobicity, solvophilicity, *etc.* This simplification step allows the computational power to cope with minimising many molecular-scale building blocks in a solvent box over extended distances and timescales – hence much larger nanoscale assembly processes can be computationally modelled. This approach can also be combined with specific and significant noncovalent interactions, such as hydrogen bonds, which may have a direct impact on fibrillar assembly. As such, insight can be gained into the precise way in which multiple molecules come together. Although such methods are quite widely used in modelling self-assembly, they have only rarely been applied to fibrillar or gel-phase materials.[112,113] Clearly there is considerable future scope for modelling methods to shed light on some of the aspects of gelation that are more challenging to understand, such as the mechanism responsible for fibre–fibre interactions and predictive insight into the rheological performance of the material on the macroscopic length scale.

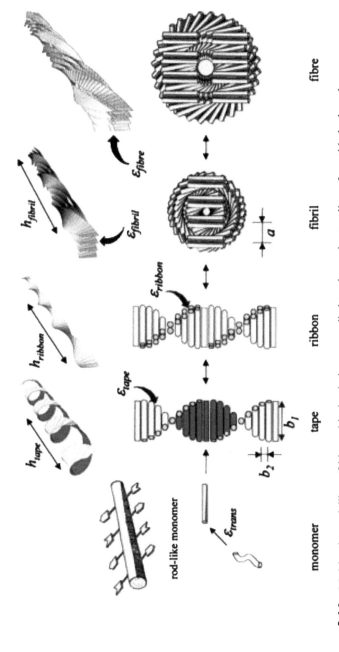

Figure 2.16 Multiscale modelling of hierarchical gelation as applied to the understanding of peptide hydrogels. Figure adapted from ref. 36 with permission of the National Academy of Science (USA).

2.6 Analysis of Chirality in Gels

2.6.1 Aspects of Chirality in Gels

As mentioned above, many gels are assembled from stereoisomeric building blocks, and in a number of cases, the molecular-scale chirality can be translated up to the nanoscale level.[37] There has also been considerable interest in how self-assembly responds to mixtures of enantiomers. There are several possible outcomes when an enantiomeric mixture of building blocks is allowed to self-assemble:

- the presence of one enantiomer *disrupts the assembly* of the other and cause a breakdown in chiral order – this indicates that there is no preference between heterochiral and homochiral interactions and the mixtures that form are unable to support gelation;[57,114]
- the enantiomers gel more effectively when mixed together, with one enantiomer enhancing the assembly of the other – reflecting the fact that heterochiral interactions are preferred to homochiral ones and that the diastereomeric complex that forms is better able to support gelation than the individual enantiomers;[82]
- the enantiomers have no impact on one another's self-assembly processes, with homochiral interactions being preferred over heterochiral interactions, and the individual enantiomeric gelators self-sorting into their own homochiral assemblies.[115–117]

Chirality at the nanoscale level can sometimes be observed by electron microscopy,[38] but there are many cases in which chiral molecules do not assemble into morphologies that are visibly chiral. Given the limitations associated with sample drying, microscopy methods are not always the most representative way of studying what is truly present within the gel-phase. Furthermore, subtle effects of chiral mixing on assembly can be difficult to probe with electron microscopy.

2.6.2 Circular Dichroism Spectroscopy

Circular dichroism spectroscopy is a solution-phase technique which is effectively UV-Vis spectroscopy using circularly polarised light – as such it can be thought of as "chiral spectroscopy".[118,119] This method can be precisely controlled and can therefore provide detailed information about the solvated native gel. As such, this is the preferred methodology for analysing the impact of chirality on self-assembly and soft materials behaviour.

An achiral molecule will exhibit no bands in its CD spectrum, whilst a chiral molecule can exhibit a signal (either positive or negative), in the same wavelength region where it has its UV-Vis absorption. In general terms, however, solvated isolated chiral molecules often have relatively small CD signals (unless they have well-organised chromophores inherent within the molecular structure). As individual molecules assemble into a chiral nanostructure, the interaction with polarised light can be significantly enhanced, and hence self-assembled nanostructures often exhibit much larger CD bands than their isolated molecular building blocks. The presence of a CD signal therefore provides good evidence for the presence of a chiral nanoscale object. To confirm this assignment, however, it is essential to carry out a variable-temperature (VT) experiment, or compare the CD spectrum with that observed in a solvent where aggregation does not take place.[115] In VT CD, the CD band should usually decrease in intensity as the temperature increases (for an enthalpy-controlled process), because the nanoscale aggregates break up into their constituent building blocks on heating (Figure 2.17). For well-characterised chromophores, it can sometimes be possible to use CD spectroscopy to predict whether they are packed in a clockwise or an anticlockwise manner within the helical assembly.[118]

Circular dichroism can provide an exquisitely detailed insight into some of the finer points of gelator self-assembly into chiral nanostructures. For example, on mixing a chiral gelator into an achiral system, it allows easy determination of whether the enantiomer is enforcing its own chiral mode of assembly onto the achiral system *via* a "sergeants and soldiers" type mechanism.[120,121] This is evidenced by the CD signal being "switched on" by the presence of a relatively small amount of the enantiomeric additive, as illustrated by Figure 2.18, in which just 1% of enantiomeric gelator (**2** or **3**), encourages achiral gelator (**1**) to assemble into a chiral nanostructure with a distinctive CD spectroscopic signature.

CD spectroscopy can also be used to provide insight into whether assembly is an isodesmic or cooperative process.[122] In isodesmic, or stepwise assembly, each molecular-scale building block adds on to the end of the growing fibre

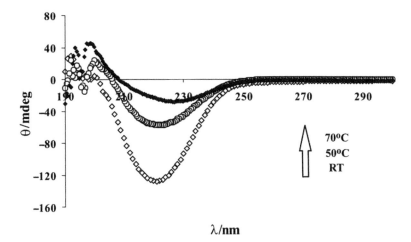

Figure 2.17 Temperature-responsive circular dichroism spectrum indicative of thermally responsive chiral self-assemblies.
Figure reproduced from ref. 115 with kind permission of Wiley-VCH.

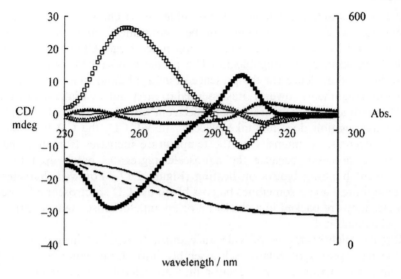

CD spectra of **1** (——), **2** (▲), **3** (△), **1/2** 99:1 (■) and **1/3** 99:1
(□). UV spectra of **1/2** 99:1 (-----) and **2** (——) in xerogel phase.

Figure 2.18   CD and UV spectra of achiral gelator (**1**) and enantiomeric gelators (**2**
and **3**). The addition of 1% of chiral additive encourages achiral gelator **1**
to assemble into chiral nanostructures with a large CD signal *via* a
sergeants and soldiers mechanism.
Figure adapted from ref. 120 with kind permission of Wiley-VCH.

with the same affinity, however, in cooperative growth, after initial nucleation
has taken place with lower binding affinity, chain elongation occurs with a
higher affinity. This gives rise to different types of network growth in each case
as can be observed by monitoring the change in the CD signal, adjusted for
concentration of chiral chromophore.[123] As noted in Section 2.5.1, other
spectroscopic techniques can also be used to achieve this (such as following the
NMR shift on increasing concentration of gelator),[21] but CD is particularly
well suited to it. By using CD to monitor self-assembly as the temperature was
slowly lowered (0.5 K/min, to suppress kinetic effects on self-assembly), it was
possible to follow the transition from monomers into helical aggregates. The
curve of ellipticity against temperature (Figure 2.19) was not sigmoidal, and
therefore the assembly could not be assigned to an isodesmic model but could
instead be understood in terms of the Oosawa–Kasai cooperative behaviour[124]
in which nonisodesmic assembly of helical structures is preceded by the isodesmic assembly of nonhelical structures.[123] Furthermore, detailed analysis
made clear that an additional process was taking place, particularly at elevated
concentrations – it was suggested this was due to lateral clustering of stacks – a
process that may ultimately cause the supramolecular polymers to give rise to
gel formation.

In addition to CD spectroscopy in the UV-visible range, it is possible to
perform CD in the infrared region.[125–128] This can give information about the

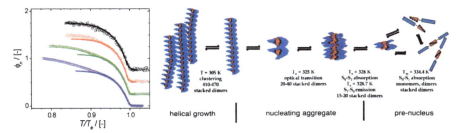

Figure 2.19 Effect of temperature on CD intensity at four different concentrations (left, data presented with a vertical offset for clarity – experimental data shown as points, modelling to Oosawa–Kasai model shown as lines) and the proposed mechanism of chiral self-assembly and disassembly (right). Figure adapted from ref. 123 with kind permission of Science.

chiral organisation of IR-active units. This is a particularly powerful way to study chiral hydrogen-bonding pathways that often drive organogelation and experience a chiral environment within the self-assembled nanofibre. This methodology can also been applied in combination with variable-temperature methods in order to monitor the effect of temperature on self-assembly.

2.7 Characterising Solvent Effects in Gels

Gel-phase materials are usually constituted of *ca.* 99% of solvent, and in some cases significantly more. Given that the solvent is the major component of a gel, it can therefore play a vital role in controlling the material's performance and properties. As such, the study of solvent effects on gelation is an important topic, as understanding how solvent and self-assembled fibres interface can provide significant insights into gelation. In many cases, researchers will try to correlate a specific experimental parameter associated with the gel, with some physical-organic characteristics of the solvent. Most commonly, authors will simply attempt to correlate (i) whether gelation occurs, (ii) the minimum gelation concentrations and/or (iii) the T_{gel} value of the soft material – however, in some cases other characteristics will be considered (such as the spectroscopic shifts, *etc.*). Many authors attempt to understand gelation in terms of bulk solvent parameters such as the dielectric constant.[82] However, such considerations do not properly account for direct interactions between solvent and gelator – which are important in mediating gelation. As such, other studies have attempted to correlate solvent properties that allow for such interactions, such as the Reichardt, Hildebrand and Hansen parameters,[43,129–133] and Kamlet–Taft parameters.[134–136] Dissolution enthalpies and entropies have also been used as a predictor for gelation in a given solvent – these are in some way, experimentally derivable surrogates for the physical organic solvent parameters but applied to a specific gelator.[137]

The use of physical-organic solvent parameters can allow authors to dissect the relative importance of solvent polarity and hydrogen-bonding effects in terms of allowing gelation to take place. In general terms, solvent–gelator interactions will hinder gelator–gelator interactions and hence prevent self-assembly from taking place and tend to give rise to isotropic solutions. However, in the absence of any solvent–gelator interactions, gelator–gelator interactions will dominate and the gelator will simply not dissolve. As such, gelation sits on a knife-edge between solubility and crystallisation[138] – parameterising this for a given gelator allows the gel properties to be controlled in a range of solvents in a predictable way.

Careful investigation of solvent effects on self-assembly processes by circular dichroism has been carried out by Meijer and coworkers using a variety of different alkanes.[123] It was noted that the chain length of the alkane solvent was having an odd–even effect on the assembly process, and it was suggested that this could be understood in terms of the solvent becoming directly involved in the ordering process. This led Meijer and coworkers to develop the theory of solvent-assisted nucleation – in which solvent is not simply passive in supporting self-assembly, but plays an explicit role in interacting with the aggregates *via* solvation and hence directing the assembly event. Stereoselective deuteration of the solvent allowed these authors to gain further insight into the directing role played by solvent molecules.[139] Other recent work supports the view that the solvent can play an explicit role in directing gel nucleation and assembly. Indeed, it has been reported that the steric demands of specific solvents can be correlated with gelation potential,[140] and it has also been noted that the chirality of the solvent can modify the performance of self-assembled chiral gelators.[141]

2.8 Kinetics of Gelation

In principle, many of the techniques described in detail above can be used to probe gelation kinetics by carrying out time-resolved experiments and monitoring how the property being observed evolves over time. There are certain key aspects of gel kinetics that should be considered when interpreting the data from this type of experiment. The Avrami model, originally developed to understand crystallisation has been successfully applied for the study of supramolecular gelation and allows researchers to determine the dimensionality of crystal growth.[142–144] This serves to emphasise the similarity between gelation and crystallisation events, as previously mentioned. As noted previously, gelation can be considered as a type of crystallisation with dimensional restrictions on the growth of the crystalline phase. The Avrami model can be expressed mathematically as:

$$1 - X(t) = \exp(-Kt^n) \tag{2.2}$$

or:

$$\ln(\ln(1/(1 - X(t)))) = \ln K + n \ln t \tag{2.3}$$

where $X(t)$ is the volume fraction of the gel phase, K is a temperature-dependent constant similar to a rate constant, n is the Avrami exponent, that reflects the type of growth leading to gelation, and t is time. This approach assumes that nucleation occurs instantaneously, and that the growth of different domains in the gel phase are independent of one another. ^{1}H NMR offers one approach by which the amount of material in the gel ($X(t)$) can be very easily quantified, but other characterisation techniques can also be applied. In terms of signal intensity, using a spectroscopic technique of choice, $X(t)$ can be expressed as follows:

$$X(t) = (I(\infty) - I(t))/(I(\infty) - I(0))$$

where $I(\infty)$ is the intensity of the signal when self-assembly has reached equilibrium and $I(0)$ is that measured at time zero (usually defined as the first measured data point, or the point at which the signal starts to change rapidly if there is a delay to assembly). Plotting $\ln(-\ln(I(\infty) - I(t))/(I(\infty) - I(0)))$ against $\ln t$ allows the determination of Avrami coefficients n and K. The Avrami coefficient n can be interpreted qualitatively in terms of the dimensionality of the object being formed. For example, an n value of 1 indicates one-dimensional growth (as would be expected for fibrils) while a value closer to 2 would indicate the growth of objects in more than one dimension. In a landmark paper,[145] rheology, fluorescence SANS and CD data were each used to probe the kinetics of self-assembly and to confirm that for the cholesterol system being investigated, the Avrami coefficient, n, was approximately 1 – consistent with one-dimensional and interface-controlled growth. The Avrami coefficient, K, represents the effective "time constant" for gelation, and as such, varies with concentration and temperature. Analysing this coefficient in more detail can give rise to estimations of "activation energies" for gelation, which can shed light on the "rate-limiting" step in the assembly process.

In a number of cases, authors have used more complex fractal growth models for network formation,[146,147] which include further branching steps in addition to the initial nucleation and growth pathways as discussed in the simple Avrami mechanistic analysis above. The growth of such fractal patterns can also be fully mathematically modelled, for example by using the Dickinson model.[148] In cases where a high degree of branching occurs, more open and well-established 3D networks with smaller mesh sizes are obtained.

2.9 Conclusions

In summary, gels are fascinating hierarchical materials, and as such must be investigated using a wide range of experimental methods. This can give insight into a wide range of different aspects such as:

- mode of self-assembly;
- morphology of nanostructure;
- materials performance;
- thermodynamics of self-assembly;
- kinetics of self-assembly.

Of course, it is not always necessary to apply every single technique, and depending on the problem being addressed, different combinations of techniques will be desirable/required. However, it is evident that for modern investigations into gel-phase soft materials, it is no longer sufficient to simply make visual observation that a gel has formed in a particular solvent. It is vital that we continue to gain more detailed insights into the mechanism of gelation and the way in which information translates from molecular-scale to nanoscale and to macroscopic behaviour. Only by achieving this degree of predictive and fundamental insight will we be able to successfully develop these materials for the next generation of sophisticated and tunable high-tech applications.[149]

References

1. P. Terech and R. G. Weiss, ed. *Molecular Gels: Materials with Self-Assembled Fibrillar Networks*, Springer, Dordrecht, The Netherlands, 2006.
2. M. George and R. G. Weiss, *Acc. Chem. Res.*, 2006, **39**, 489.
3. F. M. Menger and K. L. Caran, *J. Am. Chem. Soc.*, 2000, **122**, 11679.
4. J. H. van Esch, *Langmuir*, 2009, **25**, 8392.
5. I. A. Coates and D. K. Smith, *Chem. Eur. J*, 2009, **15**, 6340.
6. L. Yan, Y. Xue, G. Gao, J. Lan, F. Yang, X. Su and J. You, *Chem. Eur. J.*, 2010, **16**, 2250.
7. G. Cravotto and P. Cintas, *Chem. Soc. Rev.*, 2009, **38**, 2684.
8. T. Naota and H. Koori, *J. Am. Chem. Soc.*, 2005, **127**, 9324.
9. J. Wu, T. Yi, Q. Xia, Y. Zou, F. Liu, J. Dong, T. Shu, F. Li and C. Huang, *Chem. Eur. J.*, 2009, **15**, 6234.
10. X. Yu, Q. Liu, J. Wu, M. Zhang, X. Cao, S. Zhang, Q. Wang, L. Chen and T. Yi, *Chem. Eur. J.*, 2010, **16**, 9099.
11. M. Suzuki, Y. Nakajima, M. Yumoto, M. Kimura, H. Shirai and K. Hanabusa, *Org. Biomol. Chem.*, 2004, **2**, 1155.
12. U. K. Das, D. R. Trivedi, N. N. Adarsh and P. Dastidar, *J. Org. Chem.*, 2009, **74**, 7111.
13. Y. Wang, L. Tang and J. Yu, *Cryst. Growth Des.*, 2008, **8**, 884.
14. J. Cui, Z. Shen and X. Wan, *Langmuir*, 2010, **26**, 97.
15. F. Rodríguez-Llansola, J. F. Miravet and B. Escuder, *Chem. Commun.*, 2009, 209.
16. M. M. Smith and D. K. Smith, *Soft Matter*, 2011, **7**, 4856.
17. S. R. Raghavan and B. H. Cipriano, 'Gel Formation: Phase Diagrams using Tabletop Rheology and Calorimetry', Chapter 8 in *Molecular Gels, Materials with Self-Assembled Fibrillar Networks*, ed. R. G. Weiss and P. Terech, Springer, Dordrecht, Netherlands, 2006.
18. A. Takahashi, M. Sakai and T. Kato, *Polym. J.*, 1980, **12**, 335.
19. A. R. Hirst, D. K. Smith, M. C. Feiters, H. P. M. Geurts and A. C. Wright, *J. Am. Chem. Soc.*, 2003, **125**, 9010.
20. K. Murata, M. Aoki, T. Suzuki, T. Harada, H. Kawabata, T. Komori, F. Ohseto, K. Ueda and S. Shinkai, *J. Am. Chem. Soc.*, 1994, **116**, 6664.

21. A. R. Hirst, I. A. Coates, T. Boucheteau, J. F. Miravet, B. Escuder, V. Castelletto, I. W. Hamley and D. K. Smith, *J. Am. Chem. Soc.*, 2008, **130**, 9113.
22. G. W. H. Höhne, W. F. Hemminger, and H.-J. Flammersheim, *Differential Scanning Calorimetry*, Springer, Heidelberg, 2010.
23. S. K. Samanta, A. Pal and S. Bhattacharya, *Langmuir*, 2009, **25**, 8567.
24. S. K. Samanta, A. Pal, S. Bhattacharya and C. N. R. Rao, *J. Mater. Chem.*, 2010, **20**, 6881.
25. P. Sollich, 'Soft Glassy Rheology', Chapter 5 in *Molecular Gels, Materials with Self-Assembled Fibrillar Networks*, ed. R. G. Weiss and P. Terech, Springer, Dordrecht, Netherlands, 2006.
26. H. F. Mark, *Encyclopedia of Polymer Science and Technology*, John Wiley & Sons, New York, 2004.
27. P. Terech, C. Rossat and F. Volino, *J. Colloid Interface Sci.*, 2000, **227**, 363.
28. M. O. M. Piepenbrock, N. Clarke and J. W. Steed, *Langmuir*, 2009, **25**, 8451.
29. M. O. M. Piepenbrock, G. O. Lloyd, N. Clarke and J. W. Steed, *Chem. Commun.*, 2008, 2644.
30. M. Lescanne, A. Colin, O. Mondain-Monval, F. Fages and J. L. Pozzo, *Langmuir*, 2003, **19**, 2013.
31. P. Terech and S. Friol, *Tetrahedron*, 2007, **63**, 7366.
32. C. A. Schalley, *Analytical Methods in Supramolecular Chemistry*, Wiley-VCH, Weinheim, 2012.
33. Z.-T. Li and X. Zhao, 'Scanning Electron Microscopy' in *Supramolecular Chemistry: From Molecules to Nanomaterials, Vol. 2 – Techniques*, ed. P. A. Gale and J. W. Steed, John Wiley & Sons, Chichester, 2012, pp. 619–631.
34. J.-i. Kikuchi and K. Yasuhara, 'Transmission Electron Microscopy (TEM)' in *Supramolecular Chemistry: From Molecules to Nanomaterials, Vol. 2 – Techniques*, ed. P. A. Gale and J. W. Steed, John Wiley & Sons, Chichester, 2012, pp. 633–645.
35. J. G. Hardy, A. R. Hirst, I. Ashworth, C. Brennan and D. K. Smith, *Tetrahedron*, 2007, **63**, 7397.
36. A. Aggeli, I. A. Nyrkova, M. Bell, R. Hardig, L. Carrick, T. C. B. McLeish, A. N. Semenov and N. Boden, *Proc. Natl. Acad. Sci. USA*, 2001, **98**, 11857.
37. D. K. Smith, *Chem. Soc. Rev.*, 2009, **38**, 684.
38. T. Tachibana and H. Kambara, *J. Am. Chem. Soc.*, 1965, **87**, 3015.
39. D. K. Smith, 'Molecular Gels – Nanostructured Soft Materials' in *Organic Nanostructures*, ed. J. L. Atwood and J. W. Steed, Wiley-VCH, Weinheim, 2008.
40. R. H. Wade, P. Terech, E. A. Hewat, R. Ramasseul and F. Volino, *J. Colloid Interface Sci.*, 1986, **114**, 442.
41. U. J. Kim, J. Y. Park, C. M. Li, H. J. Jin, R. Valluzzi and D. L. Kaplan, *Biomacromolecules*, 2004, **5**, 786.

42. C. A. Lagadec and D. K. Smith, *Chem. Commun.*, 2012, **48**, 7817.
43. A. R. Hirst and D. K. Smith, *Langmuir*, 2004, **20**, 10851.
44. E. Krieg, E. Shirman, H. Weissman, E. Shimoni, S. G. Wolf, I. Pinkas and B. Rybtchinski, *J. Am. Chem. Soc.*, 2009, **131**, 14365.
45. D. J. Abdallah, S. A. Sirchio and R. G. Weiss, *Langmuir*, 2000, **16**, 7558.
46. H. S. Ashbaugh, A. Radulescu, D. Schwahn, D. Richter and L. J. Fetters, *Macromolecules*, 2002, **35**, 7044.
47. R. Schmidt, F. B. Adam, M. Michel, M. Schmutz, G. Decher and P. J. Mésini, *Tetrahedron Lett.*, 2003, **44**, 3171.
48. A. R. Hirst, D. K. Smith and J. P. Harrington, *Chem. Eur. J*, 2005, **11**, 6552.
49. Y. Zhou, T. Yi, T. Li, Z. Zhou, F. Li, W. Huang and C. Huang, *Chem. Mater.*, 2006, **18**, 2974.
50. J. H. Jung, Y. Ono, K. Sakurai, M. Sano and S. Shinkai, *J. Am. Chem. Soc.*, 2000, **122**, 8648.
51. N. S. S. Kumar, S. Varghese, G. Narayan and S. Das, *Angew. Chem. Int. Ed.*, 2006, **45**, 6317.
52. V. Lozano, R. Hernández, C. Mijangos and M. J. Pérez-Pérez, *Org. Biomol. Chem.*, 2009, **7**, 364.
53. C. B. Minkenberg, W. E. Hendriksen, F. Li, E. Mendes, R. Eelkema and J. H. van Esch, *Chem. Commun.*, 2012, **48**, 9837.
54. A. M. Donald, *Curr. Opin. Colloid Interface Sci*, 1998, **3**, 143.
55. J. T. Hyotyla and R. Y. H. Lim, 'Atomic Force Microscopy (AFM)' in *Supramolecular Chemistry: From Molecules to Nanomaterials, Vol. 2 – Techniques*, ed. P. A. Gale and J. W. Steed, John Wiley & Sons, Chichester, 2012, pp. 659–668.
56. B. W. Messmore and S. I. Stupp, *J. Am. Chem. Soc.*, 2005, **127**, 7992.
57. T. Koga, M. Matsuoka and N. Higashi, *J. Am. Chem. Soc.*, 2005, **127**, 17596.
58. S. Cicchi, G. Ghini, L. Lascialfari, A. Brandi, F. Betti, D. Berti, P. Baglioni, L. Di Bari, G. Pescitelli, M. Mannini and A. Caneschi, *Soft Matter*, 2010, **6**, 1655.
59. J. Puigmartí-Luis, V. Laukhin, A. P. del Pino, J. Vidal-Gancedo, C. Rovira, E. Laukhina and D. B. Amabilino, *Angew. Chem. Int. Ed.*, 2007, **46**, 238.
60. J. Puigmartí-Luis, A. P. del Pino, V. Laukhin, L. N. Feldborg, C. Rovira, E. Laukhina and D. B. Amabilino, *J. Mater. Chem.*, 2010, **20**, 466.
61. M. Anne, 'X-Ray Diffraction of Poorly Organized Systems and Molecular Gels', Chapter 11 in *Molecular Gels, Materials with Self-Assembled Fibrillar Networks*, ed. R. G. Weiss and P. Terech, Springer, Dordrecht, Netherlands, 2006.
62. O. Glatter and O. Kratky (eds), *Small Angle X-ray Scattering*, Academic Press, London, 1982.
63. A. C. Toma and T. Pfohl, 'Small-Angle X-ray Scattering (SAXS) and Wide-Angle X-ray Scattering (WAXS) of Supramolecular Assemblies' in *Supramolecular Chemistry: From Molecules to Nanomaterials,*

Vol. 2 – Techniques, ed. P. A. Gale and J. W. Steed, John Wiley & Sons, Chichester, 2012, pp. 437–450.

64. P. Terech, S. Dourdain, S. Bhat and U. Maitra, *J. Phys. Chem. B*, 2009, **113**, 8252.

65. B. Huang, A. R. Hirst, D. K. Smith, V. Castelletto and I. W. Hamley, *J. Am. Chem. Soc.*, 2005, **127**, 7130.

66. P. Dastidar, S. Okabe, K. Nakano, K. Iida, M. Miyata, N. Tohnai and M. Shibayama, *Chem. Mater.*, 2005, **17**, 741.

67. J. H. Jung, G. John, M. Masada, K. Yoshida, S. Shinkai and T. Shimizu, *Langmuir*, 2001, **17**, 7229.

68. D. K. Kumar, D. A. Jose and A. Das, *Langmuir*, 2004, **20**, 10413.

69. K. M. Anderson, G. M. Day, M. J. Paterson, P. Byrne, N. Clarke and J. W. Steed, *Angew. Chem. Int. Ed.*, 2008, **47**, 1058.

70. J. Braun, K. Renggli, J. Razumovitch and C. Vebert, 'Dynamic Light Scattering in Supramolecuar Materials Chemistrys' in *Supramolecular Chemistry: From Molecules to Nanomaterials, Vol. 2 – Techniques*, ed. P. A. Gale and J. W. Steed, John Wiley & Sons, Chichester, 2012, pp. 411–424.

71. M. Shibayama and T. Norisuye, *Bull. Chem. Soc. Jpn.*, 2002, **75**, 641.

72. S. Richter, *Macromol. Chem. Phys.*, 2007, **208**, 1495.

73. For examples see: (a) W. Weng, J. B. Beck, A. M. Jamieson and S. J. Rowan, *J. Am. Chem. Soc.*, 2006, **128**, 11663; (b) Z. Li, L. E. Buerkle, M. R. Orseno, K. A. Streletzky, S. Seifert, A. M. Jamieson and S. J. Rowan, *Langmuir*, 2010, **26**, 10093; (c) Q. Chen, Y. Feng, D. Zhang, G. Zhang, Q. Fan, S. Sun and D. Zhu, *Adv. Funct. Mater.*, 2010, **20**, 36.

74. P. Dastidar, S. Okabe, K. Nakano, K. Iida, M. Miyata, N. Tohnai and M. Shibayama, *Chem. Mater.*, 2005, **17**, 741.

75. S. Okabe, K. Ando, K. Hanabusa and M. Shibayama, *J. Polym. Sci. B*, 2004, **42**, 1841.

76. S. K. Kundu, N. Osaka, T. Matsunaga, M. Yoshida and M. Shibayama, *J. Phys. Chem. B*, 2008, **112**, 16469.

77. B. Escuder, M. Llusar and J. F. Miravet, *J. Org. Chem.*, 2006, **71**, 7747.

78. Y. E. Shapiro, *Prog. Polym. Sci.*, 2011, **36**, 1184.

79. (a) A. R. Hirst, J. F. Miravet, B. Escuder, L. Noirez, V. Castelletto, I. W. Hamley and D. K. Smith, *Chem. Eur. J.*, 2009, **15**, 372; (b) W. Edwards and D. K. Smith, *J. Am. Chem. Soc.*, 2013, **135**, 5911–5920.

80. V. J. Nebot, J. Armengol, J. Smets, S. F. Prieto, B. Escuder and J. F. Miravet, *Chem. Eur. J.*, 2012, **18**, 4063.

81. X. Salvatella and E. Giralt, *Chem. Soc. Rev.*, 2003, **32**, 365.

82. J. Makarević, M. Jokić, Z. Raza, Z. Stefanić, B. Kojić-Prodić and M. Žinić, *Chem. Eur. J*, 2003, **9**, 5567.

83. B. Escuder, J. F. Miravet and J. A. Sáez, *Org. Biomol. Chem.*, 2008, **6**, 4378.

84. J. A. Sáez, B. Escuder and J. F. Miravet, *Chem. Commun.*, 2010, **46**, 7996.

85. G. Lippens, M. Bourdonneau, C. Dhalluin, R. Warrass, T. Richert, C. Seetharaman, C. Boutillon and M. Piotto, *Curr. Org. Chem.*, 1999, **3**, 147.

86. R. Warrass, J.-M. Wieruszeski and G. Lippens, *J. Am. Chem. Soc.*, 1999, **121**, 3787.
87. R. Warrass, J.-M. Wieruszeski, C. Boutillon and G. Lippens, *J. Am. Chem. Soc.*, 2000, **122**, 1789.
88. S. Iqbal, F. Rodríguez-Llansola, B. Escuder, J.-F. Miravet, J. Verbruggen and R. Willem, *Soft Matter*, 2010, **6**, 1875.
89. M. Mayer and B. Meyer, *Angew. Chem. Int. Ed.*, 1999, **38**, 1784.
90. V. Lozano, R. Hernández, A. Ardá, J. Jiménez-Barbero, C. Mijangos and M.-J. Pérez-Pérez, *J. Mater. Chem.*, 2011, **21**, 8862.
91. M. Suzuki, M. Nanbu, M. Yumoto, H. Shirai and K. Hanabusa, *New J. Chem.*, 2005, **29**, 1439.
92. A. M. Pierce, P. J. Maslanka, A. J. Carr and K. S. McCain, *Appl. Spectrosc.*, 2007, **61**, 379.
93. D. J. Abdallah, L. D. Lu and R. G. Weiss, *Chem. Mater.*, 1999, **11**, 2907.
94. W. Edwards and D. K. Smith, *Chem. Commun.*, 2012, **48**, 2767.
95. U. Maitra, P. V. Kumar, N. Chandra, L. J. D'Souza, M. D. Prasanna and A. R. Raju, *Chem. Commun.*, 1999, 595.
96. P. Babu, N. M. Sangeetha, P. Vijaykumar, U. Maitra, K. Rissanen and A. R. Raju, *Chem. Eur. J*, 2003, **9**, 1922.
97. B. G. Bag, G. C. Maity and S. K. Dinda, *Org. Lett.*, 2006, **8**, 5457.
98. J. R. Moffat and D. K. Smith, *Chem. Commun.*, 2008, 2248.
99. X. J. Wang, L. B. Xing, W. N. Cao, X. B. Li, B. Chen, C. H. Tung and L.-Z. Wu, *Langmuir*, 2011, **27**, 774.
100. B. Xing, C.-W. Yu, K.-H. Chow, P.-L. Ho, D. Fu and B. Xu, *J. Am. Chem. Soc.*, 2002, **124**, 14846.
101. Y. Kamikawa and T. Kato, *Langmuir*, 2007, **23**, 274.
102. I. A. Coates, A. R. Hirst and D. K. Smith, *J. Org. Chem.*, 2007, **72**, 3937.
103. M. Suzuki, M. Yumoto, M. Kimura, H. Shirai and K. Hanabusa, *Chem. Commun.*, 2002, 884.
104. U. Maitra, S. Mukhopadhyay, A. Sarkar, P. Rao and S. S. Indi, *Angew. Chem. Int. Ed.*, 2001, **40**, 2281.
105. R. Sheehan and P. J. Cragg, 'Computational Techniques (DFT, MM, TD-DFT, PCM)' in *Supramolecular Chemistry: From Molecules to Nanomaterials, Vol. 2 – Techniques*, ed. P. A. Gale and J. W. Steed, John Wiley & Sons, Chichester, 2012, pp. 689–707.
106. D. J. Adams, K. Morris, L. Chen, L. C. Serpell, J. Basca and G. M. Day, *Soft Matter*, 2010, **6**, 4144.
107. T. Ide, D. Takeuchi and K. Osakada, *Chem. Commun.*, 2012, **48**, 278.
108. K. A. Houton, K. L. Morris, L. Chen, M. Schmidtmann, J. T. A. Jones, L. C. Serpell, G. O. Lloyd and D. J. Adams, *Langmuir*, 2012, **28**, 9797.
109. X. Mu, K. M. Eckes, M. M. Nguyen, L. J. Suggs and P. Ren, *Biomacromolecules*, 2012, **13**, 3562.
110. S. E. Paramanov, H.-W. Jun and J. D. Hartgerink, *J. Am. Chem. Soc.*, 2006, **128**, 7291.
111. K. Binder, *Monte Carlo and Molecular Dynamics Simulations in Polymer Science*; Oxford University Press, Inc., New York, 1995.

112. Y. S. Velichko, S. I. Stupp and M. O. de la Cruz, *J. Phys. Chem. B*, 2008, **112**, 2326.
113. O.-S. Lee, V. Cho and G. C. Schatz, *Nano Lett.*, 2012, **12**, 4907.
114. A. Brizard, R. Oda and I. Huc, *Top. Curr. Chem.*, 2005, **256**, 167.
115. A. R. Hirst, V. Castelletto, I. W. Hamley and D. K. Smith, *Chem. Eur. J*, 2007, **13**, 2180.
116. S. Cicchi, G. Ghini, L. Lascialfari, A. Brandi, F. Betti, D. Berti, P. Baglioni, L. Ḍ. Bari, G. Pescitelli, M. Mannini and A. Caneschi, *Soft Matter*, 2010, **6**, 1655.
117. B. Adhikari, J. Nanda and A. Banerjee, *Soft Matter*, 2011, **7**, 8913.
118. N. Berova, K. Nakanishi and R. W. Woody, *Circular Dichroism: Principles and Applications* 2nd edn, Wiley-VCH, Weinheim, 1994.
119. M. A. Mateos-Timoneda, M. Crego-Calama and D. N. Reinhoudt, *Chem. Soc. Rev.*, 2004, **33**, 363.
120. S. R. Nam, H. Y. Lee and J.-I. Hong, *Chem. Eur. J*, 2008, **14**, 6040.
121. R. K. Das, R. Kandanelli, J. Linnanto, K. Bose and U. Maitra, *Langmuir*, 2010, **26**, 16141.
122. A. Kentsis and K. L. B. Borden, *Curr. Prot. Peptide Sci.*, 2004, **5**, 125.
123. P. Jonkheijm, P. van der Schoot, A. P. H. J. Schenning and E. W. Meijer, *Science*, 2006, **313**, 80.
124. F. Oosawa, M. Kasai and S. Asakura, *J. Mol. Biol.*, 1962, **4**, 10.
125. V. Kral, S Pataridis, V. Setnicka, K. Zaruba, M. Urbanova and K. Volka, *Tetrahedron*, 2005, **61**, 5499.
126. V. Setnicka, J. Novy, S. Böhm, N. Sreenivasachary, M. Urbanova and K. Volka, *Langmuir*, 2008, **24**, 7520.
127. P. Iavicoli, H. Xu, L. N. Feldborg, M. Linares, M. Paradinas, S. Stafstrom, C. Ocal, B. L. Nieto-Ortega, J. Casado, J. T. L. Navarrete, R. Lazzaroni, S. De Feyter and D. B. Amabilino, *J. Am. Chem. Soc.*, 2010, **132**, 9350.
128. B. Nieto-Ortega, V. J. Nebot, J. F. Miravet, B. Escuder, J. T. L. Navarrete, J. Casado and F. J. Ramirez, *J. Phys. Chem. Lett.*, 2012, **3**, 2120.
129. K. Hanabusa, M. Matsumoto, M. Kimura, A. Kakehi and H. Shirai, *J. Colloid Interface Sci.*, 2000, **224**, 231.
130. G. Zhu and J. S. Dordick, *Chem. Mater.*, 2006, **18**, 5988.
131. M. Bielejewski, A. Łapiński, R. Luboradzki and J. Tritt-Goc, *Langmuir*, 2009, **25**, 8274.
132. M. A. Rogers and A. G. Marangoni, *Langmuir*, 2009, **25**, 8556.
133. M. Raynal and L. Bouteiller, *Chem. Commun.*, 2011, **47**, 8271.
134. K. S. Partridge, D. K. Smith, G. M. Dykes and P. T. McGrail, *Chem. Commun.*, 2001, 319.
135. W. Edwards, C. A. Lagadec and D. K. Smith, *Soft Matter*, 2011, **7**, 110.
136. J. G. Hardy, A. R. Hirst and D. K. Smith, *Soft Matter*, 2012, **8**, 3399.
137. M. L. Muro-Small, J. Chen and A. J. McNeil, *Langmuir*, 2011, **27**, 13248.
138. D. K. Smith, *Tetrahedron*, 2007, **63**, 7283.

139. S. Cantekin, Y. Nakano, J. C. Everts, P. van der Schoot, E. W. Meijer and A. R. A. Palmans, *Chem. Commun.*, 2012, **48**, 3803.
140. T. Pinault, B. Isare and L. Boutellier, *ChemPhysChem*, 2006, **7**, 816.
141. M. Mukai, H. Minamikawa, M. Aoyagi, M. Asakawa, T. Shimizu and M. Kogiso, *Soft Matter*, 2012, **8**, 11979.
142. M. Avrami, *J. Chem. Phys.*, 1939, **7**, 103.
143. M. Avrami, *J. Chem. Phys.*, 1940, **8**, 212.
144. M. Avrami, *J. Chem. Phys.*, 1941, **9**, 177.
145. X. Huang, P. Terech, S. R. Raghavan and R. G. Weiss, *J. Am. Chem. Soc.*, 2005, **127**, 4336.
146. X. Y. Liu and P. D. Sawant, *Adv. Mater.*, 2002, **14**, 421.
147. X. Huang, S. R. Raghavan, P. Terech and R. G. Weiss, *J. Am. Chem. Soc.*, 2006, **128**, 15341.
148. E. Dickinson, *J. Chem. Soc., Faraday Trans.*, 1997, **93**, 111.
149. A. R. Hirst, B. Escuder, J. F. Miravet and D. K. Smith, *Angew. Chem. Int. Ed.*, 2008, **47**, 8002.

CHAPTER 3

Molecular Gels Responsive to Physical and Chemical Stimuli

MING XIONG,[a] CHENG WANG,*[a] GUANXIN ZHANG[b] AND DEQING ZHANG*[b]

[a] College of Chemistry and Molecular Sciences, Wuhan University, Wuhan 430072, China; [b] Beijing National Laboratory for Molecular Sciences, Organic Solids Laboratory, Institute of Chemistry, Chinese Academy of Sciences, Beijing 100190, China
*Email: dqzhang@iccas.ac.cn; chengwang@whu.edu.cn

3.1 Introduction

Apart from studies on the rational design of LMWGs, extensive efforts have been made to prepare stimuli-responsive molecular gels.[1-3] The gel–solution transition is thermally reversible for molecular gels. Besides heating and cooling the gel–solution transition can also be induced by other physical and chemical stimuli for gels with appropriate LMWGs. The physicochemical properties are varied accompanying the gel–solution transition (Figure 3.1). These stimuli-responsive molecular gels are promising soft materials for a diverse range of applications.

In principle, external stimuli can be classified (Figure 3.1) as physical and chemical ones. Heat, mechanical forces, ultrasound waves, and UV-vis light can be included as physical stimuli, whereas acid–base reagents, salts with appropriate anions and cations, neutral molecules, redox reagents and enzymes can be used as chemical stimuli. In this chapter, we will highlight recent progress on molecular gels responsive to physical and chemical stimuli. First, the

RSC Soft Matter No. 1
Functional Molecular Gels
Edited by Beatriu Escuder and Juan F. Miravet
© The Royal Society of Chemistry 2014
Published by the Royal Society of Chemistry, www.rsc.org

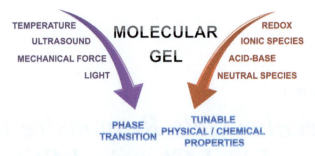

Figure 3.1   The illustration of stimuli-responsive molecular gel.

physical-stimuli responsive gels are discussed, followed by the introduction of chemical-stimuli responsive gels. We will describe representative examples of stimuli-responsive gels based on LMWGs with responsive groups to demonstrate the molecular design principle. Moreover, the potential applications of these stimuli-responsive molecular gels are also discussed. However, we would exclude enzymes-triggered gels, which will be treated in Chapter 4.

3.2   Physical-Stimuli-Responsive Gels

Molecular gels responding to physical stimuli including temperature, mechanical stress, ultrasound and light have been reported in recent years. In this section, we will introduce representative examples of physical stimuli-responsive gels and highlight how the properties of molecular gels can be tuned after applying physical stimuli.

3.2.1   Temperature-Responsive Gels

Since the association of the network of the supramolecular gel system is usually enthalpy driven, an increase in temperature will shift the equilibrium to the nonaggregated state, leading to a gel–solution phase transition. The temperature at which the gel turns into solution is called the gelation temperature, T_{gel}. Therefore, temperature responsiveness is intrinsic to molecular gels. Apart from such a gel–solution transition, unusual temperature responsiveness has also been disclosed for molecular gels.

Hamachi and coworkers reported[4] a novel hydrogelator **1** (Figure 3.2) based on glycosylated amino acid. The hydrogel behaves remarkably different with normal heat-set gels, as it shows dramatic shrinking and swelling behavior in response to temperature changes (Figure 3.3) instead of undergoing a gel–solution phase transition. Furthermore, this gel can be used to release various water-soluble drugs and related molecules trapped in the gel matrix in a thermally controlled manner. For instance, more than 90% of entrapped DNA can be released from the gel by increasing temperature, but hydrophobic substances still remain entrapped within the gel.

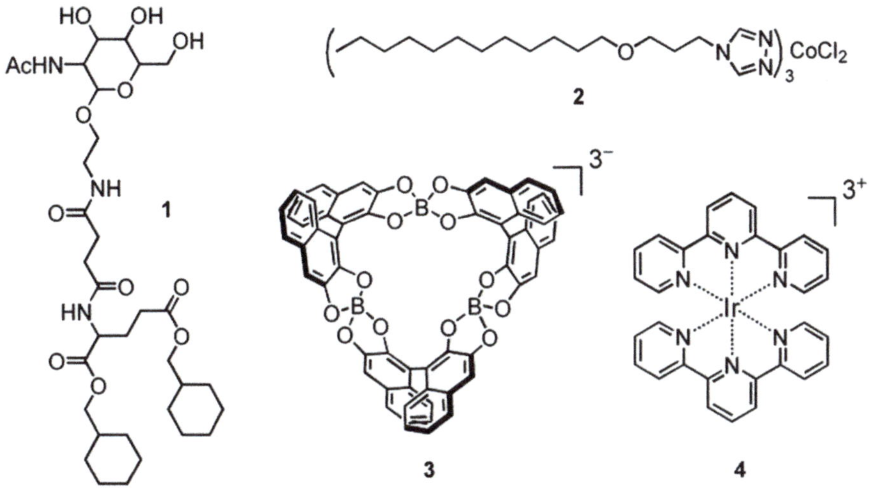

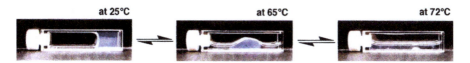

Figure 3.2   Molecular structures of compounds **1–4**.

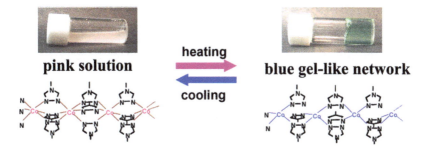

Figure 3.3   Shrinking and swelling behavior of hydrogel **1** in response to temperature changes.
Reprinted with permission from ref. 4. Copyright (2002) American Chemical Society.

Figure 3.4   Thermally induced gelation of **2** and schematic molecular packing of the corresponding polymeric tetrahedral and octahedral complexes.
Reprinted with permission from ref. 5. Copyright (2004) American Chemical Society.

Kimizuka and coworkers described[5] a class of gel-like networks formed from the lipophilic Co(II)-1,2,4-triazole complexes **2** (Figure 3.2). This system shows a complete reversal of the behavior of all other reported gels for the first time (Figure 3.4). When LMWG **2** is dissolved in chloroform, a blue gel-like phase is formed at room temperature. However, the gel turns into a pale pink solution on cooling to 0 °C rather than heating. This gel–solution transition is completely reversible compared to normal molecular gels. This is attributed to the flexible coordination geometry of Co(II), which is transformed from the tetrahedral polymeric aggregate upon gelation into the octahedral complex of cobalt (II) in solution.

Yamaguchi and coworkers reported[6] another unusual example of temperature responsive gels. Bilateral cyclophanes **3** (Figure 3.2), which have wide cavities on both sides of the symmetry plane, could form host–guest complexes with **4** *via* π–π stacking and electrostatic interactions. Interestingly, when a solution of **3** and **4** in N,N,N′,N′,N″,N″-hexamethylphosphoric triamide (HMPA) was heated above 79 °C, a phase transition occurred and an organogel was obtained. Upon cooling to ambient temperature, the gel reverted to the solution within 15 min.

3.2.2 Mechanical-Responsive Gels

One of the most prominent features of gels is their viscoelastic properties because of the incorporation of a large amount of solvent. As such, most of physical gel systems are sensitive to the mechanical stress. If the mechanical stress is applied, a gel can be deformed or destroyed, depending on the magnitude of the applied stress. For most systems, the gel state can only be restored by a heating–cooling cycle. However, some of the gels are thixotropic, which means that the gel state is spontaneously restored if the stress is removed. For instance, van Esch and coworkers reported[7] the rheological properties of the cyclohexane bisureas **5** (Figure 3.5) in various primary alcohols. The gel disintegrates above the yield stress to give a viscous fluid of submillimeter gel particles. Interestingly, the gel state can be restored after the stress is removed and the system is left at rest. Even more remarkable is that during steady-shear viscosity measurements, the viscosity increases faster if higher shear rates are applied.

Fang and coworkers reported[8] a thixotropic gel system for cholesterol-based gelator **6** (Figure 3.5). Gelation tests demonstrate that LWMG **6** can

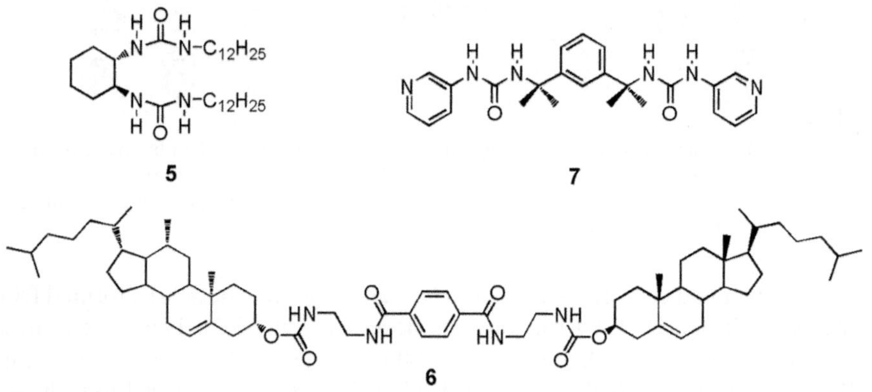

Figure 3.5   Molecular structures of compounds 5–7.

spontaneously form organogels in various solvents at room temperature. More importantly, after shaking the gels, a phase transition from a gel state to a solution state occurs. Once the shaking is stopped, the gel state recovers.

Alternatively, the shaking can also induce gelation. Steed and coworkers reported[9] the unique behavior of the pyridyl-urea **7**/CuBr$_2$ system recently. The complex system can form (Figure 3.6) unstable organogel in MeOH. However, after a rapid shaking of the samples by hand, a transformation from these weak gel-like materials to robust gels occurs. They hypothesized that the firm mechanical shaking of this dynamic system results in a significant increase of the number and connectivity of the nodes within the gel structure.

3.2.3 Ultrasound-Induced Gels

Thermotreatment is considered as a common method for the formation of a physical gel. However, it is not always a necessary one. In recent years, some gelation, which cannot occur under ordinary heating/cooling process, can be triggered[10] by ultrasound. The first ultrasound induced gel was reported[11] by Naota and Koori. The dinuclear palladium(II) complex **8** (Figure 3.7), which is stabilized by intramolecular π-stacking interactions, can instantly gelatinize (Figure 3.8) a variety of organic solvents upon a very brief presonication. Gels thus obtained were thermoreversible and solutions were obtained upon heating at above T_{gel}. Uniquely, the duration of sonication allows control of the gel strength. It should be mentioned that other external stimuli, such as vigorous

Figure 3.6   Shear-induced gelation was observed for **7** in MeOH in the presence of 0.1, 0.2, 0.3, 0.4, and 0.5 equivalents of CuBr$_2$ (from left to right in each picture).
Reproduced from ref. 9.

8   **9**

Figure 3.7   Molecular structures of compounds **8** and **9**.

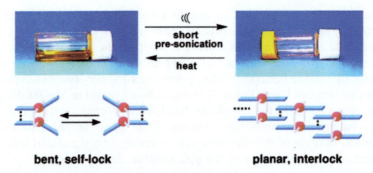

bent, self-lock   **planar, interlock**

Figure 3.8   Illustration of ultrasound-induced gelation of **8** with schematic molecular
packing.
Reprinted with permission from ref. 11. Copyright (2005) American
Chemical Society.

shaking, quick heating/cooling, and microwave irradiation, were unable to
induce gelation.

They also reported[12] another ultrasound-induced gelation system based on
palladium-bound dipeptides. Intramolecular H-bonding involving the chloro
ligand prevents this dipeptide from intermolecular self-assembly. As such, after
cooling of the hot solution, no gel formed. Ultrasound irradiation releases this
self-lock and induces the formation of semistable initial aggregates that further
assemble into gels. More importantly, increasing the sonication time will increase the concentration of β-sheet type dipeptide aggregates, thus accelerating
gelation and generating higher-order nanostructures with heat-resistant
properties.

In another example, Ratcliffe and coworkers described[13] a new gelation
mechanism based on dipeptide **9** (Figure 3.7). After cooling the hot homogeneous alkane solutions of dipeptide **9**, gelation does not occur. However, a
complete gel is formed after sonication the solution at room temperature for a
few minutes. More importantly, ultrasound modifies the morphology of the
material from sheet-like particles into three-dimensional networks of fibers or
ribbons, which are responsible for gelation.

3.2.4   Light-Responsive Gels

Light is one of the ideal stimuli due to its clean, fast, and controllable features.
As such, photoresponsive gels have gained much attention in the last decade.
By incorporation of a photochromic moiety into the LMWGs, it is reasonable
to design photoresponsive gel systems. Simultaneously, the gel properties can
be tuned after light irradiation by photochemical reactions that will transform
the gelator molecules into different species.

Azobenzene moieties are known for their photoresponsive properties and
have been incorporated (Figure 3.9) into LMWGs to generate photoresponsive
gels. Shinkai and coworkers initially designed[14] LMWG **10** bearing an

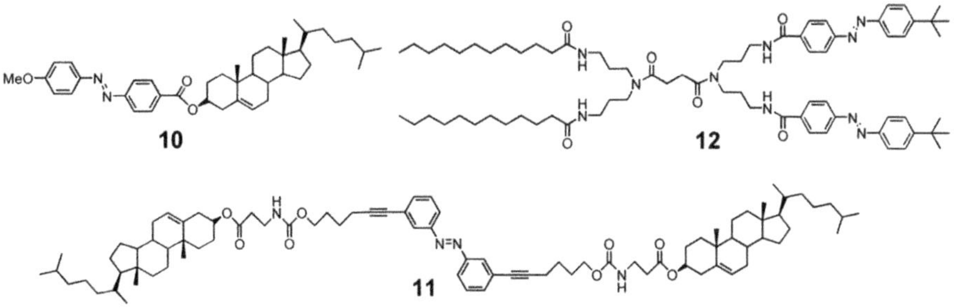

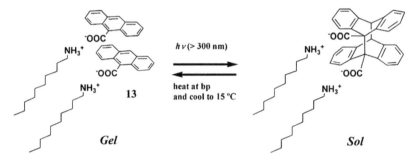

Figure 3.9 Molecular structures of compounds **10**–**12**.

Figure 3.10 Illustration of gel–solution transition based on the photodimerization and dissociation of **13** under UV-light irradiation and heating.

azobenzene segment. The gel–solution transition for the corresponding organogels can be reversibly tuned by light irradiation as a result of *trans-cis* isomerization of the azobenzene unit. Later, Koumura and coworkers reported[15] another azobenzene-based compound **11** with two urethane moieties linked to two cholesteryl ester units. After photoirradiation, the gel–solution transition can also occur. In these systems, it usually takes a long time to complete the phase transition. However, Kim and coworkers reported[16] a gelator **12** consisting of an azobenzene dendron that could undergo a rapid gel–solution transition. The collapse of the gel occurred in 2.0 min upon UV-light irradiation and gel regeneration was achieved by subsequent irradiation of visible light in just 5.0 s.

Photodimerization of anthracene may occur when intermolecular arrangements are appropriate. Shinkai and coworkers reported[17] the binary gelator **13** composed of alkylammonium and anthracene-9-carboxylate. The gels formed in cyclohexane were converted (Figure 3.10) to the solutions after UV-light irradiation because of the generation of photodimers. The gel phase can be reformed by heating the solution and further cooling. This finding suggests that the photoinduced monomer–dimer transformation eventually destroys the gel superstructure. However, the monomer–dimer conversions occur with side products and as a result multiple cycles of photoinduced gel–solution phase transitions cannot be achieved.

Due to their unique photochemistry, stilbenes have also been incorporated into LMWGs to generate photoresponsive gels. As an example, compound **14** (Figure 3.11) is composed of a *cis*-stilbene group and one oxamide moiety,[18] which can be dissolved in ethanol. Interestingly, the clear solution of **14** turned into an opaque gel after light irradiation because of the photoisomerization of *cis*-stilbenes to *trans*-stilbenes. However, further light irradiation of *trans*-isomer cannot regenerate the *cis*-isomer. Thus, the gel–solution transition cannot be induced by further light irradiation.

Dithienylethene have been extensively investigated as a photochromic switch for its excellent fatigue resistance and thermal stability. Gelators containing dithienylethene units have also been reported.[19,20] For example, Huang and coworkers reported[21] LWMG **15** (Figure 3.12) containing a dithienylethene moiety. After UV-light irradiation, the pale-brown gel was converted to a blue gel and the fluorescence of the gel was efficiently quenched. Since only small geometrical changes upon photochemical conversion of the open-form isomer to the closed-form isomer, the gel–solution transition could not be triggered by light irradiation. Further visible-light irradiation led to the reformation of the pale-brown gel.

Spiropyrans (SP) are another type of photochromic molecules and the respective closed form and open form (merocyanine form) can be reversibly interconverted by photoirradiation. Zhang and coworkers reported[22] a photoresponsive hydrogelator **16** (Figure 3.12) featuring spiropyran moiety. After visible-light irradiation, the gel is transformed into solution and the gel phase is

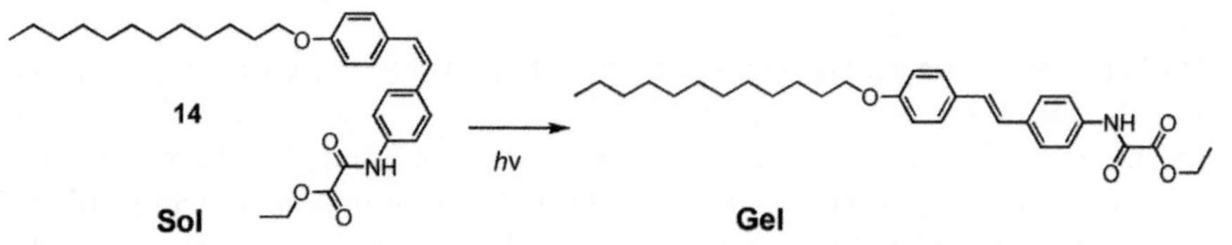

Figure 3.11 Photoinduced gelation of **14** under UV-light irradiation.

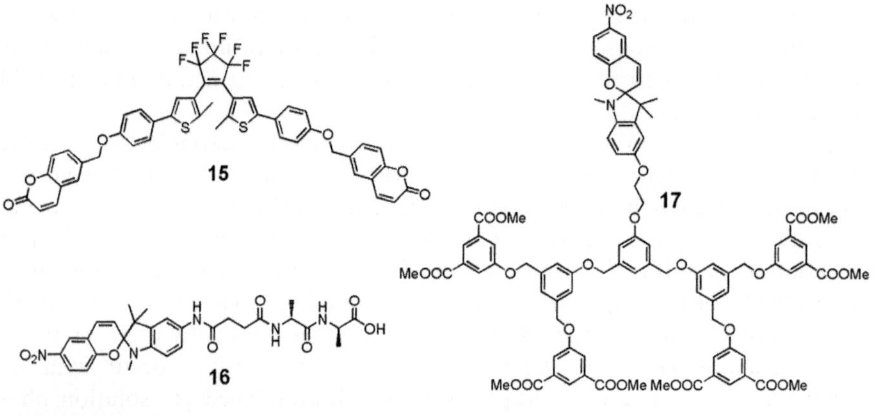

Figure 3.12 Molecular structures of compounds **15–17**.

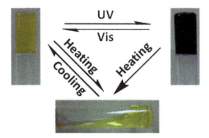

Figure 3.13   Illustration of the gel-to-gel transformation of **17** after heating, cooling, UV and visible-light irradiations.
Reprinted with permission from ref. 23. Copyright (2010) John Wiley & Sons, Inc.

reformed again after further UV irradiation. This light-triggered gel–solution transition is due to the fact that merocyanine moiety has a strong tendency to form π–π stacking in comparison with closed form.

Some of us have reported[23] a spiropyran-functionalized dendron **17** (Figure 3.12) that can form organogels in toluene and benzene (Figure 3.13). Absorption spectral studies clearly demonstrate that the photochromic reaction of the spiropyran unit in **17** can take place in both solution and gel phase. However, after UV-light irradiation, the gel phase is not destroyed, but the yellow gel is transformed into the purple-blue gel. It is very interesting to find that the purple-blue gel exhibits relatively strong red fluorescence. Further visible-light irradiation of the purple-blue gel leads to the yellow gel again, which is nonfluorescent.

3.3   Chemical Stimuli-Responsive Molecular Gels

A number of molecular gels responding to chemical stimuli have been disclosed in recent years. These chemical stimuli-responsive gels can be categorized as follows:

- pH-responsive gels;
- anion-responsive gels;
- metal-ion-responsive gels;
- redox-responsive gels;
- gels responding to neutral chemical species.

Furthermore, the combination of several different stimuli may lead to multiresponsive gels. In the following, we will present and discuss examples of each of these chemical stimuli-responsive gels.

3.3.1   pH-Responsive Gels

The pH-sensitive materials are interesting for applications in drug delivery and biomedical systems. As such, pH-responsive gels, which can be generated by

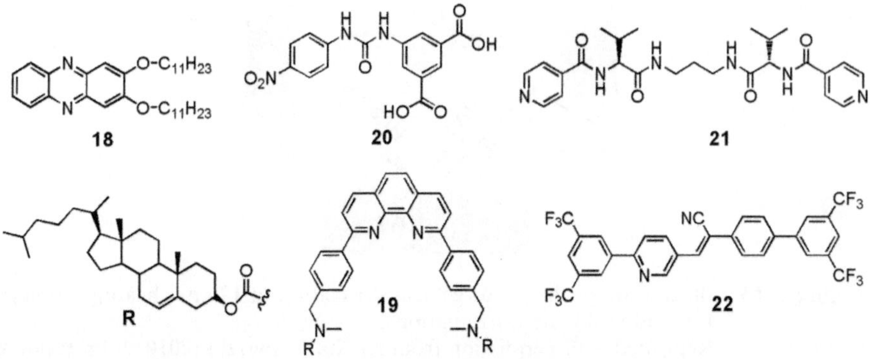

Figure 3.14   Molecular structures of compounds **18–22**.

incorporation acidic/basic moieties into LMWGs, have been studied intensively. For instance, Pozzo and coworkers reported[24] LWMG **18** (Figure 3.14) entailing a 2,3-di-N-alkoxyphenazine group. The protonation of the basic sites of the phenazine moiety was exploited to make pH-responsive gels. Addition of trifluoroacetic acid to gels led to monoprotonation and the gelling stability was enhanced. The reversibility of the process was demonstrated by the recovery of the initial thermal stability of the gel when ammonia was bubbled through the system.

Shinkai and coworkers reported[25] the 1,10-phenanthroline-appended cholesterol-based gelator **19** (Figure 3.14). It can gelate alcohols, polar aprotic solvents, organic acids, and triethylamine. Upon addition of acetic acid, the gel phases are still kept. However, the emission color of the gel is changed from purple to yellow. This result shows that the protonation of the 1,10-phenantroline in the gel state has a big influence on the emissive property of the gel phase. Furthermore, the emission intensity of **19•H$^+$** is particularly strong in the gel phase.

The gel–solution transition can also be regulated by pH changes. Hayes and coworkers reported[26] a pH-responsive hydrogelator **20** (Figure 3.14) that is derived from isophthalic acid and contains a urea functional group. The gelator **20** is completely soluble in basic media; however, after the addition of HCl solution, it can be converted into a hydrogel. On the other hand, Miravet and Escuder reported[27] another pH-responsive hydrogelator **21** (Figure 3.14) bearing a pyridine group. Gelator **21** is soluble in acidic water (HCl, 0.1 M), but it forms a transparent hydrogel after exposure to concentrated ammonia vapors in a few minutes. Moreover, the gelation is favored with increasing ionic strength of the solution (NaCl, 0.1 M).

Park and coworkers reported[28] a highly fluorescent organogel system based on a functional molecule with aggregation-induced enhanced emission (AIEE) properties. The gelator **22** (Figure 3.14) bearing a pyridine unit can form gels in several nonpolar and polar protic solvents. As a result of the presence of the pyridine unit, the highly fluorescent gel changes (Figure 3.15) to nonfluorescent solution after the addition of a photoacid generator and light irradiation.

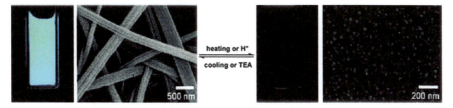

Figure 3.15 Photographs and SEM images of the reversible phase transition between gel (left) and protonated solution state (right) of **22** in 1,2- dichloroethane.
Reprinted with permission from ref. 28. Copyright (2005) American Chemical Society.

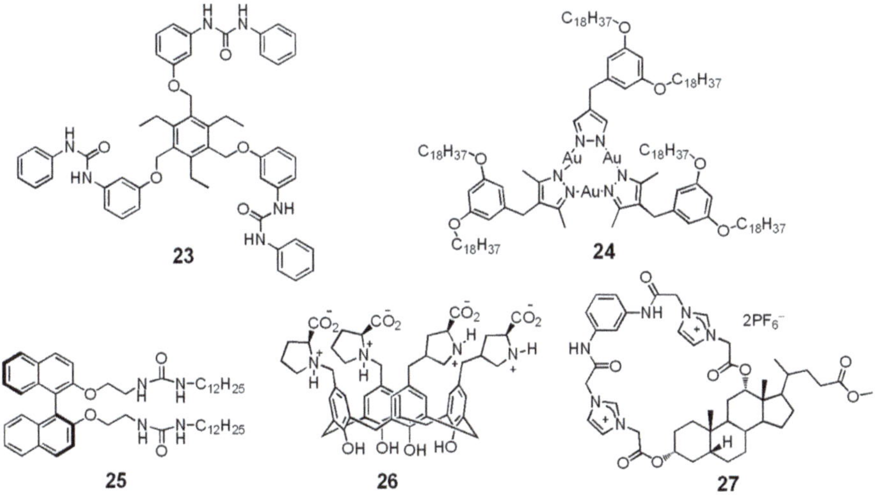

Figure 3.16 Molecular structures of compounds **23**–**27**.

Moreover, they demonstrated selective spatial patterning by using the gel–solution transition in this organogel system.

3.3.2 Anion-Responsive Gels

The field of anion binding by supramolecular hosts is well established and reasonably well understood. As a result, there is a rapidly growing interest in anion-responsive gels.[29,30] Examples of anion-triggered gel–solution transitions as well as of the opposite situations, where the presence of an anion leads to a gel formation, are known.

In urea-based gels, urea groups tend to form intermolecular H-bonds leading to extended structures. As such, the competition between the anion–gelator binding and the self-assembly of gelator molecules can be used to tune the gel properties. For example, LWMG **23** (Figure 3.16) contains multiple urea moieties forms gels in polar solvents.[31] The gels in acetone of **23** respond to

Figure 3.17 Photographs of acetone gel of **23** showing the transitions between gel state and liquid state upon addition of chemical stimuli.
Reprinted with permission from ref. 31. Copyright © 2007, Elsevier.

chemical stimuli in the form of anions, resulting in homogeneous solutions (Figure 3.17). As a result of the existence of three urea units, three equivalents of the respective anions are required for the complete gel–solution transition. Reversibility of the dissolution of the gel is demonstrated by the addition of the Lewis acid $BF_3 \cdot OEt_2$ to a solution of **23** containing tetrabutylammonium fluoride (TBAF).

We reported[32] a chiral LMWG **24** (Figure 3.16) based on the axially chiral binaphthalene with two urea moieties. LMWG **24** can form a transparent gel in cyclohexane. Interestingly, after gel formation, modulation of the CD spectrum of **24** was observed that could be reversed by alternate heating and cooling. Furthermore, the gel phase could be destroyed by addition of F^- due to the disruption of intermolecular H-bonds.

The ability of chloride anion to remove silver ion has been utilized in gels formed[33] by trinuclear Au(I) pyrazolate complex **25** (Figure 3.16). LMWG **25** forms a red-luminescent organogel in hexane *via* a Au(I)–Au(I) metallophilic interaction. On addition of a small amount of Ag^+, the gel becomes blue emissive without disruption of the gel. After removing doped Ag^+ with cetyl-trimethylammonium chloride, the original red-luminescent gel is completely restored. This study demonstrates the application of anions for tuning the physical property of molecular gels.

Alternatively, the presence of an anion can also lead to a gel formation. For instance, Ogden and coworkers reported[34] a calix[4]arene-based compound **26** (Figure 3.16) that forms hydrogel only in the presence of specific anions, such as NO_3^- and Br^-. Tripathi and Pandey reported[35] another anion-triggered gel-formation system based on bile acid derivative **27** (Figure 3.16) with imidazolium and amide moieties. In solution, compound **27** shows a high selectivity and affinity for Cl^-. However, it can form a stable gel in a $CHCl_3/DMSO$ mixture in the presence of anions. It is found that among the tested anions (F^-, Cl^-, Br^-, AcO^-, $H_2PO_4^-$, and HSO_4^-), only the HSO_4^- was effective in inducing the gelation. The system can thus be utilized for the selective naked-eye detection of HSO_4^-.

3.3.3 Metal-Ion-Responsive Gels

Metal-ion-responsive gels are attractive for smart materials. Thus, the use of metal cations to tune gel properties has received increasing interest.[29] Shinkai

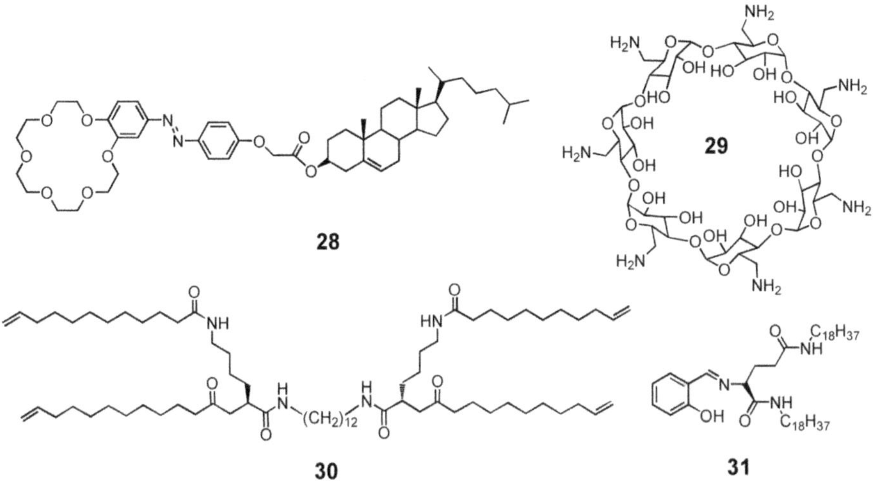

Figure 3.18 Molecular structures of compounds **28–31**.

and coworkers reported one of the early examples for which cation binding has an influence on gel rheology. LWMG **28** (Figure 3.18) containing a benzocrown moiety and an azobenzene linker can gel[36] a selection of hydrocarbons and alcohols. As the cations can bind the crownether fragment, the gel stabilities can be tuned by introducing certain cations to the systems. In the presence of Li^+, Na^+, K^+, Rb^+, and NH_4^+, T_{gel} of the gel first increases and then gradually decreases on increasing concentration of metal ions. On the other hand, Cs^+ lowers the T_{gel} and inhibits gelation completely at higher concentration. They assume that the formation of a 1:2 metal cation-crown sandwich complex may disrupt the helical stacking and lead to gel–solution transition.

Deng and Thompson reported[37] cation-responsive supramolecular gels formed with the amine-modified cyclodextrins **29** (Figure 3.18) in DMSO. Direct cooling of hot solution enables the gel formation over a few days. In order to accelerate the gelation process, a gel is yielded after sonication for 5.0 min. The gel responds quickly to the addition of metal ions (Co^{3+}, Ni^{2+}, Cu^{2+} and Ag^+), namely, the gel–solution transition occurs quickly upon addition of these metal ions. This is presumably due to the coordination of amine groups in **29** with metal ions. Interestingly, the gel can be restored again when the metal ions are precipitated out. In comparison, the addition of Na^+, K^+ and Ca^{2+} cannot induce the gel–solution transition.

Edwards and Smith reported[38] the first example of the Ag^+–alkene interaction as the driving force for the gel–solution transition. The gelator **30** (Figure 3.18) with L-lysine head groups and peripheral alkenes could form organogels (Figure 3.19) in ethyl acetate. The gels respond to Ag^+ and Li^+ by undergoing a gel–solution transition. However, Na^+ or K^+ cannot induce the gel–solution transition. Moreover, the gel–solution transition induced by Ag^+ is much faster than that induced by Li^+. Thus, Ag^+–alkene interactions play an important role in mediating the gel–solution transition.

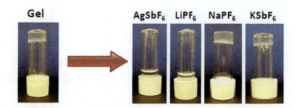

Figure 3.19   Response of gels of **30** in ethyl acetate to solutions of Ag^+ and Li^+. Reproduced from ref. 38.

Apart from tuning rheological properties, other physical properties of gels can also be modulated after addition of metal ions. Liu and coworkers reported[39] an amphiphilic gelator **31** (Figure 3.18) entailing both Schiff base and L-glutamide moieties. LMWG **31** can form thermotropic gels in many organic solvents. Interestingly, the introduction of Cu^{2+} ions causes a chiral twist in the nanofiber, whilst the addition of Mg^{2+} ions leads to a significant enhancement of the fluorescence of the gel. More importantly, the presence of Mg^{2+} enables the gel to be able to exhibit chiral recognition towards chiral molecules such as tartaric acid.

3.3.4   Redox-Responsive Gels

The redox stimulus is important for the construction of the electromechanical soft materials such as artificial muscles and electrorheological fluids. In this respect, redox-responsive gels have attracted wide interest. Functional groups that can undergo reversible oxidation–reduction reactions have been incorporated into LMWGs to render the gel responding to redox reactions.

It is known that tetrathiafulvalene (TTF) and its derivatives can be reversibly transformed into the respective radical cation ($TTF^{\bullet+}$) and dication (TTF^{2+}) states by either chemical or electrochemical redox reactions. Conversion of TTF group to either $TTF^{\bullet+}$ or TTF^{2+} would largely modulate the interaction of adjacent TTF units, therefore, the gel may be destroyed. By taking advantage of this unique feature, we and other groups have reported[40–43] the TTF-based gelators and the resulting gels that show responsiveness to redox reactions. For instance, we reported[44] TTF-derived LMWG **32** (Figure 3.20), which can gel (Figure 3.21) several solvents including cyclohexane and 1,2-dichloroethane. Either addition of Fe^{3+} or application of an oxidation potential to the gel can lead to gel–solution transition. Interestingly, the gel state can be restored by electrochemical reduction, followed by heating and cooling (Figure 3.21). Alternatively, the gel phase of LMWG **32** can also be modulated by reaction with tetracyanoquinodimethane (TCNQ), which is a strong electron acceptor. Addition of TCNQ to the gel of **32** from 1,2-dichloroethane results in the destruction of the gel state, but the reaction of **32** with TCNQ in cyclohexane leads to a dark-green gel. This may be attributed to the fact that 1,2-dichloroethane is more polar than cyclohexane and as a result, the TTF unit in LMWG **32** interacts more strongly with TCNQ in 1,2-dichloroethane.

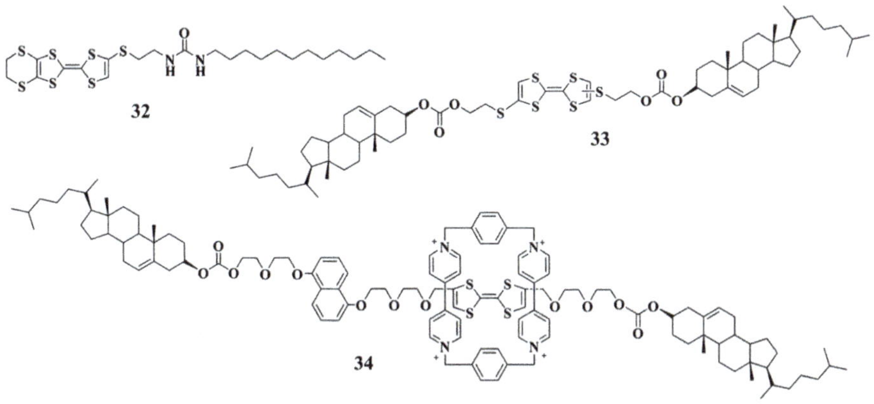

Figure 3.20 Molecular structures of compounds **32–34**.

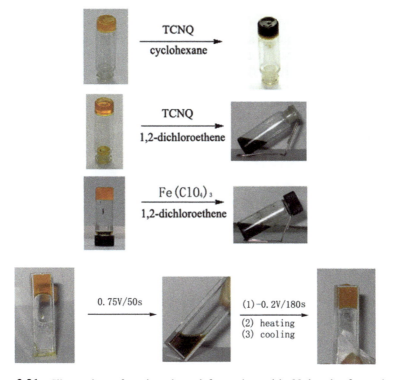

Figure 3.21 Illustration of tuning the gel formation with **32** by the formation of charge-transfer complexes and chemical or electrochemical oxidation and reduction.
Reprinted with permission from ref. 44. Copyright (2005) American Chemical Society.

We also reported another TTF-based gelator **33** and the gel formation could be tuned by addition of iodine.

Stoddart and coworkers built[45–47] a fascinating chemistry of TTF-based rotaxanes/catenanes that can be incorporated into molecular electronics devices. They reported[48] a rotaxane **34** (Figure 3.20) with TTF and 1,5-dioxynaphthalene recognition units situated in the rod and with cholesterol units as stoppers, and investigated its gelation behavior and switching properties. LMWG **34** can form an organogel in a mixture of CH_2Cl_2/ MeOH (3 : 2, v/v). Moreover, the organogel can be transformed into the solution after oxidation with $Fe(ClO_4)_3$. This is likely due to the oxidation of the TTF unit and in turn the cyclobis(paraquat-*p*-phenylene) ($CBPQT^{4+}$) ring is switched from the TTF station to 1,5-dioxynaphthalene station. Accordingly, the translational motion of the $CBPQT^{4+}$ ring affects the bulk properties of the aggregate.

Apart from TTF, there are other electroactive units such as ferrocene and thiophene, which can also be reversibly transformed into the respective cations by either chemical or electrochemical stimuli. Thus, the corresponding gels with these electroactive segments may also show responsiveness to redox reactions. For instance, Fang and coworkers studied[49] LMWG **35** (Figure 3.22) bearing a ferrocene segment that can gel cyclohexane. When the ferrocenyl group was oxidized with $(NH_4)_2Ce(NO_3)_6$, the gel gradually turned into a dark-green suspension. Furthermore, after addition of hydrazine as a reducing agent, the gel can be restored and a heating-free solution–gel phase transition is achieved.

Shinkai and coworkers reported[50] a series of redox-stimuli-responsive gels based on thiophene units. For instance, LMWG **36** (Figure 3.22) entails a redox-active sexithiophene segment can be used as excellent organogelator for various organic solvents. Addition of $FeCl_3$ as an oxidizing reagent leads to the transformation of the gels into the respective solutions. The EPR studies clearly

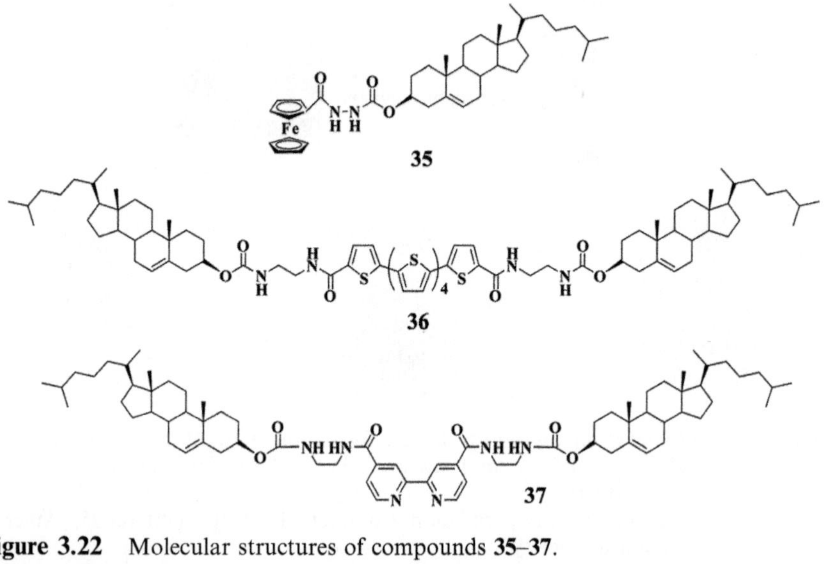

Figure 3.22 Molecular structures of compounds **35–37**.

indicate the formation of radical cations of the sexithiophene segment after introducing $FeCl_3$. Furthermore, the gel state can be restored by addition of ascorbic acid, which can reduce the radical cation of sexithiophene segment to the neutral species. As expected, such a gel–solution transition can also be implemented by electrochemical redox reactions.

By taking advantage of the redox switching behavior of Cu(I)/Cu(II) complexes, Shinkai and coworkers reported[51] another example of a reversible gel–solution phase transition that can be induced by a redox stimulus. The coordination complex of Cu(I) with **37** (Figure 3.22) can gelate some organic solvents. The gels can be destroyed after transformation of Cu(I) complex to the respective Cu(II) complex. Interestingly, the gel state can be reformed by further reduction of Cu(II) to Cu(I).

Interconversion between thiols and disulfides is an important reaction in many biological processes. The thiols can reversibly convert to disulfide bond by exposure to various oxidation agents. McNeil and coworkers reported[52] the use of an oxidative stimulus to convert soluble thiol **38** into gelator **39** by the construction of a disulfide bond (Figure 3.23). When TATP (triacetonetriperoxide) is added to a mixture of **38** and TsOH in MeOH, a stable gel is formed within 30 min. The authors further optimized the system for the potential application as a portable naked-eye detector for peroxide-based explosive TATP.

3.3.5 Gels Responding to Neutral Chemical Species

The exposure of gels or solutions to neutral chemical species may tune the gel–solution transition in two ways: i) the neutral chemical species will react with LMWGs and thus gels will be transformed into the respective solutions; ii) gels are formed after activation of pro-gelators by reactions with certain neutral chemical species.

Miravet and Escuder reported[53,54] one of the first examples of reactive molecular gels. Amino acid derived LMWG **40** (Figure 3.24) can form a weak gel in acetonitrile with T_{gel} below 50 °C. The LMWG **40**, bearing *p*-nitrophenyl carbamate fragments, can easily react with amines in a diffusion-controlled process to give bisureas. After the reaction of **40** with alkylamine, the resulting gel[53] has a different rheological behavior with T_{gel} above 80 °C. However, after

Figure 3.23 Illustration of interconversion of compounds **38** and **39**.

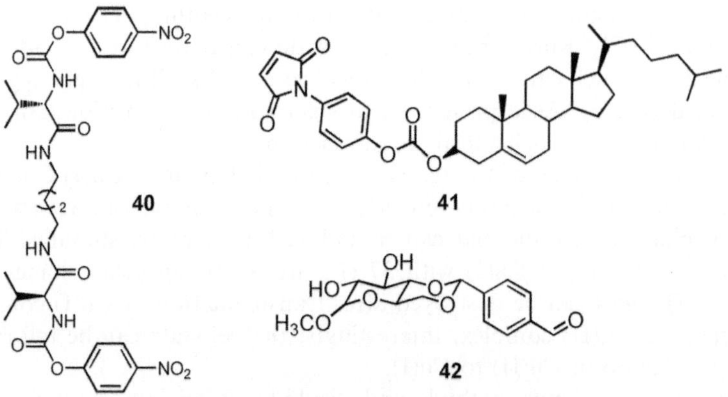

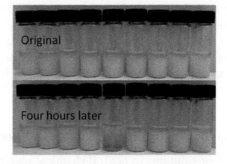

Figure 3.24   Molecular structures of compounds **40–42**.

Figure 3.25   Hydrogels with **42** covered with the aqueous solutions of different amino
acids (from left to right: aspartic acid, lysine, alanine, serine, cysteine,
histidine, arginine, proline, and phenylalanine).
Reprinted with permission from ref. 56. Copyright (2010) American
Chemical Society.

the addition of (1*R*,2*R*)-(+)-1,2-diphenyl- 1,2-ethanediamine, the gel is completely disassembled[54] and gel–solution transition occurs.

   Some of us have reported[55] a cholesterol-based gelator **41** (Figure 3.24) with
maleimide unit that can form organogels in several solvents. It is interesting to
note that the gels can be gradually transformed into the solution after addition
of n-hexylthiol and triethylamine. Such a gel–solution transition is attributed to
the Michael reaction between the maleimide unit in **41** and thiol under basic
conditions. However, if n-hexylthiol is added alone, the gel–solution transition
cannot occur even after several days. This is probably because the Michael
reaction between maleimide and thiol is slow in the absence of bases.

   Some of us have also reported[56] a saccharide-derived hydrogelator **42**
(Figure 3.24) containing an aldehyde group. Figure 3.25 displays the variation
of hydrogels of **42** after the addition of aqueous solution of amino acids. The
hydrogel containing cysteine was completely transformed into solution,
whereas the others were unaffected. Such selective responsiveness of the

hydrogel toward cysteine should be related to the specific reaction of aldehyde with cysteine that will form a thiazolidine derivative. The aldehyde group cannot react with other amino acids including the acidic and basic ones under the same conditions.

Hanabusa and coworkers reported[57] the *in situ* gelation at room temperature. By simply mixing solutions of highly reactive isocyanates and alkylamines, urea-based gels (Figure 3.26) are formed spontaneously within seconds. In these systems, the formation of the gel phase is the output for the presence of a chemical species that can be utilized for the naked-eye detection of certain chemicals. It provides a unique approach to produce molecular gels for which gelators are generated *in situ*.

Alkylamines are generally nongelators or inefficient ones. However, George and Weiss demonstrated[58,59] that the amines can react with carbon dioxide to form ammonium carbamates, which are very potent gelators for several solvents – the gels can be formed within 30 s after the introduction of CO_2. Interestingly, mild heating of the gels can revert the ammonium carbamates to the corresponding amines and CO_2, and the gel will transform to the solution. In this manner, reversible formation of organogels is realized by employing "latent" gelators (amines) and CO_2.

Alternatively, the properties of molecular gels can be tuned after addition of neutral chemical species *via* noncovalent intermolecular interactions. Several reports indicate that it is possible to tune the gel stabilities by introducing certain molecules to the systems. For instance, Shinkai and coworkers reported[60] a Zn(II) porphyrin-based gelators **43** (Figure 3.27) with cholesterol segments and the gelation ability can be tuned by addition of C_{60}. LMWG **43** can form organogels in various aromatic liquids. Interestingly, thermal stabilities of the gels gradually increase after addition of more C_{60}. This stabilizing effect is attributed to the formation of 2:1 sandwich complexes between the Zn(II) porphyrins and C_{60}.

Ajayaghosh and coworkers have intensively investigated[61,62] the self-assembly and gel formation of oligo(p-phenylenevinylene)s (OPVs). For instance, LMWG **44** (Figure 3.27) with two –OH groups can gel[63] several

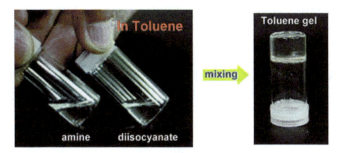

In situ organogelation

Figure 3.26 Typical procedure of *in situ* organogelation.
Reproduced from ref. 57.

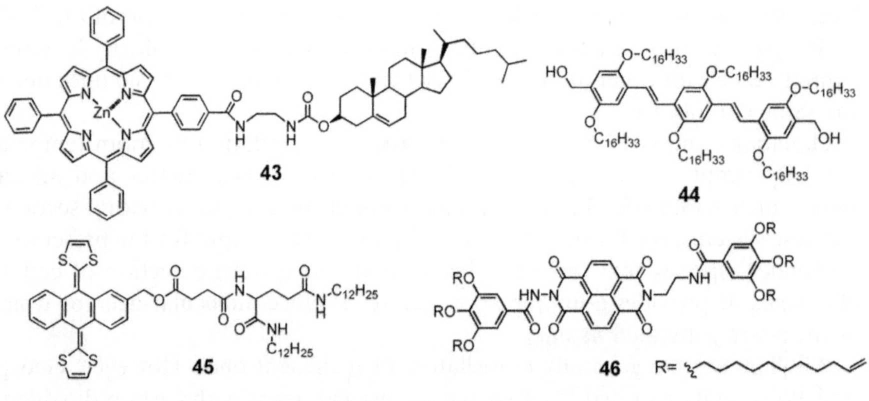

Figure 3.27   Molecular structures of compounds **43–46**.

nonpolar solvents and they found that the gels of **44** in relatively polar solvents, *e.g.* toluene, are thixotropic. It is interesting to note that the self-assembly of OPV molecules is accelerated in the presence of carbon nanotubes (CNTs) through physical interactions of OPV molecules and CNTs. In this way, CNTs are dispersed in the solvent, which reinforces the self-assembly processes and leads to the formation of hybrid gels. Furthermore, CNTs may act as physical crosslinks between these self-assembled structures, thus enhancing the gel stability.

9,10-Bis(1,3-dithiol-2-ylidene)-9,10-dihydroanthracene, which is usually referred to as ex-TTF, is unique in terms of its structure and electrochemical behavior. Martin and coworkers published[64,65] a series of reports with regard to the binding of ex-TTF with fullerene and assembly of ex-TTF-fullerene derivatives into nanostructures. Some of us have reported[66] an ex-TTF derived LMWG **45** (Figure 3.27) with *L*-glutamide-derived lipid segment that can facilitate the gelation through hydrophobic and H-bonding interactions. Notably, the gelation ability of **45** can be enhanced after addition of C_{60}. The T_{gel} increases gradually and reaches the maximum after addition of 0.5 eq. of C_{60}.

Shinkai and coworkers reported[67] molecular recognition within molecular gels with LMWG **46** entailing an electron accepting naphthalenediimide moiety (Figure 3.27). Gels of **46** are able to interact with dihydroxynaphthalene (DHN) by intercalation between consecutive naphthalenediimide moieties. As a result, different colored gels are generated with different isomers of DHN. Thus, naked-eye differentiation of several positional isomers of DHN becomes possible. The authors also mentioned that the gel superstructure was extremely sensitive to the amount of inducing DHN compounds. For example, when more than 1.2 equiv of DHN was added, the gel matrix can readily transform into the solution phase.

Harada and coworkers described[68] a host–guest based chemically responsive hydrogel formed by a gelator based on β-cyclodextrin (β-CD) substituted with a cinnamoyl-trinitrophenyl unit. The modified β-CD forms supramolecular fibrils through host–guest interactions, and the hydrogen bonds between the

CDs result in crosslinking of the fibrils to give the hydrogel. Addition of competitive guests for β-CD such as 1-adamantane carboxylic acid, urea, or methyl orange, leads to destruction of the hydrogel.

3.3.6 Multistimuli-Responsive Gels

Molecular gels responding to multistimuli have gained increasing attention in recent years as such gels may find practical applications. For instance, as discussed above the gel–solution transition of LMWG **35** can be reversibly tuned by redox reaction. In addition, it can also be affected by shear stress and sonication, and as a result the gels of **35** are multistimuli responsive.

It is expected that multistimuli-responsive gels can be yielded by incorporating several stimuli-responsive moieties into LMWGs. Huang and coworkers reported[69] a crownether-appended gelator **47** (Figure 3.28) with excellent gelation properties. As LMWG **47** possesses one crownether moiety that can remarkably bind a wide range of cations, the reversible gel–solution transitions are easily achieved by adding and removing organic salts. Furthermore, the reversible gel–solution transitions can also be achieved (Figure 3.29) by changing pH because of the presence of organic ammonium salt unit in **47**. They also claimed that the solution-gel transition can be further tuned by changing the counteranions of the secondary ammonium salt unit in **47**. Thus, gels with LMWG **47** can be considered to be multistimuli responsive.

Pozzo and coworkers reported[70] LMWG **48** (Figure 3.28) featuring *2H-*chromene moiety that can be reversibly transformed between the closed and open forms. After UV-light irradiation, the corresponding gels became rapidly colored and transformed into the solutions. These studies clearly demonstrate that the transformation of the closed form into the open form of the 2H-chromene moieties in **48** strongly affects the intermolecular H-bonding interactions. The gel phases can be restored by heating and further cooling the solutions.

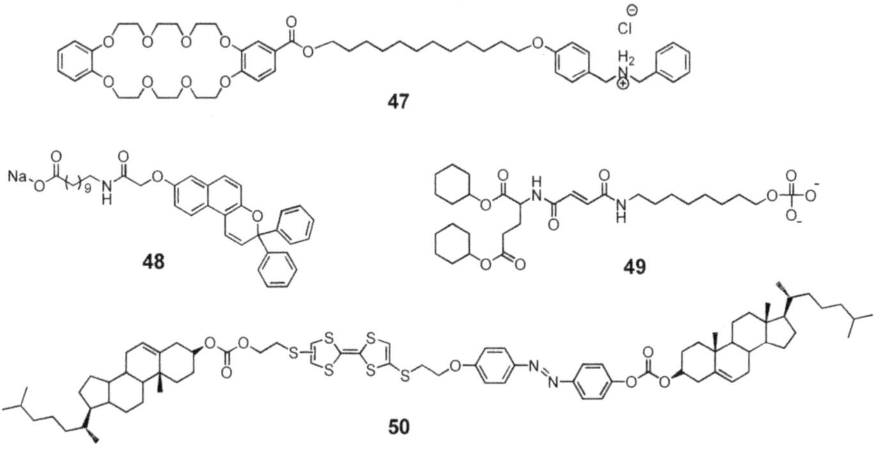

Figure 3.28 Molecular structures of compounds **47–50**.

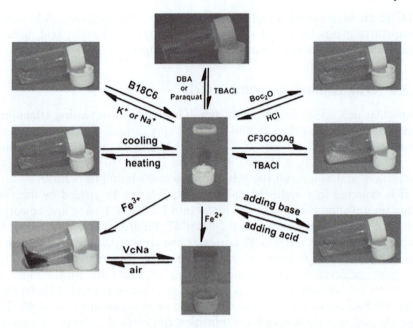

Figure 3.29   The reversible gel–solution transitions of supramolecular gel with **47** in acetonitrile triggered by a variety of stimuli.
Reprinted with permission from ref. 69. Copyright (2012) John Wiley & Sons, Inc.

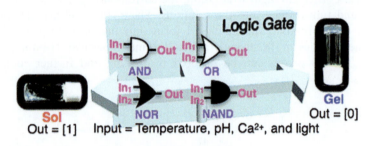

Figure 3.30   Gel-based supramolecular logic gates.
Reprinted with permission from ref. 71. Copyright (2009) American Chemical Society.

Hamachi and coworkers reported[71] a hydrogel comprising phosphate-type hydrogelator **49** (Figure 3.28). The hydrogel exhibits (Figure 3.30) macroscopic gel–solution behavior in response to four distinct input stimuli: temperature, pH, Ca^{2+}, and light. On the basis of its multistimuli responsiveness, gel-based supramolecular logic gates displaying AND, OR, NAND, and NOR functions are successfully constructed. These gel-based supramolecular logic gates are capable of holding and releasing bioactive substances in response to logic triggers. In addition, by combining the supramolecular gel-based AND logic

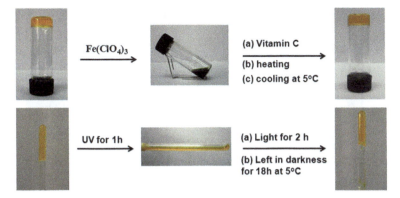

Figure 3.31 Reversible tuning of the gel formation of **50** (up) by chemical oxidation and reduction and (down) by UV- and visible-light irradiation.
Reprinted with permission from ref. 72. Copyright (2010) American Chemical Society.

gate with a photoresponsive supramolecular gel, the release rate of the bioactive substance can be modulated.

We have also described[72] LMWG **50** (Figure 3.28) featuring both electroactive TTF and photoresponsive azobenzene units with the aim of generating multistimuli-responsive gels. As anticipated, by manipulating the redox state of the electroactive TTF group, the gel–solution transition could be reversibly tuned by either chemical or electrochemical oxidation/reduction reactions (Figure 3.31). Furthermore, the *trans–cis* photoisomerization of the azobenzene group in **50** can also trigger the gel–solution transition. Therefore, the gel–solution transition for gels of LMWG **50** induced by redox reactions and light irradiations can be operated separately.

3.4 Applications

It is anticipated that stimuli-responsive gels are highly promising candidates for a diverse range of applications. Some of these stimuli-responsive gels have already demonstrated their applications in different areas. For instance, gel-based supramolecular logic gates[71] are capable of holding and releasing bioactive substances in response to logic triggers. As such, in this section we will merely introduce their applications with a few representative examples. Readers may find detailed examples in the following chapters.

Most of the gel systems are responsive to thermal stimuli and as a result, the gel will transform into solution and the viscosity of the medium can be tuned. By making use of this feature, we established[73] a thermodriven molecular fluorescence switch by studying the fluorescence for bispyrene molecules in organogel system. The results showed that the reversible modulation of the monomer/excimer emission for bispyrene molecules can be achieved through the solution–gel phase transition. Such modulation of the monomer/excimer

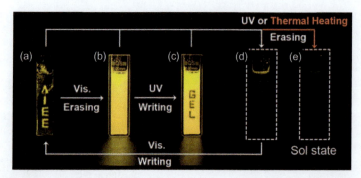

Figure 3.32 Reversible fluorescence image of organogel system in a quartz cell. (a) Writing, (b) erasing, (c) rewriting, (d) re-erasing, and (e) erasing by heat–solution state. The dark region represents the irradiated area. Reprinted with permission from ref. 74. Copyright (2009) John Wiley & Sons, Inc.

emission is due to the restriction of the intramolecular conformational change (in the excited state) of bispyrene molecules in the gel phase.

With the aim to develop fluorescent optical memory devices, fluorescence-switching organogel systems activated by external stimuli were investigated. For instance, Park and coworkers reported[74] a switchable fluorescent organogel system based on the AIEE phenomenon (Figure 3.32). The fluorescence of the organogel containing a dithienylethene compound was quenched after UV irradiations, and the fluorescence was restored after further visible-light irradiation. Such fluorescence modulation is based on the photochromic transformation of dithienylethene and the intermolecular energy transfer. Notably, this fluorescence switching is reproducible. They claimed that such unique fluorescence switching behavior is potentially useful for high-density optical memory devices.

Gels have been utilized as media for the growth of molecular crystals for a long time. Compared to conventional covalent gels, the advantage of using supramolecular gels as the crystallization medium is that the reversible nature of the supramolecular gels allows facile release of crystals. Recently, Steed and coworkers reported[75] the use of low molecular weight supramolecular gels as media for the growth of molecular crystals (see Chapter 8). Growth of a range of crystals of organic compounds, including pharmaceuticals, was achieved in bis(urea) gels without cocrystal formation. More importantly, after addition of acetate anions gels are then transformed into solutions. Crystals can be conveniently recovered by filtration, and there was no visible degradation of the crystalline sample.

Hydrogels are generally regarded as biocompatible materials. Their porous structure, along with their water content, is extremely suitable to accommodate water-soluble molecules. Thus, hydrogels are promising media for the controlled release of molecules, in particular for drug delivery. For example, Escuder and coworkers reported[76] a highly biocompatible hydrogelator bearing a nucleophilic reactive site. The hydrogels are sensitive to aldehydes that are biomarkers of certain diseases. Moreover, the response rate of hydrogels toward aldehydes is found to be dependent on the chemical structures of aldehydes. The authors successfully demonstrated the application of such hydrogels for the controlled release of model drugs entrapped within the hydrogel network.

3.5 Conclusion and Perspectives

In summary, molecular gels that can be addressed by chemical and physical stimuli have experienced rapid development in recent years. These responsive gels as intelligent materials have shown potential applications in a number of areas. Therefore, further expansion of this fascinating area can be imagined and the following issues deserve more attention: 1) it is still challenging to design LMWGs leading to stimuli-responsive gels, since the incorporation of functional groups into LMWGs may perturb the self-assembly processes in a destructive way. Additional studies will be required, especially the detailed mechanism of the gel formation. These studies may enable the rational design of LWMGs; 2) molecular gels responsive to other physical stimuli such as magnetic fields and X-ray radiation are unexplored. These investigations will enrich the scope of molecular gels and provide new application opportunities.

Further studies in this attractive area need the endeavors of scientists from different areas including organic chemistry, physical chemistry, nanoscience and materials science. Apart from the rational design of LMWGs, the scientists should focus on how to develop the practical applications of these responsive materials. In this aspect, issues such as the stability of the supramolecular gels in a prolonged use or enhancement of the mechanical strength of the systems, still need to be solved.

References

1. X. Yang, G. Zhang and D. Zhang, *J. Mater. Chem.*, 2012, **22**, 38–50.
2. H. Svobodová, V. Noponen, E. Kolehmainen and E. Sievänen, *RSC Adv.*, 2012, **2**, 4985–5007.
3. M. D. Segarra-Maset, V. J. Nebot, J. F. Miravet and B. Escuder, *Chem. Soc. Rev.*, 2013, **42**, 7086–7098.
4. S. Kiyonaka, K. Sugiyasu, S. Shinkai and I. Hamachi, *J. Am. Chem. Soc.*, 2002, **124**, 10954–10955.
5. K. Kuroiwa, T. Shibata, A. Takada, N. Nemoto and N. Kimizuka, *J. Am. Chem. Soc.*, 2004, **126**, 2016–2021.
6. H. Danjo, K. Hirata, S. Yoshigai, I. Azumaya and K. Yamaguchi, *J. Am. Chem. Soc.*, 2009, **131**, 1638–1639.
7. J. Brinksma, B. L. Feringa, R. M. Kellogg, R. Vreeker and J. van Esch, *Langmuir*, 2000, **16**, 9249–9255.
8. M. Xue, D. Gao, X. Chen, K. Liu and Y. Fang, *J. Colloid Interface Sci.*, 2011, **361**, 556–564.

9. M.-O. M. Piepenbrock, N. Clarke and J. W. Steed, *Soft Matter*, 2010, **6**, 3541–3547.

10. G. Cravotto and P. Cintas, *Chem. Soc. Rev.*, 2009, **38**, 2684–2697.

11. T. Naota and H. Koori, *J. Am. Chem. Soc.*, 2005, **127**, 9324–9325.

12. K. Isozaki, H. Takaya and T. Naota, *Angew. Chem., Int. Ed.*, 2007, **46**, 2855–2857.

13. D. Bardelang, F. Camerel, J. C. Margeson, D. M. Leek, M. Schmutz, M. B. Zaman, K. Yu, D. V. Soldatov, R. Ziessel, C. I. Ratcliffe and J. A. Ripmeester, *J. Am. Chem. Soc.*, 2008, **130**, 3313–3315.

14. K. Murata, M. Aoki, T. Suzuki, T. Harada, H. Kawabata, T. Komori, F. Ohseto, K. Ueda and S. Shinkai, *J. Am. Chem. Soc.*, 1994, **116**, 6664–6676.

15. N. Koumura, M. Kudo and N. Tamaoki, *Langmuir*, 2004, **20**, 9897–9900.

16. J. H. Kim, M. Seo, Y. J. Kim and S. Y. Kim, *Langmuir*, 2009, **25**, 1761–1766.

17. M. Ayabe, T. Kishida, N. Fujita, K. Sada and S. Shinkai, *Org. Biomol. Chem.*, 2003, **1**, 2744–2747.

18. S. Miljanić, L. Frkanec, Z. Meić and M. Žinić, *Langmuir*, 2005, **21**, 2754–2760.

19. J. J. D. de Jong, L. N. Lucas, R. M. Kellogg, J. H. van Esch and B. L. Feringa, *Science*, 2004, **304**, 278–281.

20. S. Wang, W. Shen, Y. L. Feng and H. Tian, *Chem. Commun.*, 2006, 1497–1499.

21. S. H. Xiao, T. Yi, F. Y. Li and C. H. Huang, *Tetrahedron Lett.*, 2005, **46**, 9009–9012.

22. Z. J. Qiu, H. T. Yu, J. B. Li, Y. Wang and Y. Zhang, *Chem. Commun.*, 2009, 3342–3344.

23. Q. Chen, Y. Feng, D. Zhang, G. Zhang, Q. Fan, S. Sun and D. Zhu, *Adv. Funct. Mater.*, 2010, **20**, 36–42.

24. J. L. Pozzo, G. M. Clavier and J. Desvergne, *J. Mater. Chem.*, 1998, **8**, 2575–2577.

25. K. Sugiyasu, N. Fujita, M. Takeuchi, S. Yamada and S. Shinkai, *Org. Biomol. Chem.*, 2003, **1**, 895–899.

26. F. Rodríguez-Llansola, B. Escuder, J. F. Miravet, D. Hermida-Merino, I. W. Hamley, C. J. Cardin and W. Hayes, *Chem. Commun.*, 2010, **46**, 7960–7962.

27. J. F. Miravet and B. Escuder, *Chem. Commun.*, 2005, 5796–5798.

28. J. W. Chung, B.-K. An and S. Y. Park, *Chem. Mater.*, 2008, **20**, 6750–6755.

29. M.-O. M. Piepenbrock, G. O. Lloyd, N. Clarke and J. W. Steed, *Chem. Rev.*, 2010, **110**, 1960–2004.

30. J. W. Steed, *Chem. Soc. Rev.*, 2010, **39**, 3688–3699.

31. M. Yamanaka, T. Nakamura, T. Nakagawa and H. Itagaki, *Tetrahedron Lett.*, 2007, **48**, 8990–8993.

32. C. Wang, D. Zhang and D. Zhu, *Langmuir*, 2007, **23**, 1478–1482.

33. A. Kishimura, T. Yamashita and T. Aida, *J. Am. Chem. Soc.*, 2005, **127**, 179–183.

34. T. Becker, C. Y. Goh, F. Jones, M. J. McIldowie, M. Mocerino and M. I. Ogden, *Chem. Commun.*, 2008, 3900–3902.
35. A. Tripathi and P. S. Pandey, *Tetrahedron Lett.*, 2011, **52**, 3558–3560.
36. K. Murata, M. Aoki, T. Nishi, A. Ikeda and S. Shinkai, *J. Chem. Soc., Chem. Commun.*, 1991, 1715–1718.
37. W. Deng and D. H. Thompson, *Soft Matter*, 2010, **6**, 1884–1887.
38. W. Edwards and D. K. Smith, *Chem. Commun.*, 2012, **48**, 2767–2769.
39. Q. Jin, L. Zhang, X. Zhu, P. Duan and M. Liu, *Chem.-Eur. J.*, 2012, **18**, 4916–4922.
40. T. Kitamura, S. Nakaso, N. Mizoshita, Y. Tochigi, T. Shimomura, M. Moriyama, K. Ito and T. Kato, *J. Am. Chem. Soc.*, 2005, **127**, 14769–14775.
41. T. Kitahara, M. Shirakawa, S. Kawano, U. Beginn, N. Fujita and S. Shinkai, *J. Am. Chem. Soc.*, 2005, **127**, 14980–14981.
42. J. Puigmartí-Luis, V. Laukhin, A. P. del Pino, J. Vidal-Gancedo, C. Rovira, E. Laukhina and D. B. Amabilino, *Angew. Chem., Int. Ed.*, 2007, **46**, 238–241.
43. T. Akutagawa, K. Kakiuchi, T. Hasegawa, S. Noro, T. Nakamura, H. Hasegawa, S. Mashiko and J. Becher, *Angew. Chem., Int. Ed.*, 2005, **44**, 7283–7287.
44. C. Wang, D. Q. Zhang and D. B. Zhu, *J. Am. Chem. Soc.*, 2005, **127**, 16372–16373.
45. J. F. Stoddart, *Chem. Soc. Rev.*, 2009, **38**, 1802–1820.
46. C. Wang, M. A. Olson, L. Fang, D. Benitez, E. Tkatchouk, S. Basu, A. N. Basuray, D. Zhang, D. Zhu, W. A. Goddard and J. F. Stoddart, *Proc. Natl. Acad. Sci. USA*, 2010, **107**, 13991–13996.
47. A. Coskun, J. M. Spruell, G. Barin, W. R. Dichtel, A. H. Flood, Y. Y. Botros and J. F. Stoddart, *Chem. Soc. Rev.*, 2012, **41**, 4827–4859.
48. Y.-L. Zhao, I. Aprahamian, A. Trabolsi, N. Erina and J. F. Stoddart, *J. Am. Chem. Soc.*, 2008, **130**, 6348–6350.
49. J. Liu, P. He, J. Yan, X. Fang, J. Peng, K. Liu and Y. Fang, *Adv. Mater.*, 2008, **20**, 2508–2511.
50. S. Kawano, N. Fujita and S. Shinkai, *Chem.-Eur. J.*, 2005, **11**, 4735–4742.
51. S. Kawano, N. Fujita and S. Shinkai, *J. Am. Chem. Soc.*, 2004, **126**, 8592–8593.
52. J. Chen, W. Wu and A. J. McNeil, *Chem. Commun.*, 2012, **48**, 7310–7312.
53. J. F. Miravet and B. Escuder, *Org. Lett.*, 2005, **7**, 4791–4794.
54. J. F. Miravet and B. Escuder, *Tetrahedron*, 2007, **63**, 7321–7325.
55. Q. Chen, D. Zhang, G. Zhang and D. Zhu, *Langmuir*, 2009, **25**, 11436–11441.
56. Q. Chen, Y. Lv, D. Zhang, G. Zhang, C. Liu and D. Zhu, *Langmuir*, 2010, **26**, 3165–3168.
57. M. Suzuki, Y. Nakajima, M. Yumoto, M. Kimura, H. Shirai and K. Hanabusa, *Org. Biomol. Chem.*, 2004, **2**, 1155–1159.
58. M. George and R. G. Weiss, *Langmuir*, 2002, **18**, 7124–7135.
59. M. George and R. G. Weiss, *J. Am. Chem. Soc.*, 2001, **123**, 10393–10394.

60. T. Ishi-i, R. Iguchi, E. Snip, M. Ikeda and S. Shinkai, *Langmuir*, 2001, **17**, 5825–5833.
61. A. Ajayaghosh and V. K. Praveen, *Acc. Chem. Res.*, 2007, **40**, 644–656.
62. A. Ajayaghosh, V. K. Praveen and C. Vijayakumar, *Chem. Soc. Rev.*, 2008, **37**, 109–122.
63. S. Srinivasan, S. S. Babu, V. K. Praveen and A. Ajayaghosh, *Angew. Chem., Int. Ed.*, 2008, **47**, 5746–5749.
64. N. Martín, L. Sánchez, M. Á. Herranz, B. Illescas and D. M. Guldi, *Acc. Chem. Res.*, 2007, **40**, 1015–1024.
65. E. M. Perez and N. Martín, *Chem. Soc. Rev.*, 2008, **37**, 1512–1519.
66. X. Yang, G. Zhang, D. Zhang and D. Zhu, *Langmuir*, 2010, **26**, 11720–11725.
67. P. Mukhopadhyay, Y. Iwashita, M. Shirakawa, S. Kawano, N. Fujita and S. Shinkai, *Angew. Chem., Int. Ed.*, 2006, **45**, 1592–1595.
68. W. Deng, H. Yamaguchi, Y. Takashima and A. Harada, *Angew. Chem., Int. Ed.*, 2007, **46**, 5144–5147.
69. S. Dong, B. Zheng, D. Xu, X. Yan, M. Zhang and F. Huang, *Adv. Mater.*, 2012, **24**, 3191–3195.
70. S. A. Ahmed, X. Sallenave, F. Fages, G. Mieden-Gundert, W. M. Müller, U. Müller, F. Vögtle and J. L. Pozzo, *Langmuir*, 2002, **18**, 7096–7101.
71. H. Komatsu, S. Matsumoto, S.-i. Tamaru, K. Kaneko, M. Ikeda and I. Hamachi, *J. Am. Chem. Soc.*, 2009, **131**, 5580–5585.
72. C. Wang, Q. Chen, F. Sun, D. Zhang, G. Zhang, Y. Huang, R. Zhao and D. Zhu, *J. Am. Chem. Soc.*, 2010, **132**, 3092–3096.
73. C. Wang, Z. Wang, D. Zhang and D. Zhu, *Chem. Phys. Lett.*, 2006, **428**, 130–133.
74. J. W. Chung, S.-J. Yoon, S.-J. Lim, B.-K. An and S. Y. Park, *Angew. Chem., Int. Ed.*, 2009, **48**, 7030–7034.
75. J. A. Foster, M.-O. M. Piepenbrock, G. O. Lloyd, N. Clarke, J. A. K. Howard and J. W. Steed, *Nature Chem.*, 2010, **2**, 1037–1043.
76. F. Rodríguez-Llansola, J. F. Miravet and B. Escuder, *Chem. Commun.*, 2011, **47**, 4706–4708.

CHAPTER 4

Enzyme-Responsive Molecular Gels

SISIR DEBNATH AND REIN V. ULIJN*

WestCHEM, Department of Pure and Applied Chemistry, University of
Strathclyde, 295 Cathedral Street, Glasgow, G1 1XL UK
*Email: rein.ulijn@strath.ac.uk

4.1 Introduction

The last century has seen extensive development in the field of chemical science
that was primarily focused on understanding the behaviour of the molecules
and their constituent atoms. During the last two decades, chemists have become
increasingly focused on interactions at the molecular level and behaviour of
mixtures, rather than pure components, *i.e.* supramolecular chemistry.[1,2] Inspiration in this area often comes from the highly specific intermolecular processes in nature that are responsible for emergent properties and complexity in
living systems. While traditional chemistry focuses on covalent bonds, supramolecular chemistry examines the reversible noncovalent interaction between
molecules.[3,4] In the last few years, this area has seen a shift in emphasis towards
the study of increasingly complex mixtures of molecules and on how their
interactions change over time, an area known as systems chemistry.[5,6] One of
the main branches of the supramolecular chemistry and also emerging as a
subfield in systems chemistry, is related to the materials science and specifically
molecular gelation. In these systems, the solid component of a hydrogel consists
of a supramolecular structure derived from the self-assembly of oligomers or
nonpolymeric molecules.[7,8] These materials have a range of emerging applications in biomedicine and nanotechnology.[9]

RSC Soft Matter No. 1
Functional Molecular Gels
Edited by Beatriu Escuder and Juan F. Miravet
© The Royal Society of Chemistry 2014
Published by the Royal Society of Chemistry, www.rsc.org

Enzymatic reactions are increasingly recognised as powerful means to control and direct supramolecular interactions, by catalytically converting non-assembling precursors to self-assembling building blocks.[10–13] Enzyme triggered self-assembly and gelation of the hydrogelators is advantageous for biomedical applications due to the highly efficient catalytic efficiency and selectivity of enzymes in mild conditions, the localised action of enzymes that allows for space/time control and the inherent biocompatibility associated with the enzymes. Due to the role of the enzyme in the catalytic self-assembly process, it is possible to control the kinetics of the process, giving rise to supramolecular hydrogels with tunable properties (as discussed in a later section).

Although some examples of non-gelling systems exist, in this book chapter we will specifically concentrate on biocatalytic self-assembly of gel materials whereby large numbers of building blocks form extended noncovalent molecular systems that lead to supramolecular hydrogels. We will first discuss examples of biocatalytic hydrogelation in biological systems that are commonly used as a source of inspiration, followed by a review of recent man-made systems, their design rules and opportunities for nonequilibrium assembly. Finally, we will review the emerging applications of (intracellular) biosensing, controlled release and cell instructive materials that control and direct cell fate.

4.2 Biocatalytic Self-Assembly in Biology

Inspiration for development of enzyme-catalysed supramolecular gels comes from living systems, *e.g.* enzyme-controlled formation and degradation of collagen fibrils, actin filaments and microtubules.

Collagen forms stiff gel-like materials by entangled networks of collagen fibrils. This gel surrounds cells to give a scaffold to support the formation of fibrous tissues such as tendon, ligament and skin, cartilage, bone, blood vessels, and intervertebral disc, *etc.* Formation of collagen gel is controlled by an enzymatic process as follows. Procollagens (nonassembling precursors) are produced within the cells in the shape of triple helical structures and "loose ends", sensitive to the enzymes collagen peptidases, at both sides. After secretion into the extracellular space, membrane bound enzymes, collagen peptidases, remove these "loose ends" of the procollagen molecules to form tropocollagen molecules that self-assemble to form longer fibres. These are subsequently covalently crosslinked between the triple helixes by another enzyme, lysyl oxidase, forming collagen fibrils that give rise to gel-phase materials with highly tunable stiffness.[14–16]

Another example of enzymatic self-assembly is the formation of actin filaments (F-actin) that drive the locomotion of the single-cell organisms and provide the structural support mechanism for movement and cell shape changes *via* cytoskeleton reorganisation in eukaryotic cells.[17,18] These are the 1D fibres of the monomeric subunit; globular actin (G-actin) that undergoes continuous self-assembly and disassembly, initiated by the enzymatic hydrolysis of the biological fuel molecules adenosine triphosphate (ATP) to the diphosphate analogue (ADP) by ATPase. Likewise, the formation of microtubules, which are responsible for transport within the cell as well as processes of cellular reorganisation relevant to cell division, is enzymatically regulated. Here, guanosine triphosphate (GTP) is hydrolysed to guanosine diphosphate (GDP) by the enzyme GTPase to control the corresponding assembly and disassembly of the monomeric proteins in microtubules.[17,19] In summary, enzymatic self-assembly and hydrogelation is a common mechanism to control these natural self-assembly processes and provides a unique opportunity to develop new dynamic materials based on these same principles.

4.3 Biocatalytic Self-Assembly in Designed Biomaterials

A number of approaches have been used to trigger the self-assembly of small molecules into supramolecular hydrogels. They include a variety of chemical and physical means, which can broadly be divided into two types, depending on whether the stimulus applied is a bulk or localised event. In the first type, self-assembly is initiated by an overall change in environmental conditions, such as pH or ionic strength.[20,21] The other type includes the locally applied stimulus such as light or catalytic action of the enzymes that involves conversion to the self-assembly molecule by a triggered change in chemical structure.[22,23] The exploitation of the enzymes as selective biological stimuli to trigger the self-assembly of small molecules has some unique advantages. These systems always operate under mild and constant conditions and provide better control over a self-assembly process and its kinetics due to the unique role of the enzyme, which locally amplifies the concentration of the self-assembly building blocks. The selectivity means that, in principle, a range of different triggers (responding to different enzymes) can be applied within single systems, something that is very difficult to achieve using the more conventional stimuli. In addition, the difference between metabolic profiles in health and disease commonly comes down to different expression levels of certain key enzymes. If these enzymes can be exploited as triggers to control assembly or disassembly, new opportunities emerge to dynamically respond to the onset of disease.

4.4 Design of Biocatalytic Gelators

4.4.1 Basic Building Blocks

The idea of biocatalytic self-assembly is based on the fact that self-assembling molecules are always amphiphilic in nature, and the balance between hydrophobic and hydrophilic elements dictates their self-assembly propensity. By either cleaving or adding hydrophobic or hydrophilic elements from/to these molecular structures, the tendency to assembly can be modified.

Most examples of the biocatalytic self-assembly are based on peptides and their derivatives; these will be the focus of the chapter. There are 20 gene coding amino acids in nature (Figure 4.1). Peptides generated from the combination of these amino acids have different properties, depending on the nature of their

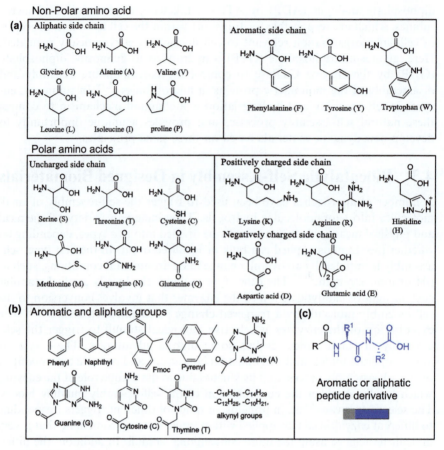

Figure 4.1 The structure of (a) 20 natural L amino acids and (b) aromatic and aliphatic residues that gives rise to (c) aromatic short peptide derivatives, which form supramolecular assembly through hydrogen-bonding and π-stacking interactions.

side-chain substitution. There exists an enormous number of possible sequences to give rise to peptides (20^5/3.2 million sequence possible for a pentapeptide). Considerable progress has been made in elucidation of design rules of peptides capable of forming well-defined self-assembled nanostructures.[24–32] Biocatalytic peptide hydrogelators are comprised of three components, i) an assembly directing unit, a component that directs the noncovalent interaction responsible for self-assembling of the precursor ii) an enzyme-recognition site (peptide sequence based on enzyme's substrate specificity), and iii) a molecular switch component that initiates self-assembly upon enzyme action.

4.4.1.1 Assembly Directing Unit

The simplest and most versatile peptide self-assembly paradigm consists of short peptide sequences (typically less than five amino acids, and often only two) functionalised at the N-terminus with aromatic moieties. These systems are known to give rise to a range of self-assembling structures, depending on the balance between the aromatic π-stacking and hydrogen bonding inter-actions between the peptide moieties. One proposed structure are the highly stable π-π-interlocked β sheets (or π-β structures) that have been observed for a number of these systems.[33,34] Some of the aromatic residues that have been used in this context are phenyl, naphthyl, pyrene, 9-fluorenyl as well as the nucleobases (Figure 4.1).

4.4.1.2 Enzyme Recognition Site

A range of enzymes have been utilised to initiate self-assembly including transglutaminase,[35] α-chymotrypsin,[13] thermolysin,[36–41] subtilisin,[42,43] phos-phatase,[44–46] *trans*-acylase,[47] penicillin G amidase,[48] and (a combination of) kinase/phosphatase.[49] Herein, we will concentrate on systems that are based on the widely used aromatic peptide amphiphiles focusing on two main categories of chemical reactions like i) Hydrolysis ("breaking" bonds, by enzymes such as phosphatase, subtilisin and β-lactamase) and ii) Condensation/transacylation ("making" bonds, by enzymes such as thermolysin and α-chymotrypsin).

4.4.1.3 Molecular Switch Component

As mentioned, the enzyme-triggered self-assembly and associated rebalancing of the hydrophobic/-philic parts of building blocks usually involves either making, or breaking of covalent chemical bonds (Figure 4.2). The resulting hydrogelator molecules often rapidly reach their critical aggregation concen-tration at the site of catalysis, and start to self-assemble into nanofibres that eventually entangle to form the higher-order network structure of a supra-molecular hydrogel. For route a, the formation of the supramolecular assembly relies on reactions that are favourable in water (hydrolysis of amides and (phosphate) esters), giving rise to a largely irreversible system. The structural details of these chemical conversions are shown in Figure 4.3. In route b, the self-assembly may take place under reversible conditions if the condensation

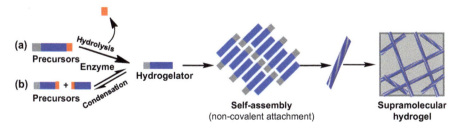

Figure 4.2   Proposed mechanism for enzyme triggered self-assembly (a) formation of supramolecular assembly *via* bond cleavage (hydrolysis) (b) formation of supramolecular assemblies *via* bond formation (condensation).

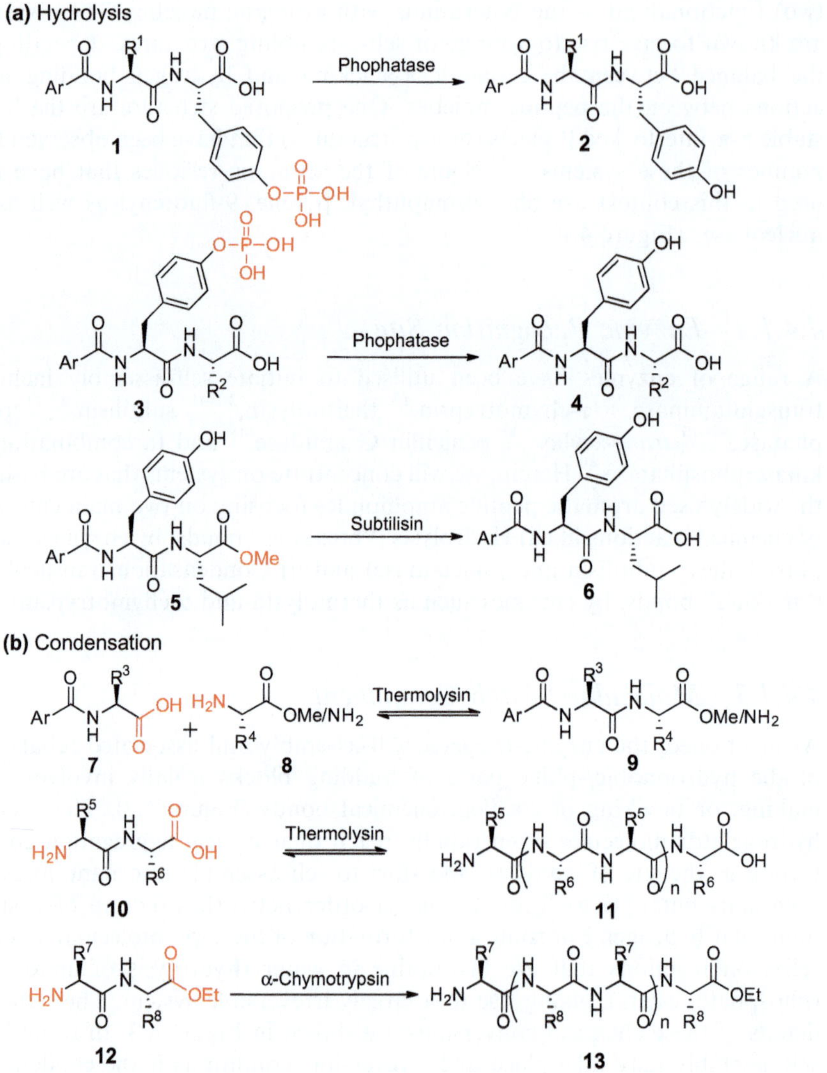

(a) Hydrolysis

(b) Condensation

Figure 4.3 Examples of enzymatic synthesis of supramolecular assemblies of aromatic short peptides *via* (a) hydrolysis by alkaline phosphatase and esterase (*e.g.* subtilisin) (b) condensation using protease (*e.g.* thermolysin, α-chymotrypsin).

reaction itself is thermodynamically unfavoured but facilitated by the free-energy contribution of the self-assembly step.

Perhaps the most commonly used approach based on hydrolysis of a non-assembling precursor takes advantage of phosphatase, which dephosphorylates serine, threonine or tyrosine residues, thereby changing the charge balance of the molecule. The enzyme is of particular interest as it forms one half of the kinase/phosphatase switches that regulate the protein activity. In a typical example, the enzyme would dephosphorylate a tyrosine residue on an aromatic peptide phosphate derivative (**1**, **3**) to form the self-assembling aromatic peptide amphiphile (**2**, **4**).[10,44] It has been found that the precursors in some cases form supramolecular micelles, which transform to fibres during enzyme-triggered dephosphorylation that ultimately causes the hydrogelation. Such supramolecular transformations may find uses in controlled release or even mimics of the dynamic fibres found in natural systems as discussed in the first section.

A second common route that takes advantage of the hydrolysis of precursors, makes use of esterase activity. Typically, subtilisin is used to cleave the methyl ester of the short Fmoc-di/tri peptide derivatives to generate the supramolecular assembly. For example, it has been found that subtilisin hydrolyses Ar-FY-OMe (**5**) to corresponding acid Ar-FY-OH (**6**) at physiological pH to generate nanofibres and ultimately gels. In the case of Fmoc-LL-OMe and Fmoc-LLL-OMe, the formation of nanotubes was described,[38] the latter were subsequently studied for their charge-transport capabilities, for potential uses in bio/electronic interfacing.[39]

In route b, a protease, thermolysin, which is commonly known to catalyse the hydrolysis of peptide bond, has been used to synthesise the peptide bond between precursors (**7** and **8**). In the case of aromatic peptide amphiphiles, the condensation reaction can involve an amino acid functionalised with an aromatic at the N terminus and a free carboxylic acid group, which in itself is not a gelator. Upon enzymatic condensation of this acid with an amino acid ester or amide, this may form an aromatic peptide amphiphile (**9**) that is capable of molecular self-assembly (Figure 4.3). Alternatively, the reaction can involve dipeptides (**10**) where the formed peptide (**11**) sequence may adopt a common beta sheet type configuration (typically alternating hydrophilic/phobic residues).[36] In a related system, peptide bond formation has also been activated by kinetically controlled hydrogelation where condensation reaction occurs *via* oligomerisation of KL-OEt (**12**) by the α-chymotrypsin to form (KL)$_4$ (**13**) hydrogelator.[13]

4.5 Mechanistic Insight into Enzyme-Catalysed Hydrogelation

As briefly explained above, biocatalytic self-assembly may take place under kinetic or thermodynamic control. The self-assembly and gelation event involves reduction in mobility of self-assembling molecules – in effect locking the system into a metastable state. If the enzymatic reaction is essentially irreversible (as is the case for route a), the rate of reaction will dictate the rate of formation of the self-assembling structure. It is reasonable to expect this to give rise to different supramolecular organisation, depending on the biocatalytic rate. On the other hand, if the self-assembling building blocks can be converted back to their precursors, it is reasonable to assume that, over time, a thermodynamically preferred supramolecular organisation results (route b).

4.5.1 Nonequilibrium Self-Assembly

Self-assembly through bond cleavage operates under kinetic control and it is possible to access different kinetically trapped supramolecular structures by simply changing the amount of enzyme present in the system. For example, enzymatic dephosphorylation of β-peptide derivative, NapFFY*p* (**14**) (Figure 4.4) to form NapFFY (**15**) hydrogelator, demonstrates the ability to control gel properties by altering the amount of catalyst present and hence tuning kinetics. The elastic modulus could be varied from 300 to 4000 Pa (Table 4.1) with increasing amount of the enzyme phophatase and hence the faster formation of the assemblies indicates that the ratio of enzyme to precursor plays a significant role to constructs the supramolecular assemblies.[46] It has found that the morphology of the nanostructures formed is directly related to the amount of the enzyme and hence to the kinetics of the formation of the assemblies. The slower the formation of the nanostructures leading to the thinner and more uniform the fibres of the hydrogel.

By exploiting the localised nucleation and growth mechanism of biocatalysed systems, a high level of control can be achieved over supramolecular order for the self-assembly of aromatic short peptide amphiphiles. For this purpose Fmoc-YL-OMe (**16**) was hydrolysed with subtilisin to form the dipeptide Fmoc-YL (**17**) that undergo molecular self-assembly (Figure 4.5a).[43] It was shown that the properties of the self-assembled structure could be affected by the enzyme concentrations since the nucleation and early-stage structure growth are spatially confined at the site of enzyme (Figure 4.5b). This was confirmed by the early-stage AFM image where we can see the nucleation and

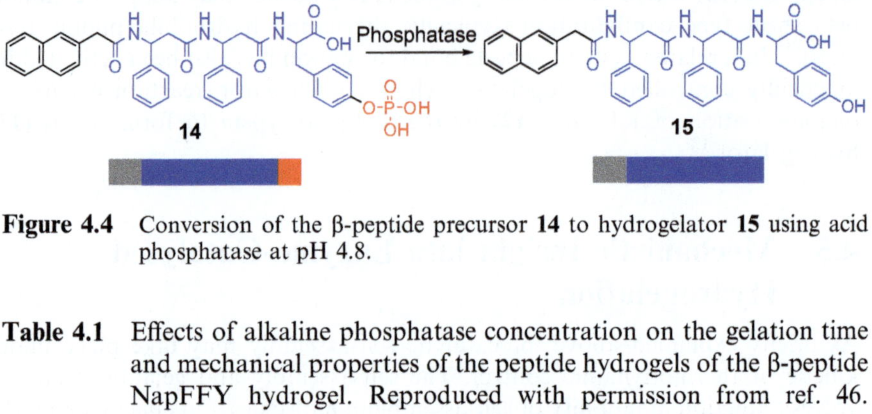

Figure 4.4   Conversion of the β-peptide precursor **14** to hydrogelator **15** using acid phosphatase at pH 4.8.

Table 4.1   Effects of alkaline phosphatase concentration on the gelation time and mechanical properties of the peptide hydrogels of the β-peptide NapFFY hydrogel. Reproduced with permission from ref. 46. Copyright 2007 Wiley-VCH Verlag GmbH & Co. KGaA.

Entry	[Enzyme] (U mL^{-1})	Time required for hydrogelation (min)	G′ (Pa)
1	5.88	2	4000
2	2.94	10	900
3	1.47	30	300

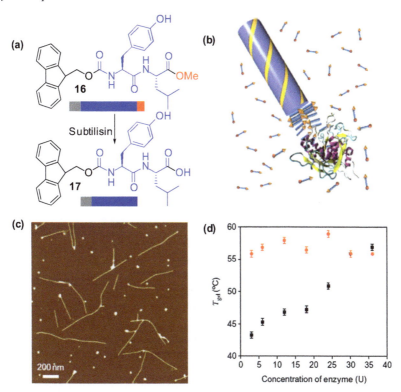

Figure 4.5 (a) Chemical structure of Fmoc-YL-OMe and their subtilisin-catalysed hydrolysis to Fmoc-YL-OH gelator. (b) Schematic of nucleation and growth mechanism of self-assembly controlled by subtilisin. (c) AFM analysis of initial stages of the self-assembly process. (d) Melting temperature (T_{gel}) of gels formed catalytically (black) and by a heating–cooling cycle (red) at different enzyme concentrations. Modified from ref. 43.

initiation of fibre growth starts from the cluster of enzymes (Figure 4.5c). The rate of catalytic self-assembly and hence the properties of the self-assembled structures could be tuned simply by changing the amount of enzyme present in the system. The resulting materials were structurally different, as was immediately evident from a comparison of the melting behaviour of these gels, which showed that higher enzyme concentrations gave rise to increasingly stable gels with increasing melting temperatures (Figure 4.5d). The supramolecular organisation was systematically different and AFM images also demonstrated clearly that the self-assembled network was directly controlled by the amount of enzyme with higher enzyme concentrations giving rise to longer and more bundled fibres. Thus, the enzymatic conversion in these reactions was dictated by its concentration leading to higher-ordered structure formed more quickly at high enzyme concentration. It was proposed that the increasing enzyme concentration controls the size of the biocatalytic cluster and activity at the aggregation nuclei, thus allowing locking of the structure under kinetic control, which corresponds to the local minima in the free-energy landscape. Thus, structurally diverse materials can be accessed, based on a single gelator molecule, which are inaccessible *via* conventional self-assembly.

4.5.2 Equilibrium-Driven Self-Assembly

As mentioned, self-assembly through condensation may operate under thermodynamic control because of the reversible nature of building block formation/hydrolysis. Such systems are highly dynamic with attractive features, including thermodynamic defect correction (which may be of interest in self-healing electronics) and self-selection of the most stable structures in the mixtures as a method for discovery of stable self-assembling systems from mixtures.

Peptide condensation/hydrolysis reactions are ideally suited to produce reversible systems because the free-energy change of amide synthesis/hydrolysis is small (4.0 kJ mole^{-1})[50] and hence can easily be reversed. This can be achieved by formation of a self-assembled structure to give the thermodynamic driving force for peptide synthesis. We have demonstrated this approach using model systems based on Fmoc-amino acids and dipeptides/ amino acid amides. The ability to sequentially access increasingly stable structures (*e.g.*, when comparing Fmoc-TL-OMe and Fmoc-TF-OMe)[50] left no doubt that these systems will self-select the most stable self-assembling structures from component mixtures.

The fully reversible behavior of the enzyme triggered self-assembly can be utilised to form competing building blocks, where the most stable structure is expected to be formed preferentially. The system generated by continuous interconversion between the building blocks to reach an equilibrium distribution is known as a dynamic combinatorial library (DCL).[5] This approach has been successfully used in the discovery of stable supramolecular assemblies from mixtures. The discovery of peptide nanomaterials through DCL is helpful as the testing of all possible peptide and with different sequences would be a major synthetic and analytical challenge if one wants to go for it by classical synthesis since the combination of twenty natural amino acid can give million structurally diverse species.

The approach was used to identify the most stable dipeptide sequences from a range of amino acid building blocks[37] that eventually led to the discovery of a new two-dimensional self-assembly structure, based on Fmoc-SF-OMe.[34] Specifically, evolution of peptide-based nanostructures has been investigated for the reaction of Fmoc-S and Fmoc-T (**18**) with the nucleophiles (**19**) (leucine, L; phenylalanine, F; tyrosine, Y; valine, V; glycine, G and alanine, A amino-acid methyl esters) in presence of thermolysin to produce Fmoc–dipeptide esters, Fmoc–XY–OMe (**20**) (Figures 4.6a and b).[37] Individually, the reaction of Fmoc-T with the mentioned amino acids methyl ester give 96% of Fmoc–TF–OMe and 84% of Fmoc–TL–OMe to give self-supporting hydrogels, while the yields of the other Fmoc–XY–OMe products are relatively poor. Similar results were also obtained for the individually reaction of Fmoc-S with the

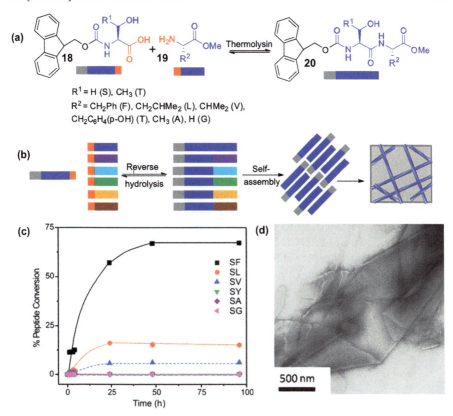

Figure 4.6 (a) Reversible amide synthesis/hydrolysis catalysed by thermolysin. (b) Schematic representation of peptide library where the most stable sequence is preferentially produced, forming a self-supporting gel. (c) Time course for Fmoc–SX–OMe library. (d) TEM image of nanosheets of Fmoc-SF-OMe obtained after 24 h.
Modified from refs. 34 and 37.

amino acid esters. Here again, the yield of Fmoc–SF–OMe, Fmoc–SL–OMe and Fmoc–SV–OMe are 96, 85 and 56%, respectively, (Figure 4.6c) leading to hydrogelation, whereas the yield for the other esters are poor. The high yield together with hydrogelation of some specific peptide suggests that these peptides have thermodynamically favoured self-assembly whereas the low yields of the other peptides suggest the lack of it. Now, when the mixture of the methyl esters of the aforementioned amino acids are separately exposed to Fmoc-S and Fmoc-T in the presence of thermolysin, Fmoc–SF–OMe and Fmoc–TF–OMe, respectively, in the two reaction mixtures dominated (>80% of the total peptide yield) over time indicating that this peptide ester represents the lowest free-energy well in the free-energy hypersurface. When studied in isolation, it was found that the Fmoc-SF-OMe structure forms remarkably stable two-dimensional sheet-like structures that may have interesting electronic properties due to the continuously π-stacked arrangement of the fluorenyl groups

(Figure 4.6d).[34] Clearly, the approach enables discovery of new nanostructures from component mixtures.

Recently, we have shown that using donor–acceptor (DA) interaction of one of the components in the DCL can be significantly amplified (Figures 4.7a and b).[51] Here, the DCL was prepared by the reaction of Nap-Y (**21**) with the nucleophile, XNH$_2$, (**22**) (leucine, L; phenylalanine, F; tyrosine, Y; valine, V; glycine, G and alanine, A amino-acid amides) to form Nap-YXNH$_2$ (**23**). It was found that after 48 h **YF** was preferentially produced in 52%, corresponding to 77% of the total peptide yield obtained, Conversely, YL was formed in 23%, while all the other derivatives were formed in negligible amounts though in individual experiments the yield of YF, YL and YV were 89, 72 and 81%, respectively. These results indicate that the YF derivative is the lowest free energy product among the mixtures. Now, the yield of YF derivative was significantly increased up to 82% (corresponding to 95% of the total peptide yield obtained), in the DCL by using an acceptor molecule, dansyl-β-alanine, owing to donor–acceptor interaction between naphthalene and dansyl chromophores (Figure 4.7c). Conversely YL was formed in low yield (8%) and all

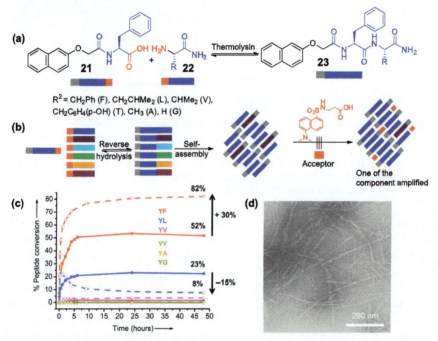

Figure 4.7   (a) Reversible amide synthesis/hydrolysis catalysed by thermolysin. (b) Schematic representation of peptide library where the most stable sequence is preferentially amplified by acceptor molecule (dansyl derivative). (c) Time course of the percentage conversion in DCL system by HPLC as measured in the absence (solid traces) and presence (dashed traces) of DA library. (d) TEM image of presence of **DA** molecules in DCL system. Modified from ref. 51.

the other derivatives were formed in negligible amounts. The donor–acceptor interaction was confirmed by efficient energy transfer between them. TEM images showed the presence of entangled nanofibres of up to several micrometres in length in the library in the presence of acceptor (Figure 4.7d). This result indicates that the suitable donor–acceptor interaction can amplify one of the products in the DCL.

4.6 Application of Biocatalytic Self-Assembly and Gelation

These nanoscale materials may find applications in several directions of biomedicine that take advantage of the ability to measure and respond to differences in enzyme levels that are often associated with certain metabolic states of cells, including disease states. Recent progress in the development of applications is highlighted in the following.

4.6.1 Controlling Cell Fate

Bing Xu and his team designed a precursor molecule (**24**) that does not self-assemble when dissolved in water. When this molecule is exposed to the cells, it enters into the cell by diffusion and being an esterase substrate, it undergoes hydrolysis by an endogeneous esterase enzyme to form a gelator (**25**), which self-assemble into nanostructures that leads to hydrogelation inside the cell (Figure 4.8).[52] This gelation induces an abrupt change in the viscosity of the cell cytoplasm and causes cell death, which depends on enzyme expression levels in

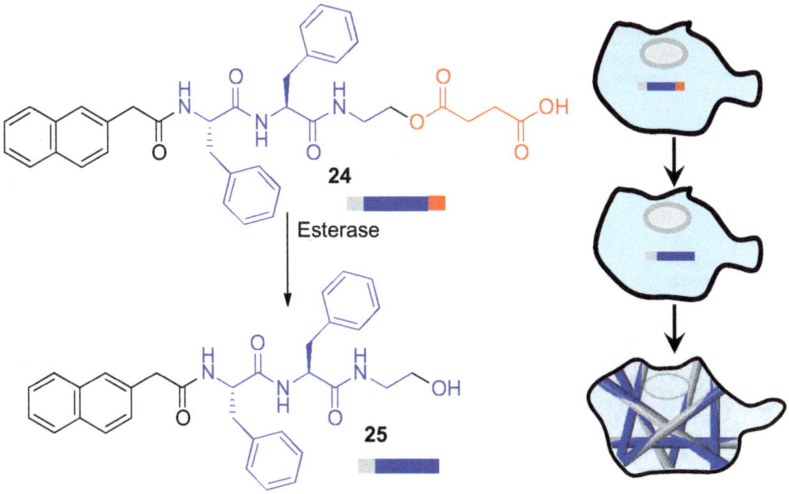

Figure 4.8 Chemical structures and schematic representation of intracellular cleavage of an esterase substrate (**24**) by an endogenous esterase to the hydrogelator (**25**) and corresponding schematic representation for the formation of supramolecular assembly within the cells.

the cells under investigation. At a certain concentration of **25** (0.04 wt%), the majority of human cervical cancer derived cells died within three days, while fibroblast NIH3T3 cells remained alive and dividing under the same conditions.

4.6.2 Antimicrobial Applications

To exploit enzymatic hydrogelation as a means to manage bacterial cells, Bing Xu and his team designed a phosphatase-sensitive small precursor molecule (**26**) that remains in solution in the absence of enzyme. By using the over-expressed alkaline phosphatase in E. coli bacteria, this precursor molecule is converted to the corresponding hydrogelator (**27**) inside the bacterial cell (Figure 4.9a). The subsequent intracellular hydrogelation inhibited the bacterial growth.[53] Isopropyl-b-d-thiogalactopyranoside (IPTG) and plasmids were used to induce the overexpression of phosphatase. The intracellular concentration of the hydrogelator was found to be significantly higher than that in the culture medium when IPTG was used, confirming the successful enzymatic conversion of precursor into hydrogelator as well as its accumulation inside the bacteria (Figure 4.9b). The formation of the hydrogel after

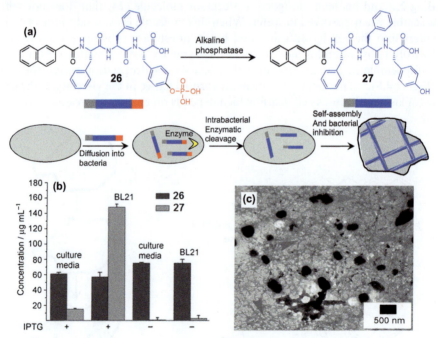

Figure 4.9   (a) On exposure to alkaline phosphatase, compound **26** is converted to com **27** which forms hydrogel. (b) Concentrations of **26** and **27** in the culture medium and within the cells (BL21, plasmid +, IPTG + or IPTG) (c) TEM images of the hydrogel formed inside the bacteria after culturing with **26** for 24 h (arrows indicate the nanofibres formed by **27**).
Reproduced with permission from ref. 53. Copyright 2007 Wiley-VCH Verlag GmbH & Co. KGaA.

lysis and the TEM image also confirmed the presence of nanofibres (Figure 4.9c). These experiments showed that intracellular enzymatic formation of supramolecular nanostructures can be used for the development of anti-microbial biomaterials.

In order to investigate whether morphological difference of nanostructures formed from different amphiphiles[42] could give different antimicrobial re-sponse, we have recently used the overexpressed enzyme alkaline phosphatase triggered self-assembly of the compounds **28–32** (Figure 4.10a) in *E. coli*, bacteria. The nonassembling precursors can be converted to self-assembling aromatic peptide amphiphiles *in vivo*, with the location of the products being directly linked to the hydrophobicity of the products – the more hydrophobic Fmoc-FY-OH products accumulating within the cells and the rest hydrophilic Fmoc-YX-OH peptides being found in the media. For all the peptides the bacterial deaths were found to be around 40% (Figure 4.10b). This response was insignificantly different regardless of the chemical structure of aromatic peptide amphiphiles, suggesting that nanoscale morphology is not a main factor in dictating antimicrobial activity of these materials.

4.6.3 Drug Delivery

Another promising field of the application of the enzymatic supramolecular assemblies is their use as localised drug-delivery systems. Issues in drug delivery include toxic effects at nontarget sites that reduce the efficiency of the drug. By exploiting locally overexpressed enzymes in dissolution of supramolecular gels it would be possible to release the drug in very specific locations. Two ways to

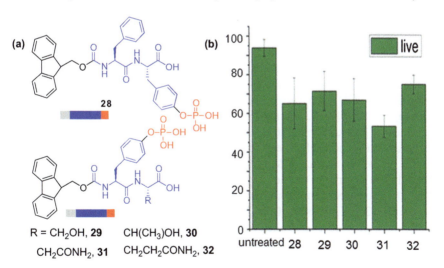

Figure 4.10 (a) Chemical structures of alkaline phosphatase responsive amphiphilic compounds **28–32**. (b) Percentage of active (live) in bacterial cultures after treatment with the phosphorylated precursors of self-assembling aromatic peptide amphiphiles **28–32**.

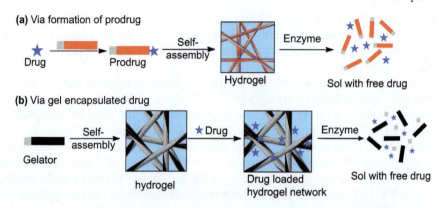

Figure 4.11   Schematic representation of drug release from (a) prodrug-based hydrogel and (b) Drug-encapsulated hydrogel.

deliver a drug from gel matrix have been investigated, *e.g.* a) *via* prodrug-based gel and b) *via* gel-encapsulated drugs (Figure 4.11).

4.6.3.1   Prodrug-Based Hydrogel

Researchers have modified (model) drug compounds as a low molecular weight hydrogel bearing an enzyme-cleavable linker, taking inspiration from the prodrug approach. Such prodrugs should encompass key functional groups that can promote self-assembly in aqueous solutions. Upon exposure to relevant enzymes, these prodrug-based hydrogels could cleave the specific bonds to release drug molecules.

A bioresponsive *cis*-diamminedichloroplatinum (II) (CDDP) derivative (**33**) was developed that in combination with self-assembling peptide amphiphile forms nanofibres that ultimately forms physical gel.[54] This prodrug contains a fatty acid with a MMP-2 sensitive peptide GTAGLIGQRGDS (Figure 4.12a). Release of CDDP derivative (**34**) from the prodrug gel was triggered by the cleavage of the MMP-2 sensitive peptide sequence that is expected to be between glycine (G) and leucine (L) in the mentioned peptide amphiphile (**33**). Enzymatic degradation of the CDDP-peptide amphiphile gel was confirmed by TEM images of the gel before and after degradation as shown in Figure 4.12b. The release of CDDP was dependent on the enzyme concentration in the medium. It was found that the increased concentration of MMP-2 increases the release of the drug (Figure 4.12c). Thus, the MMP-2-triggered CDDP release from the gel would potentially provide a controlled delivery system for targeted anticancer drugs.

Similarly Miravet and coworkers have synthesised a low molecular weight gelator (**35**) using a model drug where the drug and self-assembling units are connected through a self-immolative linker p-aminobenzoylcarbonyl that is stable under physiological conditions (Figure 4.13a).[55] This prodrug molecule forms excellent hydrogel with minimum gelator concentrations (MGC) as low as 0.1 w/v%. The gels were formed by sponge-like morphology with

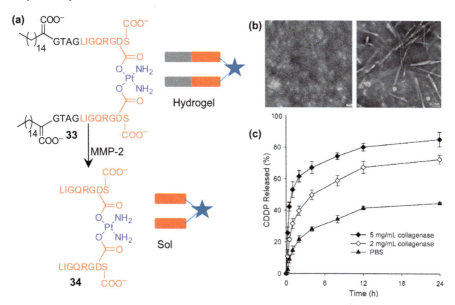

Figure 4.12 (a) Chemical structures of the prodrugs before and after the exposure of enzyme. (b) TEM images of nanofibre networks after degradation with 2 mg/mL MMP-2 enzyme. (c) Release profiles of CDDP (**34**) from CDDP-Peptide amphiphile (**33**) gels at different concentrations of type IV collagenase (MMP-2) solution.
Figure adapted from ref. 54. Copyright 2009 American Chemical Society.

micrometre-sized cavities filled by the solvent (Figure 4.13b). When the amide bond of the molecule is broken by trypsin, the amino group of p-aminobenzoylcarbonyl (**36**) becomes free and undergoes a rapid 1,6-elimination to carbamic acids that are unstable and decompose to release the model drug (**37**) (Figure 4.13c).

4.6.3.2 Drug-Encapsulated Hydrogel

Vemula *et al.* encapsulated a hydrophobic drug molecule, curcumin, in a hydrogel prepared from acetaminophen and a fatty acids based gelator (**38**) (Figure 4.14a).[56] The concentration of encapsulated curcumin is 10^5 times higher than the solubility of curcumin in water. This is because, due to the hydrophobic nature, curcumin is located at hydrophobic pockets of the gel. The formed hydrogel was degraded completely to form two nonself-assembling components (**39**, **40**) by the enzyme, lipase, while releasing the encapsulated chemopreventive hydrophobic drug curcumin that is monitored by time-dependent UV-Vis spectroscopy (Figure 4.14b). Control of the drug release rate was achieved by manipulating the enzyme concentration and temperature. It was found that the drug release was accelerated with the increase of concentration of enzyme and also with the increase of temperature.

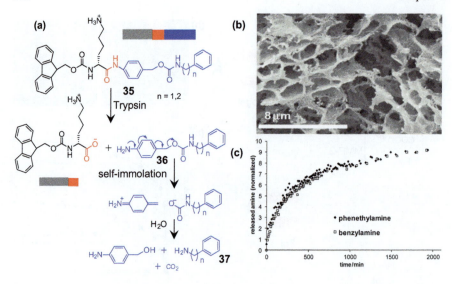

Figure 4.13 (a) Reaction of the prodrug after exposure of trypsin. (b) Cryo-SEM images of hydrogels 1. (c) Kinetic profile of the release of amines. Reproduced with permission from ref. 55.

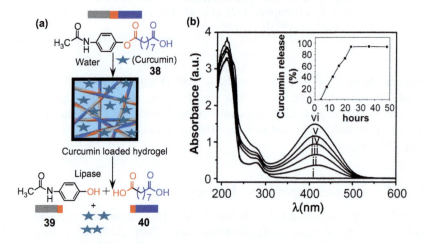

Figure 4.14 (a) Encapsulation of curcumin in hydrogel matrix followed by release of it by degradation with lipase. (b) % of curcumin release with time. Reproduced with permission from ref. 56.

4.6.4 Biocatalytic Self-Assembly for Imaging

Nanofibres, produced from self-assembly of small molecules in water, can be used as a simple system inside cells for regulating cellular processes. For this purpose it is highly desirable to visualise the self-assembly process inside the cell. In this context, Bing Xu and his team have designed a precursor molecule

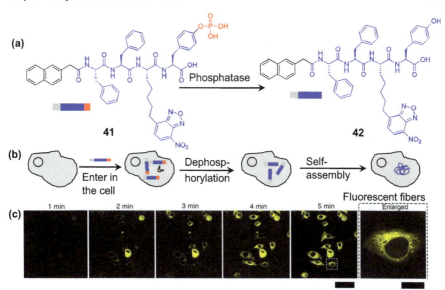

Figure 4.15 (a) On exposure to alkaline phosphatase, compound **41** is converted to compound **42** that forms a hydrogel, (b) Schematic representation of entering the precursor inside the cell and subsequent dephospholylation and self-assembly, and (c) Fluorescent confocal microscope images show the time course of fluorescence emission inside the HeLa cells incubated with 500 μM of **41** in PBS buffer (scale bar, 50 μm).
Reproduced with permission from ref. 57. Copyright Nature Publishing Group.

(**41**) (Figure 4.15) that produces the nanofibres of the hydrogelators (**42**) inside cells (Figures 4.15a and b) triggered by enzyme and visualised the process through microscopy.[57] It is expected that the precursor and the individual hydrogelators will exhibit low fluorescence, while the nanofibres of the hydrogelators display strong fluorescence. In the experiments after the addition of the precursor solution to the HeLa cells, the emission of the precursor outside the cells remains very low during the whole experiment (Figure 4.15c). The incubated cells become slightly fluorescent within 80 s, and a bright fluorescent area appears near the nucleus in one of the cells after 3 min. After 5 min, all the cells glow brighter at the centre of the cytoplasm than around the outer membrane of the cells.

4.7 Conclusions

Researchers are increasingly interested to mimic biology's approaches in an effort to enhance their control over bottom-up fabrication process. Enzymes have been recognised as a powerful means to dictate such nanofabrication process towards developing complex and highly selective next-generation nanoscale materials. Enzyme-assisted formation of supramolecular assembly have several unique features like self-assembly under constant conditions, spatiotemporal control of nucleation and structure growth, controlling mechanical properties and the defect-correcting and component-selecting abilities of systems that operate under thermodynamic control. In this way, they produce highly selective supramolecular nanostructures with fewer defects for their application in three-dimensional cell culture, antimicrobial materials, drug delivery, imaging, biosensing and supramolecular electronics.

References

1. G. M. Whitesides and B. A. Grzybowski, *Science*, 2002, **295**, 2418.
2. J. M. Lehn, *Angew. Chem. Int. Ed.*, 1990, **29**, 1304.
3. J. M. Lehn, *Supramolecular Chemistry – Concepts and Perspectives*, Wiley-VCH, Weinheim, 1995.
4. J. M. Lehn, *Polym. Int.*, 2002, **51**, 825.
5. R. F. Ludlow and S. Otto, *Chem. Soc. Rev.*, 2008, **37**, 101.
6. J. J.-P. Peyralans and S. Otto, *Curr. Opin. Chem. Biol.*, 2009, **13**, 705.
7. L. A. Estroff and A. D. Hamilton, *Chem. Rev.*, 2004, **104**, 1201.
8. Y. Zhang, Y. Kuang, Y. Gao and B. Xu, *Langmuir*, 2011, **27**, 529.
9. A. R. Hirst, B. Escuder, J. F. Miravet and D. K. Smith, *Angew. Chem. Int. Ed.*, 2008, **47**, 8002.
10. Z. Yang, H. Gu, D. Fu, P. Gao, J. K. Lam and B. Xu, *Adv. Mater.*, 2004, **16**, 1440.
11. Y. Gao, Y. Kuang, Z. F. Guo, Z. Guo, I. J. Krauss and B. Xu, *J. Am. Chem. Soc.*, 2009, **131**, 13576.
12. S. Toledano, R. J. Williams, V. Jayawarna and R. V. Ulijn, *J. Am. Chem. Soc.*, 2006, **128**, 1070.
13. X. Qin, W. Xie, S. Tian, J. Cai, H. Yuan, Z. Yu, G. L. Butterfoss, A. C. Khuonga and R. A. Gross, *Chem. Commun.*, 2013, **49**, 4839.
14. M. Colombo, R. J. Brittingham, J. F. Klement, I. Majsterek, D. E. Birk, J. Uitto and A. Fertala, *Biochemistry*, 2003, **42**, 11434.
15. A. Feratala, A. Sieron, Y. Hojima, A. Ganguli and D. J. Prockop, *J. Biol. Chem.*, 1994, **269**, 11584.
16. K. Kadler, Y. Hojima and D. J. Prockop, *J. Biol. Chem.*, 1987, **262**, 15696.
17. H. Lodish, A. Berk, P. Matsudaira, C. A. Kaiser, M. Krieger, M. P. Scott, S. L. Zipursky and J. Darnell, *Molecular Cell Biology*, 5th edn. Freeman, New York, 2003.
18. B. Alberts, A. Johnson, J. Lewis, M. Raff, K. Roberts and P. Walter, *The Cytoskeleton. Molecular Biology of the Cell*, Garland Science, New York, 2002, Chapter 16, 907.
19. A. Desai and T.J. Mitchison, *Annu. Rev. Cell. Dev. Biol.*, 1997, **13**, 83.
20. A Shome, S. Debnath and P. K. Das, *Langmuir*, 2008, **24**, 4280.
21. B. Ozbas, J. Kretsinger, K. Rajagopal, J. P. Schneider and D. J. Pochan, *Macromolecules*, 2004, **37**, 7331.
22. Y. Huang, Z. Qiu, Y. Xu, J. Shi, H. Lina and Y. Zhang, *Org. Biomol. Chem.*, 2011, **9**, 2149.

23. R. V. Ulijn, *J. Mater. Chem.*, 2006, **16**, 2217.
24. A. Dasgupta, J. H. Mondal and D Das, *RSC Adv.*, 2013, **3**, 9117.
25. S. Debnath, A. Shome, D. Das and P. K. Das, *J. Phys. Chem. B*, 2010, **114**, 4407.
26. E. T. Pashuck and S. I. Stupp, *J. Am. Chem. Soc.*, 2010, **132**, 8819.
27. K. J. Nagy, M. C. Giano, A. Jin, D. J. Pochan and J. P. Schneider, *J. Am. Chem. Soc.*, 2011, **133**, 14975.
28. M. Gungormus, M. Branco, H. Fong, J. P. Schneider, C. Tamerler and M. Sarikaya, *Biomaterials*, 2010, **31**, 7266.
29. Z. Yang, G. Liang, M. Ma, Y. Gaoa and B. Xu, *J. Mater. Chem.*, 2007, **17**, 850.
30. J. Nanda, A. Biswas and A. Banerjee, *Soft Matter.*, 2013, **9**, 4198.
31. X. Li, Y. Kuang, H. C. Lin, Y. Gao, J. Shi and B. Xu, *Angew. Chem. Int. Ed.*, 2011, **50**, 9365.
32. A. Mahler, M. Reches, M. Rechter, S. Cohen and E. Gazit, *Adv. Mater.*, 2006, **18**, 1365.
33. A. M. Smith, R. J. Williams, C. Tang, P. Coppo, R. F. Collins, M. L. Turner, A. Saiani and R.V. Ulijn, *Adv. Mater.*, 2008, **20**, 37.
34. M. Hughes, H. Xu, P.W.J.M. Frederix, A.M. Smith, N.T. Hunt, T. Tuttle, I. A. Kinloch and R.V. Ulijn, *Soft Matter.*, 2011, **7**, 10032.
35. B. H. Hu and P. B. Messersmith, *J. Am. Chem. Soc.*, 2003, **125**, 14298.
36. J. B. Guilbaud, E. Vey, S. Boothroyd, A. M. Smith, R. V. Ulijn, A. Saiani and A. F. Miller, *Langmuir*, 2010, **26**, 11297.
37. A. K. Das, A. R. Hirst and R. V. Ulijn, *Faraday Discuss.*, 2009, **143**, 293.
38. A. K. Das, R. Collins, A. Hirst and R. V. Ulijn, *Small*, 2008, **4**, 279.
39. H. Xu, A. K. Das, M. Horie, M. S. Shaik, A. M. Smith, Y. Luo, X. Lu, R. Collins, S. Y. Liem, A. Song, P. L. A. Popelier, M. L. Turner, P. Xiao, I. A. Kinloch and R. V. Ulijn, *Nanoscale*, 2010, **2**, 960.
40. M. Hughes, P. W. J. M. Frederix, J. Raeburn, L. S. Birchall, J. Sadownik, F. C. Coomer, I. H. Lin, E. J. Cussen, N. T. Hunt, T. Tuttle, S. J. Webb, D. J. Adams and R. V. Ulijn, *Soft Matter.*, 2012, **8**, 5595.
41. R. J. Williams, T. E. Hall, V. Glattauer, J. White, P.J. Pasic, A. B. Sorensen, L. Waddington, K. M. McLean, P. D. Currie and P. G. Hartley, *Biomaterials*, 2011, **32**, 5304.
42. M. Hughes, L. S. Birchall, K. Zuberi, L. A. Aitken, S. Debnath, N. Javid and R. V. Ulijn, *Soft Matter*, 2012, **8**, 11565.
43. A. R. Hirst, S. Roy, M. Arora, A. K. Das, N. Hodson, P. Murray, S. Marshall, N. Javid, J. Sefcik, J. Boekhoven, J. H. van Esch, S. Santabarbara, N. T. Hunt and R. V. Ulijn, *Nature Chem.*, 2010, **2**, 1089.
44. J. W. Sadownik, J. Leckie and R. V. Ulijn, *Chem. Commun.*, 2011, **47**, 728.
45. Z. Yang, G. Liang and B. Xu, *Acc. Chem. Res.*, 2008, **41**, 315.
46. Z. M. Yang, G. L. Liang, M. L. Ma, Y. Gao and B. Xu, *Small*, 2007, **3**, 558.
47. S. D. Santos, A. Chandravarkar, B. Mandal, R. Mimna, K. Murat, L. Saucede, P. Tella, G. Tuchscherer and M. Mutter, *J. Am. Chem. Soc.*, 2005, **127**, 11888.

48. L. A. Abramovich, R. Perry, A. Sagi, E. Gazit and D. Shabat, *ChemBio-Chem.*, 2007, **8**, 859.

49. Z. M. Yang, G. L. Liang, L. Wang and B. Xu, *J. Am. Chem. Soc.*, 2006, **128**, 3038.

50. R. J. Williams, S. M. Smith, R. Collins, N. Hodson, A. K. Das and R. V. Ulijn, *Nature Nanotechnol.*, 2009, **4**, 19.

51. S. K. M. Nalluri and R. V. Ulijn, Chem. Sci. DOI: 10.1039/C3SC51036K.

52. Z. M. Yang, K.M. Xu, Z. F. Guo, Z. H. Guo and B. Xu, *Adv. Mater.*, 2007, **19**, 3152.

53. Z. Yang, G. Liang, Z. Guo, Z. Guo and B. Xu, *Angew. Chem. Int. Ed.*, 2007, **46**, 8216.

54. J. K. Kim, J. Anderson, H. W. Jun, M. A. Repka and S. Jo, *Molec. Pharmac*, 2009, **6**, 978.

55. J. A. Saez, B. Escuder and J. F. Miravet, *Tetrahedron*, 2010, **66**, 2614.

56. P. K. Vemula, G. A. Cruikshank, J. M. Karp and G. John, *Biomaterials*, 2009, **30**, 383.

57. Y. Gao, J. Shi, D. Yuan and B. Xu, *Nature Commun.*, 2012, **3**, 1033.

CHAPTER 5

Molecular Gels as Containers for Molecular Recognition, Reactivity and Catalysis

JUAN F. MIRAVET* AND BEATRIU ESCUDER*

Departament de Química Inorgànica i Orgànica, Universitat Jaume I, 12071 Castelló, Spain
*Email: miravet@uji.es; escuder@uji.es

5.1 Introduction

The self-assembly of functional molecular building blocks into organised supramolecular architectures has been intensely studied in recent decades. The bottom-up construction of functional assemblies offers a unique opportunity for the study of reactivity and molecular recognition in an environment reminiscent of enzyme active sites in which the proximity of functional groups, steric effects and solvophobic effects, among others play a relevant role. Particularly interesting is the emergence of unusual chemical reactivities, as for instance changes in pKa values and multivalency, as a consequence of the creation of preorganised arrays of binding groups.[1,2]

The study of molecular recognition and reactivity in organised assemblies is a wide area of research in physical organic chemistry that includes, for instance, the study of chemical reactions within confined self-assembled microreactors such as micelles, vesicles, polymersomes, (nano)capsules, nanotubes as well as on organised surfaces such as monolayers or nanoparticles among others.[3–5]

RSC Soft Matter No. 1
Functional Molecular Gels
Edited by Beatriu Escuder and Juan F. Miravet
© The Royal Society of Chemistry 2014
Published by the Royal Society of Chemistry, www.rsc.org

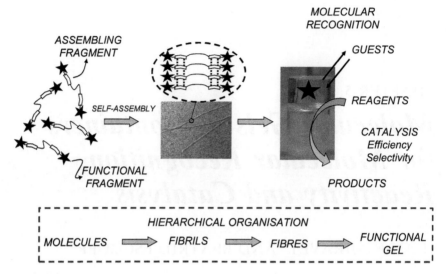

Figure 5.1   Hierarchical organisation of functional molecular gels.

In this context, the study of functional molecular gels as containers for molecular recognition and chemical reactions is quite recent, as the field of molecular gels has blossomed in the last two decades (Figure 5.1).[6,7]

5.2   The Container: Molecular Gels

Molecular gels are colloidal systems that may be described as two separate phases, a microporous solid network filled by pools of solvent (Figure 5.2). The self-assembly process of gelator molecules is determined by its solubility and the degree of cooperativity of the supramolecular polymerisation process.[8] For highly co-operative self-assembly ($K_2 < K_n$, K_2 and K_n being the dimerisation and elongation constants, respectively), the gel will consist of a network of long fibres (large aggregates) in equilibrium with a fraction of gelator monomer free in solution (determined by its solubility). For isodesmic self-assembly ($K_2 \approx K_3 \approx \ldots K_n$), the gel will consist of a mixture of oligomeric aggregates of different sizes.[9]

The composition of the gel has to be considered prior to the addition of other molecules as they will interact with all the components of the gel system (free gelator in solution, soluble oligomers and aggregates in the gel phase) and even though these components have the same molecular composition in all the cases, molecular conformations and reactivity may be different. For instance, hydrogelators bearing carboxyl groups have been described to experience an increase in apparent pKa values on going from diluted solutions to aggregates.[10,11] This phenomenon, well known to happen for ionisable groups in enzyme active sites,[12] has been attributed to the unfavourable charge–charge interactions that would occur between neighbouring carboxyl groups being deprotonated simultaneously and it is enhanced in parallel with the increase in gelator's hydrophobicity – namely, degree of aggregation.

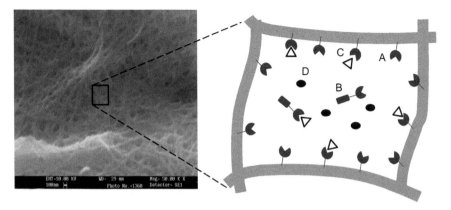

Figure 5.2   Microscopic aspect of the gel network and schematic representation of a network's compartment and guest molecules. (A: gelator molecules in the fibres; B: gelator in solution; C: interacting guest molecules; D: noninteracting guest molecules).

On the other hand, the presence of the fibrillar network introduces a barrier for diffusion of molecules through the gel that will be slow compared to diffusion in pure solvent. Quite often, gels are loaded with other compounds by adding a supernatant solution and allowing free diffusion. Even in the absence of strong intermolecular interactions between the added compounds and the gel phase, diffusion of these solutions may take several hours. For some applications this may not be the ideal case. Alternatively, additives may be entrapped during the gelation process by codissolution with the gelators (*in situ* loading). In such a case they must not interfere with the gelator self-assembly, otherwise the gel may even be destroyed.

What happens once a guest molecule has reached a solvent compartment within the gel network? As is schematised in Figure 5.2, guest molecules will be distributed between the solution and the gel phases depending on their relative affinities.

It has been shown that noninteracting guests within the solvent pools held by the gel network behave as if they were in true independent solutions. For instance, photochemical quenching rates, which measure diffusion rates in the range of nanometres, showed that molecules trapped into solvent pools of a hyaluronan gel diffuse as fast as in solution.[13] A similar behaviour has been described by Galindo and coworkers for a molecular gel.[14] On the other hand, Hanabusa and coworkers have studied organogel electrolytes and have found that the ionic conductivity is only slightly affected by the presence of the gel network.[15] Furthermore, Duncan and Whitten have studied molecular gels by NMR and reported that the line width of NMR solvent signals is not affected by gel formation, revealing that the macroscopic viscosity increase does not affect the tumbling rates of the solvent molecules.[16]

In our group we have used NMR in order to study interaction and molecular recognition of small molecules in molecular gels and we have also observed that

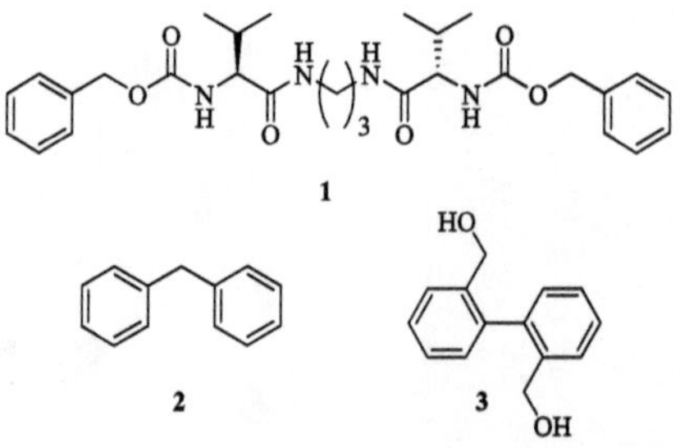

Scheme 5.1   Structures of organogelator **1** and guest compounds.

molecules that are not interacting with the gel network remain unaffected as if they were in pure solvent.[17] For instance, we studied the NMR of compounds **2**, a nonpolar molecule, and **3**, bearing two H-bonding groups, in the presence of a gel formed by **1** in C_6D_6 (Scheme 5.1). We could observe that compound **3**, capable of H-bonding interaction with the gel phase shows a decrease of T_2 values, meaning a slower tumbling rate than the corresponding solution in the absence of the gel ($T_{2sol} = 1.77$ s, $T_{2gel} = 1.20$ s). On the contrary, T_2 values for compound **2**, noninteracting with gel phase and well solvated by C_6D_6 were unaffected by the presence of the gel ($T_{2sol} = T_{2gel} = 4.95$ s).

A relevant example of noninteracting species is represented by the orthogonal self-assembly of gelators – the so-called "self-sorting" – in which non-mutually interacting molecules may develop independent networks of fibres or even other self-assembled systems such as micelles or vesicles.[18,19]

In summary, molecular gels may behave as compartmentalised sample holders that physically entrap solutions or aggregates of noninteracting molecules. However, they may be also chemically active and, as we shall see in the following sections, have a strong influence in relevant processes such as the stabilisation of proteins and drug crystals, the selective molecular recognition of guests and catalysis among others.

Here, we will discuss all these phenomena in depth through a selection of recent illustrative examples. The chapter will be divided into three sections: 1) host–guest noncovalent effects, 2) reactive gels that form covalent bonds and 3) catalytic gels.

5.3   Host–Guest Noncovalent Effects

Guest compounds loaded in the gel network may interact with gelator molecules by all kind of noncovalent interactions (H-bonding, van der Waals interactions, π-stacking, solvophobic interactions, *etc.*). As these weak interactions are the same that sustain the gel network several competitive equilibria will be established and depending on which one dominates the system will be driven to either the incorporation of the guest into the gel phase or the weakening and disassembly of the gel network. This mutual influence has to be carefully considered when a given application is sought. For instance, the incorporation of the guest into the gel phase is convenient for the use of this material as a scavenger, whereas the disassembly of the gel is relevant for the use as a chemically responsive material (see Chapter 3).

5.3.1   Noncovalent Interactions with Small Organic Molecules

Shinkai and coworkers reported one of the first examples of molecular recognition within molecular gels.[20] Naphthalenediimide derivative **4** forms organogels by using π–π stacking, hydrogen bonding and van der Waals interactions (Figure 5.3). These gels are able to interact with naphthalenediols

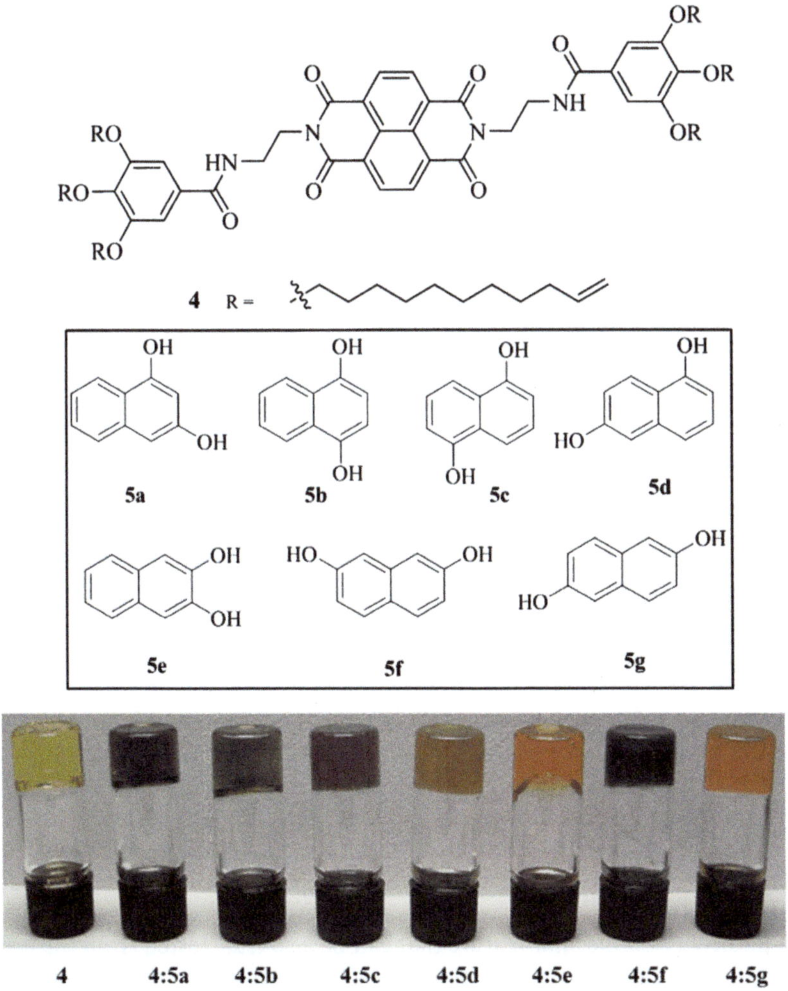

Figure 5.3   NDI-based organogelator **4** and dihydroxynaphthalene guests **5a–g**. Photograph showing the colour changes of organogel **4** upon addition of different guests.
Adapted with permission from ref. 20. Copyright 2006 John Wiley and Sons.

by intercalation between consecutive naphtalenediimide (NDI) moieties forming 1D π-stacked donor–acceptor systems. Furthermore, positional isomers (**5a–g**) may be distinguished by the naked eye due to a selective photochromic response of the gel phase at a very low concentration of compound as compared with usual sensors that operate in solution. Additionally, in a related work, they have shown that compounds **5a** and **5d** could act as "molecular adhesives" for the self-healing process of the organogel formed by compound **4**. This organogel shows thixotropic behaviour, namely, it thins under mechanical stress and it is reconstituted during resting time. However, the regenerated fibres are shorter, with a large number of active ends. The authors envisage that the addition of a donor chemical stimulus able to glue the fibre ends would enhance the rate of recovery. Indeed, after vortexing the organogel for 10 min 1,3-dihydroxynaphthalene (**5a**) was added and the organogel was recovered twice as faster as without the additive.[21]

We, together with Hayes' group, have also reported a very simple compound (**6**) that forms hydrogels by pH tuning and without the need of heating by a combination of orthogonal π–π stacking and H-bonding interactions. The gel phase presents a layered structure – similar to the structure found for single crystals grown in water:methanol mixtures – that is able to intercalate flat aromatic dyes such as methylene blue (Scheme 5.2). Moreover, the partial deprotonation of carboxyl groups in the gel allows for the selective adsorption of positively charged dyes.[22]

As mentioned before, the precise organisation of functional groups on the surface of the fibres may lead to the emergence of multivalent binding sites.[23] For instance, compound **7** forms gels in acetonitrile with an extended β-sheet-like conformation presenting an array of pyridyl groups accessible on the surface of the aggregates that are able to interact with hydroxylated aromatic compounds by H-bonding. This interaction is stronger for divalent guests and in particular for those with a distance between the two OH that matches with the disposition of the pyridyl binding sites of the assemblies – resorcinol (**8**) and 2,7-dihydroxynaphthalene (**5f**) – and causes the collapse of the gel network (Figure 5.4).[24]

The solvent also plays a relevant role both in the shaping of the gel container as well as a modulator of the noncovalent interactions. For example, compound **7**, that also forms gels in toluene, presents a completely different interaction against phenolic compounds from that in acetonitrile. In this case, the organogel is selectively disassembled in the presence of catechol (**11**) by interaction with gelator molecules by H-bonding and π-stacking (Scheme 5.3).[25]

Chirality is a molecular feature that can be also transferred from the primary structure of the gelator into the supramolecular level in molecular gels.[26] The gel formed by an enantiomerically pure gelator can be seen as a chiral vessel and therefore enantioselective interactions with guest molecules may appear. For example, Ihara and coworkers have described recently an L-glutamide-functionalised zinc porphyrin (**12**) that forms chiral gels in apolar organic solvents such as cyclohexane (Scheme 5.4). They have shown by CD that the

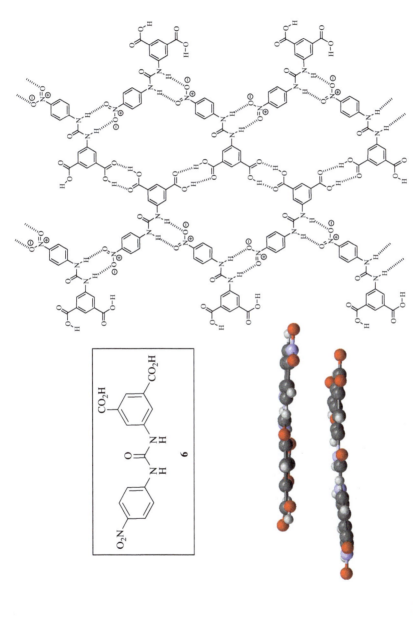

Scheme 5.2   H-bonding pattern and layered structure found in water:methanol single crystals of hydrogelator **6**.

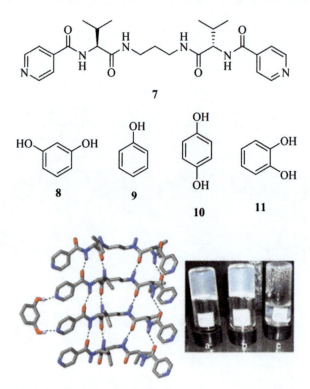

7

8   **9**

10   **11**

Figure 5.4 (A) Structures of gelator **7** and phenolic guests. (B) Molecular model of the interaction of aggregated compound **2** and resorcinol (**8**) (left) and pictures of samples prepared by gelation of **2** in acetonitrile together with phenol derivatives (right: (a) phenol (**9**), (b) hydroquinone (**10**) and (c) resorcinol (**11**)).

aggregation induces a strong supramolecular Cotton effect for the Soret band of the achiral porphyrin fragment. This system has been studied for the enantio-selective axial coordination of amino acid methyl esters and a particularly high enantioselectivity has been observed for L-histidine derivatives – imidazole resi-due binds stronger than primary amino groups to Zn porphyrins.[27]

Recently, several groups have reported on enantioselective gel collapsing as a tool for visual chiral sensing. For instance, Pu and coworkers have reported BINOL-terpy Cu(II) complex **13** that forms ultrasound-induced gels in CHCl$_3$ and can be used for the visual sensing of chiral amino alcohols.[28] For example, the addition of 0.1 eq of (S)-phenylglycinol to a gel of (R)-**13** in CHCl$_3$ caused gel collapse after 2 min of sonication whereas after the addition of 0.1 eq of (R)-phenylglycinol the gel remained stable and additional 0.1 eq of additive was necessary to break the gel. The reverse behaviour was observed for the gel of (S)-**13** (Scheme 5.4). Following the same approach Tu *et al.* described an ALS (aromatic-linker-steroid) chiral organometallic gelator with a bipyridine ligand that is coordinated to platinum (**14**) (Scheme 5.4).[29] When this gel is mixed with 0.1 eq. of bulky phosphine chiral ligands, only with the *R* enantiomers of the

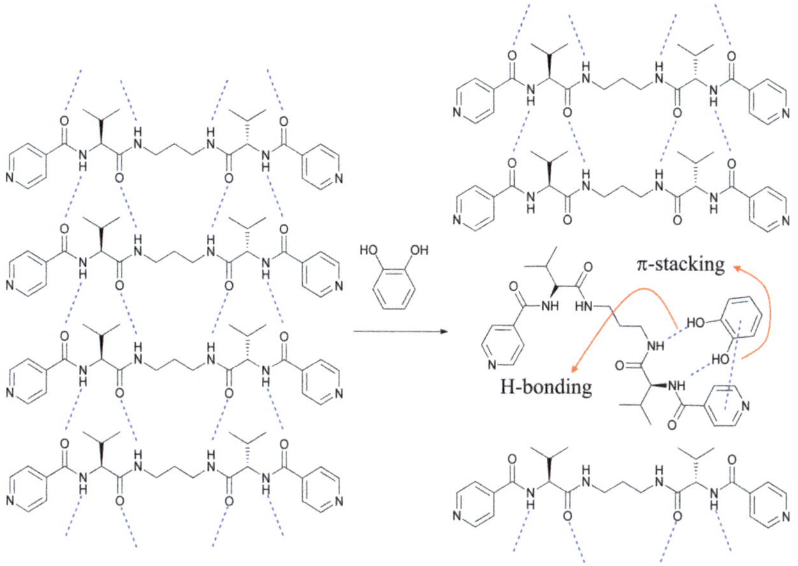

Scheme 5.3   Disruption of gel obtained from **7** in toluene due to noncovalent interactions with catechol (**11**).

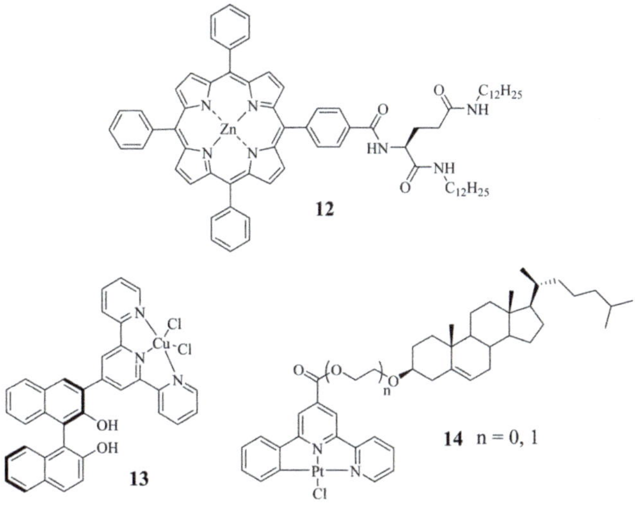

Scheme 5.4   Structures of gelators with a chiral response.

ligands does the gel remain stable. On the other hand, 0.1 eq. of the *S* enantiomer drives to the collapse of the gel. The results are widely supported by microscopy pictures as well as CD and NMR. However, the mechanism of gel disruption still remains unravelled, as when using other proportions of the ligands no difference is observed between the effects of both enantiomers. Nevertheless, this study stands out as a promising visual sensor for chiral discrimination.

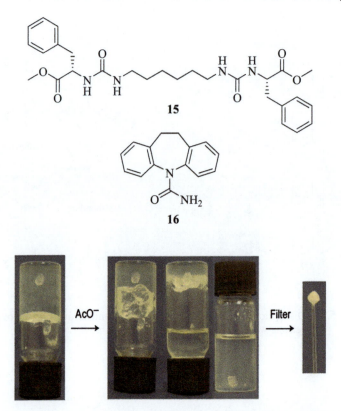

Figure 5.5 A single crystal of compound **16** is grown in an organogel of **15**. The crystal is recovered after gel disruption when adding acetate.
Reprinted with permission from Macmillan Publishers Ltd: Nature Chemistry (ref. 30), copyright (2010).

The noncovalent interaction of molecular gels and small molecules has been also explored in the field of crystal growth. For instance, Steed and coworkers have reported the relevance of gel–small-molecule interactions for the crystallisation of polymorphs of drugs within the matrix of bis(urea) organogels (**15**) (Figure 5.5). They have shown that besides the improved quality of the crystals related to the viscosity effect of the gel medium, differences in polymorphic preference appeared depending on the gelator and drug structures as well as on the solvent employed. Additionally, the presence of urea groups responsive to the addition of anions allowed for the easy recovery of the crystals by simple addition of tetrabutylammonium acetate.[30]

5.3.2 Stabilisation/Activation of Reactive Intermediates and Protein Structure

Shumburo and Biewer described the use of molecular organogels formed by compound **17** for the stabilisation of photomerocyanine species (PM) in

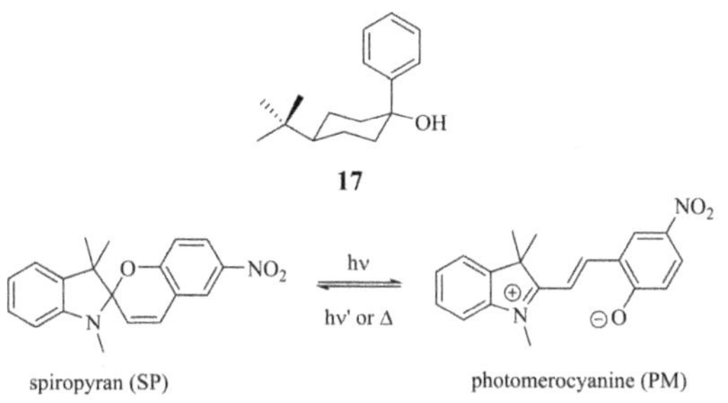

Scheme 5.5   Structure of gelator **17** and scheme of the spiropyran photochromism.

spiropyran (SP) photochromic switches (Scheme 5.5). Measurements of half-life for the PM species within gels of **17** in mineral oil reveal an enhancement of its lifetime by a factor of 195 with respect to the solution. Intriguingly, this effect is not as strong for other related SP derivatives and the authors correlate it with specific interactions between PM and the gel fibres.[31]

Molecular gels have also been employed for the confinement of proteins revealing interesting results both on protein structure and function. Hydrophobic microenvironments created within the fibrillar network of hydrogels have been described to minimise the denaturation of proteins.[32] The fibrillar network of hydrogels has been shown to mimic the cellular environment and it has been used for biomimetic processes. In this field, Xu and coworkers have reported several examples of the use of molecular hydrogels for the increase of activity and stability of enzymes in aqueous as well as in organic media.[33]

For instance, they have reported simple amino acid-based molecular hydrogels (**18** and **19**) that mimic the cellular environment of bioluminescence.[34] These materials have been used for the confinement of heme proteins (met-hemoglobin or horseradish peroxidase) and the study of the chemoluminescence (CL) of the catalytic oxidation of luminol (**20**) by H_2O_2 (Scheme 5.6A). Compared with the reaction in solution, the incorporation of the reagents in the hydrogel enhanced the quantum yield of CL by more than ten times as well as the emission half-life of the oxidised product. The authors suggest that the nanofibres of the hydrogel provide a protein-like viscous medium and hydrophobic nanoenvironment that capture the fluorophore and reduce its radiationless decay, improving the quantum yield.

They have also used the same hydrogels as the scaffold to encapsulate hemin as the prostetic group for an artificial peroxidase (Scheme 5.6B). For that purpose, a hydrogel was prepared incorporating hemin (**22**) and histidine (**23**) as Fe axial ligand. The oxidation of pyrogallol (**24**) was used as a model reaction and it was shown that the hydrogels exhibited higher activity than free hemin in the buffer solution and other encapsulated hemin systems. The authors suggested that the localisation of the hemin on the nanofibres avoids the dimerisation and inactivation of the catalyst and that the nanopores in the hydrogel facilitate the access of the substrate to the catalytic site. Nevertheless, the activity was still below that of the natural horseradish peroxidase (HPR).

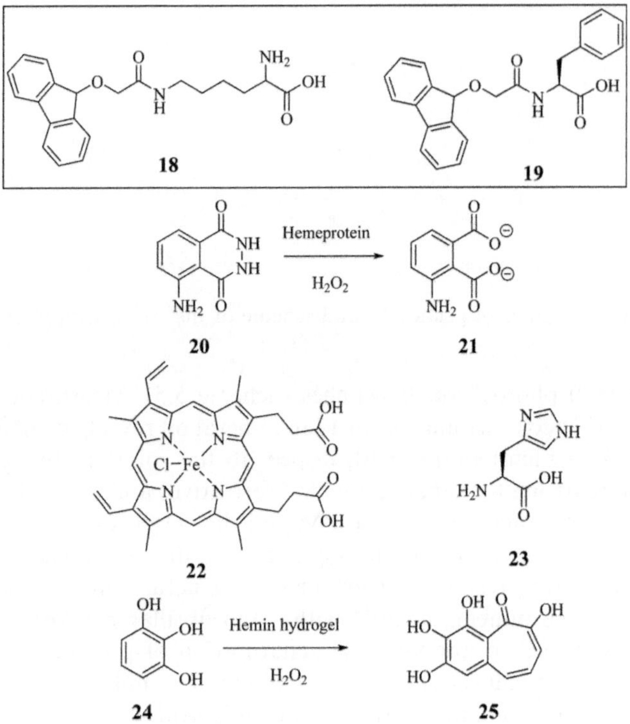

Scheme 5.6   (A) Structures of hydrogelators **18** and **19**. (B) Biocatalytic oxidation of luminol (**20**). (C) Prostetic groups of the artificial enzyme. (D) Oxidation of pyrogallol (**24**).

However, better results were obtained when this system was used in an organic solvent, toluene. In that case an increase in activity was observed compared to the system in aqueous solution as well as to free hemin in toluene, achieving about 60% of the catalytic activity of natural HRP. They ascribed this improvement to the phase equilibrium of reactants and products between the nonpolar solvent and polar regions of the artificial enzyme.[35]

Nanofibrillar aggregates formed by low molecular weight compounds have also been shown to specifically interact with dormant proteases such as procaspases, enzymes involved in biologically relevant transformations, and activate them through an allosteric mechanism that promotes autoproteolysis. Wells and coworkers have shown that compound **26** self-assembles into fibres of 2.6 nm thickness and micrometre length (Figure 5.6).[36] Procaspase-3 becomes immobilised on the surface of the fibres and generates the active caspase-3. The authors suggest that the fibrils are acting as scaffolds that concentrate procaspase-3 and then it can be processed by other enzymes or alternatively, the conformation of the proenzyme may be altered in a way that promotes intramolecular processing. They also find structural and functional analogies with the fibrous β-sheet aggregates formed by amyloid proteins,

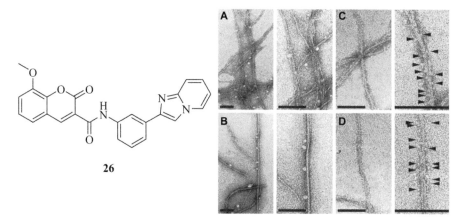

Figure 5.6 Structure of compound **26** and TEM images of its nanofibrils with and without procaspase-3. Negatively stained fibrils (A) at 25 °C and (B) at 37 °C. (C, D) Fibrils decorated with procaspase-3 (arrowheads) at 25 °C and 37 °C, respectively. Scale bars = 100 nm.
Adapted with permission from ref. 36. Copyright 2011 American Chemical Society.

suggesting the role played by nanofibrils in the activation of procaspases. Although in this case the nanofibres do not form a self-sustainable hydrogel it exemplifies the potential of the interaction between nanostructures formed by small synthetic molecules and large biomolecules with envisaged applications in nanomedicine. Other examples have been described, in which peptide amphiphiles bearing active epitopes have been used for the specific activation of biological processes (see Chapter 6) however, the example discussed here reports on a molecule of about only 400 Da.

5.3.3 Templates for Reactions in Solution: Polymer Imprinting

The nanostructured gel matrix may be used as a template for reactions taking place in solution. Relevant examples are the assistance of the gel for the controlled synthesis of nanoscale objects such as nanoparticles or the transcription of the gel network nanostructures into inorganic materials such as silica among others. In these examples, functional groups on the surface of the fibres may act as catalysts for the reaction or nucleation sites for the growth of nanoparticles. These inorganic materials are discussed in detail in Chapter 8.

Here we will focus on the use of gel nanostructures for the preparation of imprinted organic polymers. These materials can be obtained when a gel is prepared in a solvent with polymerisable groups that can be crosslinked, leading to a bulk polymer that after removal of the template nanostructures will show their mould. This strategy was first reported by two groups in 1997. Weiss and coworkers reported the use of tetraoctadecylammonium bromide (**27**) as a gelator for methyl methacrylate and styrene and the imprinting of mesoscopic

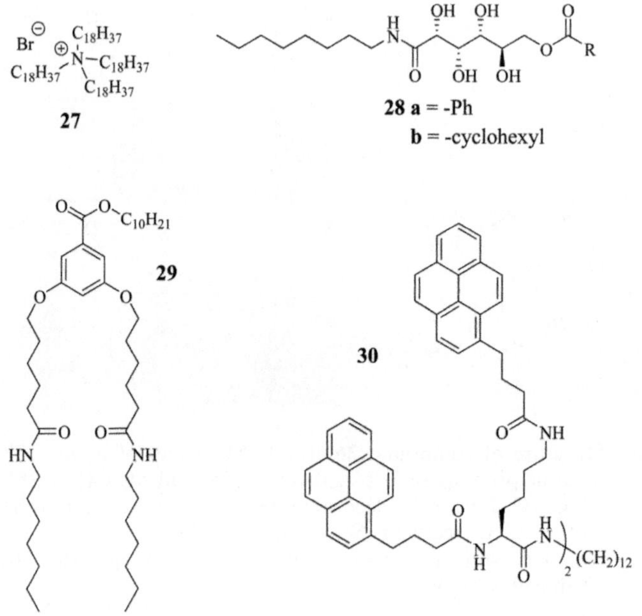

27

28 a = -Ph

b = -cyclohexyl

29

30

Scheme 5.7 Examples of gelators used as templates for polymer imprinting.

channels.[37] In parallel, Nolte and coworkers reported the transcription of the fibrillar nanostructures formed in methacrylate mixtures by n-octyl gluco-namides (**28**) after polymerisation of the solvent (Scheme 5.7). Nanopores could be observed, however, the imprinting process failed in the transcription of the supramolecular chirality observed in the original gel probably due to the shrinking of the methacrylate gel during polymerisation.[38]

More recently, Mésini and coworkers have succeeded in the transcription of supramolecular chirality from helical tapes formed by 3,5-bis(5-hexyl-carbamoylpentyloxy) benzoic acid decyl ester (**29**) in ethylene glycoldiacrylate into helical pores after photopolymerisation and extraction of the template with CH_2Cl_2 (Scheme 5.7).[39] The same helical tapes, closed into nanotubes, have been used by this group for the preparation of imprinted nanotubes with application as catalytic materials.[40]

In most of these examples the gelator is washed-out from the material after imprinting because the transcription of the nanostructured shape is the main objective. However, it may also be interesting to keep the supramolecular nanostructures embedded within the polymer especially when the gelator in-corporates a useful functionality. For example, Moffat and Smith have re-ported the preparation of fluorescent "two-faced" polymer wafers by using gels formed by pyrene-based compound **30** as templates for styrene/divinylbenzene polymerisation. They show that during the reaction some gelator molecules migrate towards one of the faces of the polymer leading to a material with different luminescent properties in both sides of the wafer.[41]

5.4 Reactive Gels: Formation of Covalent Bonds

Molecular gels formed by compounds bearing reactive functional fragments (electrophilic, nucleophilic, photosensitive) allow for the postmodification of the gel after self-assembly. These (photo)chemical reactions could be induced by an external reagent or a physical stimulus (light or heat) and would produce changes in the morphology and the rheology characteristics of the gel network. For instance, as we shall see later the photochemical cleavage of a fragment of the gelator may cause the gel weakening and disassembly or the addition of a nucleophilic reagent that reacts with an electrophilic centre in the gel phase may reinforce the network. Both results may be relevant for applications; the disassembly in response to a chemical stimulus may be useful for sensing or for the release of entrapped components, whereas covalent capture of the gel nanostructures may be useful to improve the mechanical properties of the material. In this section we will discuss some cases in which covalent bonds are being formed or broken in the gel phase, either in an irreversible or reversible way.

5.4.1 Reactions of Gelator Molecules with an Effect on Gel Rheology

Almost ten years ago, Koshima and coworkers described for the first time a photochemical reaction in the gel state (Scheme 5.8).[42] A gel of benzophenone-based compound **31** in 2-propanol was irradiated and gradually converted into a solution. Careful analysis revealed that the pinacol product **32** was obtained only in a 20% yield and as a diastereomeric mixture, showing that the photoreaction was not stereospecific. Moreover, the reaction in the gel was much slower than in solution, and other degradation products were also obtained. The authors suggested that the opaqueness of the samples and the lack of mobility in the gel phase could explain these poor results.

Closely in time, a landmark in the field was established by Feringa and coworkers by the design of a system that allows photocontrol of chirality on the molecular as well as the supramolecular level (Scheme 5.9).[43] They reported a system based on a dithienylethene photochromic unit functionalised with chiral assembling fragments (**33**) that may exist as two interconvertible open forms with *P*- and *M*- helicity that can be reversibly converted into two closed diastereoisomers (**34**) upon irradiation with UV light. Compound **33** forms gels in organic solvents such as toluene and molecular chirality is transferred into the

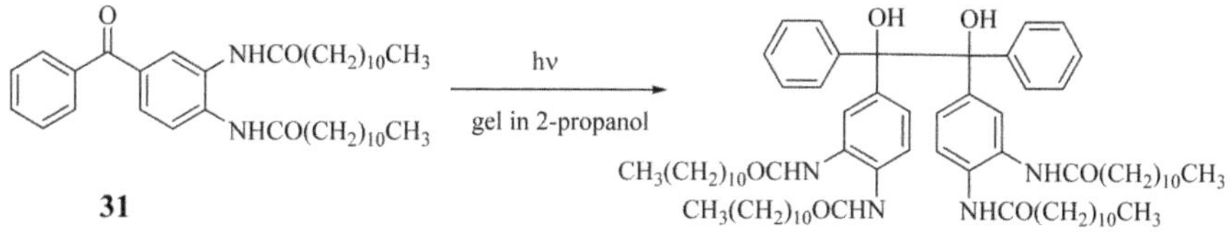

Scheme 5.8 Photoreaction of benzophenone-based gelator **31**.

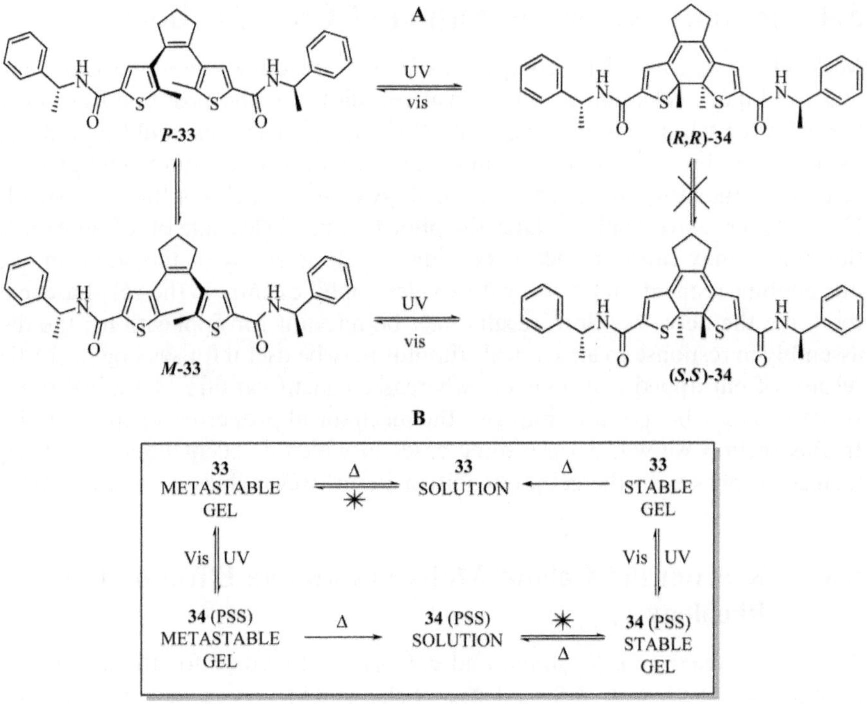

Scheme 5.9   (A) Photocontrolled dithienylethene chiral switch. (B) Aggregation and switching processes and relation between different states (PSS: photo-stationary state).

supramolecular level by the formation of helical fibres. Gels formed by the open form at low temperature lock compound **33** into one chiral conformation (*P*) in the gel and the chirality is maintained after UV irradiation (313 nm) in the closed form (***R,R*-34**) with a large diastereomeric excess (96% de). However, the gel formed by ***R,R*-34** (*P*-helicity) is metastable and after a heating–cooling cycle a thermodynamically stable gel with inverted supramolecular chirality (**S,S-34**, *M*-helicity) is formed. Irradiation of the *M*-helical gel with visible light leads to the formation of an open form gel (**33**) but with retention of the *M*-helicity. This gel is metastable and can be converted back to the *P*-helical initial gel by a heating–cooling cycle. This example reveals that the rigid environment of the gel phase preserves the supramolecular chirality during the photoreaction and allows the access to metastable aggregates of opposite chirality.

On the other hand, some years ago our group demonstrated that reactive groups on the gel phase are accessible to reactive small molecules (Scheme 5.10).[44,45] Thus, gels formed by compound **35** in acetonitrile were reacted with amines to produce bis-urea derivatives **36a–d** that formed strong gels *in situ*. Compounds **36a–d** were also prepared by conventional synthesis in solution and tested for gelation. They resulted as being highly insoluble, strong heating was required to dissolve them and the obtained gels were opaque and

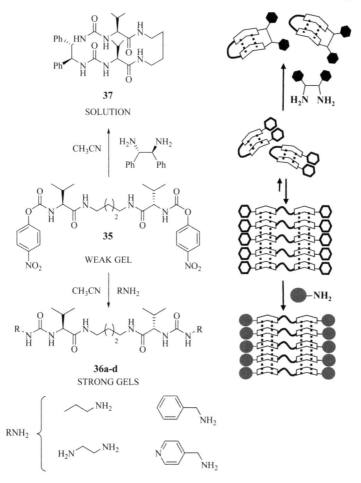

Scheme 5.10   Reactive gelator **35** and its bisurea products (left). Self-assembly and reaction scheme (right).

poorly stable, in contrast to the soft conditions required for the *in situ* reaction-gelation procedure.

An intriguing question that appeared – and should be considered in general for reactions in gels – is whether the reaction was taking place in the aggregated phase (gel-to-gel) or with the free gelator molecules in solution in equilibrium with the gel phase. The study of the reaction with diamines was used to shed light on that issue. First, the reaction of gel **35** with ethylene diamine gave a gel formed by simple products (cyclic and acyclic dimers, and short oligomers) whereas the reaction in solution phase gave a complex mixture of oligomeric compounds. This result suggested that the reaction proceeds mainly in the gel phase that is highly preorganised and supports the hypothesis that reactive sites in that phase are accessible. On the other hand, the reaction with 1,2-diphenyl-1,2-ethanediamine, a bulkier diamine, resulted in the gel disassembly and formation of a $1 + 1$ macrocyclic compound (**37**) in high yield. This compound could not be formed directly from the gel phase as the two ends of the molecule are far away in an extended conformation and should come from the reaction of free molecules in solution. Indeed, compound **35** presents a folded conformation in solution that could explain why the small macrocycle is the main product. In summary, these examples reveal that whether the reaction takes place in solution or in the gel will depend on the steric requirements of the reactive sites as well as on the dynamics of the gel-to-solution equilibrium.

Mésini and coworkers have used the direct functionalisation of gel-forming nanotubes to access functional nanotubes that were otherwise inaccessible from the free molecules in solution (Figure 5.7). Compounds **38a,b** containing alkyne and azido end groups self-assemble in alkanes into nanotubes that form gels above 0.4 wt%. These nanotubes are reacted with azides and alkynes, respectively, by the "click reaction" and the nanotubular shape is preserved after

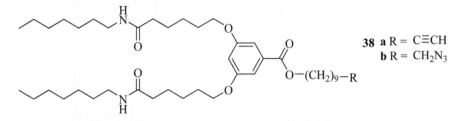

38 **a** R = C≡CH
    b R = CH$_2$N$_3$

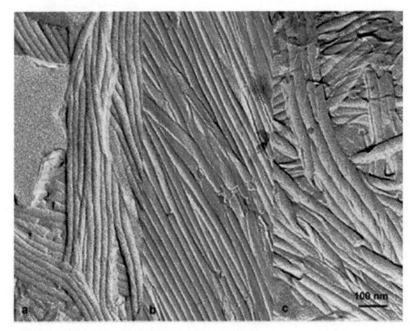

Figure 5.7 "Clickable" organogelators **38** and freeze fracture TEM of the gels of nanotubes after reaction ((a) **38a** and N$_3$-C$_{10}$H$_{21}$; (b) **38a** and N$_3$-C$_{10}$H$_{21}$-OH; (c) **38b** and CH≡C-C$_9$H$_{18}$-OH.

the reaction. The conversions of the functional groups in the gels are lower than in solution, between 55% and 61%, probably due to the inaccessibility of some of the reactive groups oriented towards the inner cavity of the tubes as well as to the heterogeneous conditions of the reaction (the Cu catalyst is sparingly soluble in alkanes).[46]

Reactions in the gel phase may have an effect not only on the product distribution and gel stability but also on the stereochemistry of the process. In this respect, Shinkai and coworkers have described anthracene-containing organogelators that can experience topotactic photodimerisation reactions in the gel phase (Scheme 5.11). In one example a two-component gel is formed by D-alanine coupled to a gallic acid assembling fragment and 2-anthracenecarboxylic acid (**39**) in cyclohexane. The preorganisation of these molecules within the one-dimensional supramolecular assemblies of the gel allows for a high degree of stereochemical control of the photodimerisation compared with the reaction in solution. In this sense, only *head-to-head* (*h–h*) photocyclodimers were formed, although with a low ee (10%).[47]

In order to improve the transfer of chirality into the photoreaction, the anthracene fragment and the gallic acid-amino acid moiety were connected covalently (**40**). The preference for the formation of *h–h* photodimers was maintained in different solvents, whereas the enantioselectivity was significantly improved. For instance, gels of **40** in cyclohexane/*n*-hexane mixtures gave a *h–h* relative yield of 93% and an ee for the major product of 33%. The enantioselectivity increased up to 45–56% by the use of glycidyl methyl ether (as a racemate or enantiopure solvent). In this case the relative yield of *h–h*

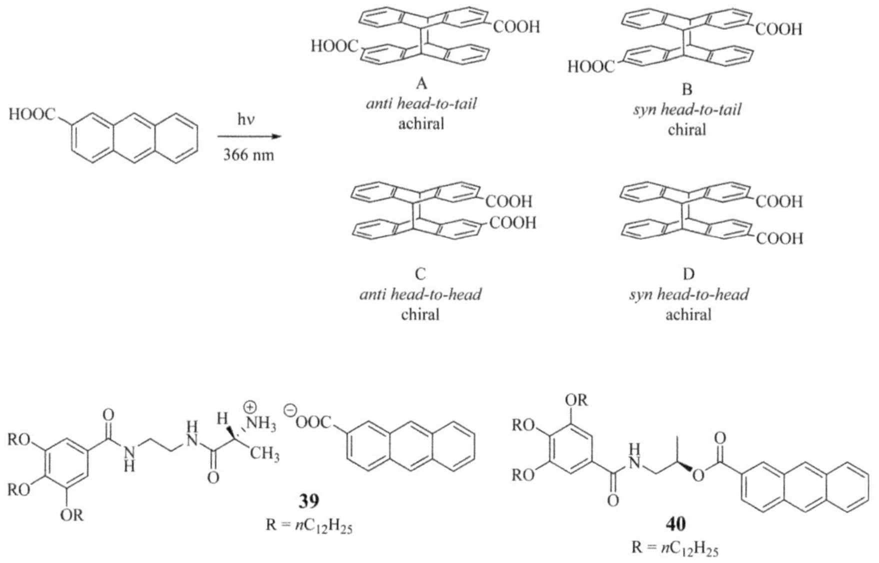

Scheme 5.11 (A) Schematic representation of [4+4] photocyclodimerisation of 2-anthracenecarboxylic acid in the free state. (B) Anthracene-based gelators **39** and **40**.

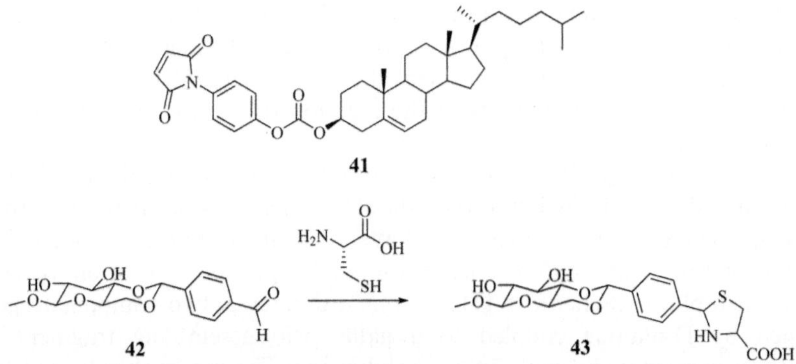

Scheme 5.12   (A) Maleimide-based gelator **41**. (B) Reaction scheme of aldehyde **42** with *L*-cysteine.

photodimers was lowered to about 55%. After a deep photophysical and structural study the authors concluded that there is a solvent-dependent correlation between molecular packing, gel strength and photoreaction outcome.[48]

The disassembly of the gel phase after a covalent bond forming reaction has been used by Zhang and coworkers to prepare chemoresponsive organogels (Scheme 5.12). They have designed a cholesterol-based gelator **41** bearing a maleimide fragment that forms gels in cyclohexane that are converted into solutions by Michael reaction with *n*-hexylthiol and triethylamine.[49] The same group has also designed cysteine responsive hydrogels. Compound **42**, bearing a reactive aldehyde functional group forms hydrogels and after addition of different amino acid solutions on top of the gels only the sample containing cysteine caused the gel disassembly. They demonstrate that the selectivity is based on the specific reaction of the aldehyde with the thiol group to form a thiazolidine derivative (**43**) that is soluble in water.[50]

In this context, our group has recently reported compound **44**, an amphiphilic proline derivative that forms hydrogels above 2 mM (Figure 5.8).[51] This compound bearing a secondary amino group is able to react with aldehydes causing the network disassembly with a rate that depends on the aldehyde hydrophobicity. This effect was studied using the release of methylene blue as the reporter signal and it was observed that the addition of a hydrophobic aldehyde such as 3-phenyl propanal caused the disassembly and the complete release of the dye in 30 min, whereas addition of hydrophilic aldehydes (acetaldehyde and propanal) only released a 11% of the dye after that time. Moreover, this hydrogel has been used for the entrapment and release of ketoprofen, a nonsteroidal anti-inflammatory drug with a low solubility in water (0.01% wt). The drug could be loaded up to 20% wt of gelator and released by the addition of 3-phenyl propanal. These results open the door for application in drug release in specific chemical environments.

The formation of reversible covalent bonds may be used to obtain dynamic gel systems. Lehn and coworkers have reported the generation of dynamic covalent libraries by reaction of guanosine hydrazide **45** with different

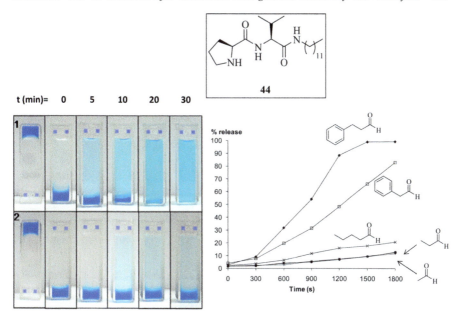

Figure 5.8   (A) Structure of hydrogelator **44**. (B) Release of methylene blue after addition of (1) 3-phenylpropanal and (2) water. (C) Percentage of released methylene blue with time after addition of different aldehydes.

aldehydes. In the presence of metal cations these derivatives form G-quartets that self-assemble into hydrogels as a consequence of hydrophobic interactions and π-stacking. Remarkably, the reversibility of the hydrazone bonds leads to the selection of the components based on the formation of the strongest hydrogel (Scheme 5.13).[52] This behaviour has been applied, for instance, for the slow release of volatile aldehydes from the hydrogel, in which the authors suggest that the different evaporation profiles of some of the tested aldehydes could be related with differences in the hydrazone stability.[53] The dynamic covalent approach has been shown to be a powerful tool even with very simple low molecular weight compounds such as lauric hydrazide and aliphatic and aromatic aldehydes.[54]

So far, we have described two kinds of reactive systems: gels that react with added compounds leading to an irreversible change in molecular structure and gel properties and dynamically reversible systems that are formed under thermodynamic equilibrium. However, reactive gels may be also used to create systems formed by dissipative self-assembly (DSA) – systems formed by non-assembling entities that, through the activation by an energy source, assemble into temporarily ordered structures that after further energy dissipation deactivate causing the collapse of the formed structures. This behaviour is found in natural self-assembling systems such as, for example, microtubule dynamic self-assembly/disassembly sequences.

In this context, van Esch and coworkers have recently reported the first example of a molecular gel that is temporarily formed by DSA and using a

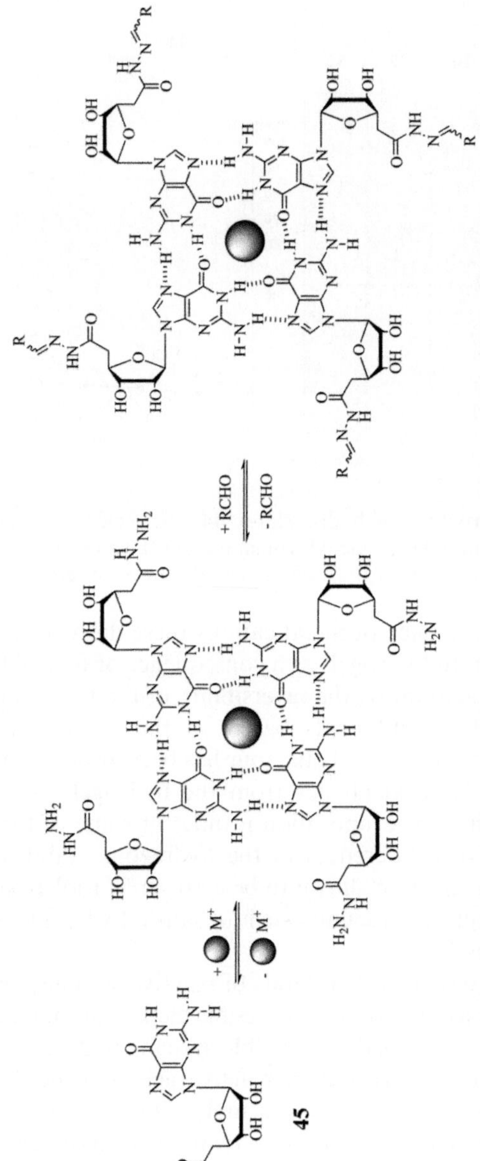

Scheme 5.13   General scheme showing the assembly of **45** into G-quartets and further reversible reaction with aldehydes leading to hydrogels.

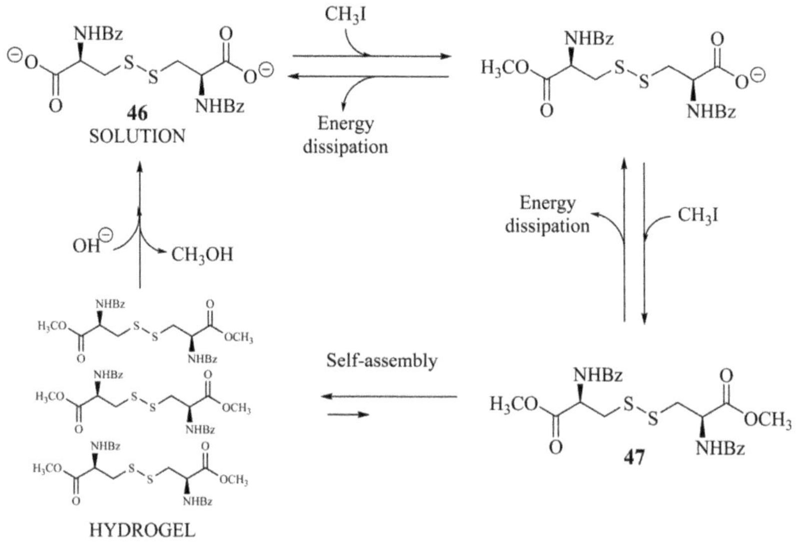

Scheme 5.14 Reaction cycle of the DSA of compound **47**.

chemical fuel.[55] The self-assembly cycle is based on dibenzoyl-L-cystine (DBC, **46**) (Scheme 5.14), a pH-responsive hydrogelator that self-assembles below its pKa (*ca.* 4.5), and DBC-dimethyl ester (**47**), which lacks the carboxylic groups and self-assembles independently of the pH. Methyl iodide (MeI) is used as the chemical fuel that slowly converts the water-soluble dicarboxylate into the dimethyl ester, activating the self-assembly process. However, the diester is prone to hydrolysis into the starting dicarboxylate DBC in an energy-dissipating step. A careful selection of the experimental parameters (concentration, pH, temperature) allows the control of the rate of the different steps and creates a transient hydrogel network that is continuously formed and destroyed as long as the chemical fuel is available. Although this is a simple example compared with the natural counterparts, it is a breakthrough towards the construction of sophisticated gel materials.

5.4.2 Covalent Capture of Gel Nanostructures

Molecular gels are built by weak noncovalent interactions and therefore are quite sensitive to environmental changes. For instance, changes in temperature, ageing or partial evaporation of the solvent can modify the rheological as well as the morphological properties of the gel network. Besides, molecular gels are too weak for some applications under shear conditions where a robust material is required. In order to improve the robustness of gels the nanoscale structures may be "captured" by the formation of covalent bonds between gelator molecules. For that purpose, polymerisable functionalities are introduced such as alkenes and alkynes in the gelator structure that will be reacted in the gel phase. Several examples are shown in Scheme 5.15. In particular, in the case of

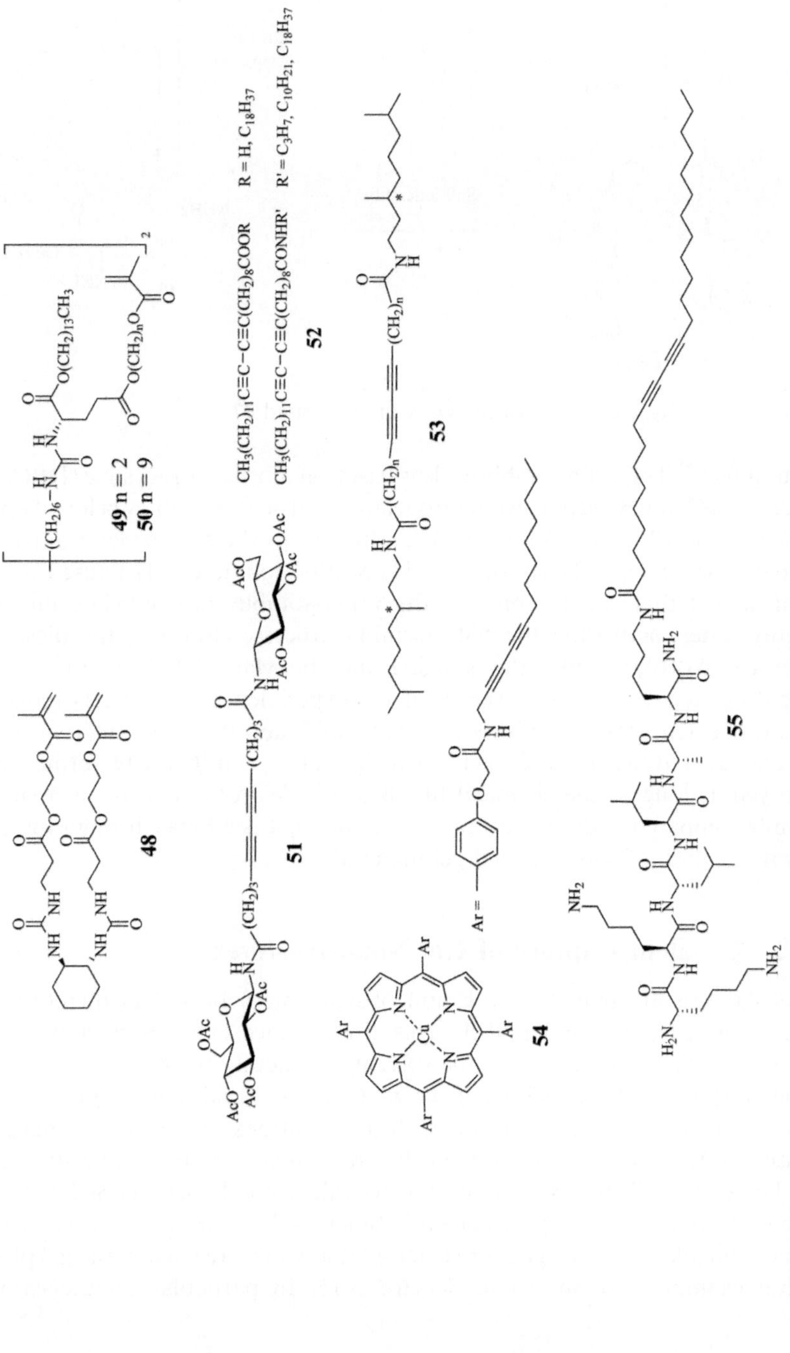

Scheme 5.15 Examples of gelators bearing photopolimerisable alkene and alkyne groups.

alkynes, the preorganisation of the reactive groups within the nanostructures leads to highly controlled topotactic reactions. Additionally, the use of UV light as an initiator avoids the addition of other compounds that could affect the self-assembly process.[56–62]

Alkene metathesis has also been used to capture gel nanostructures by Smith and coworkers. For that purpose, solutions of second-generation Grubbs' catalyst were added on top of the gels and left to diffuse and react over a 24-h period. As an example, they used this methodology in order to capture meta-stable spherical nanostructures obtained for compound **56** being able to ef-fectively capture more than 75% of the material. (Scheme 5.16A).[63,64]

"Click" chemistry has been also used for the crosslinking and stabilisation of gel nanostructure (Scheme 5.16B). Finn and coworkers were the first to de-scribe the use of the Cu(I)-catalysed cycloaddition reaction of alkynes and azides for that purpose.[65] Gelators bearing the *trans*-1,2-diaminocyclohexane assembling scaffold and "clickable" terminal groups (**57, 58**) were prepared. Gels could be formed in acetonitrile in the presence of the additives required to perform the cycloaddition reaction at optimised ratios and an improvement on rheological and thermal properties was observed after crosslinking of the gel nanostructures. This strategy has been also used by Torres and coworkers for the fabrication of stable photoactive phthalocyanine-based organogels.[66]

Finally, some examples have also been reported in which covalent capture of the gelator is combined with the imprinting of the nanostructures by poly-merisation of the solvent (**59**, Scheme 5.16C). These new polymeric nano-composites would retain the functionality of the low molecular weight gelators within a robust polymeric matrix.[67]

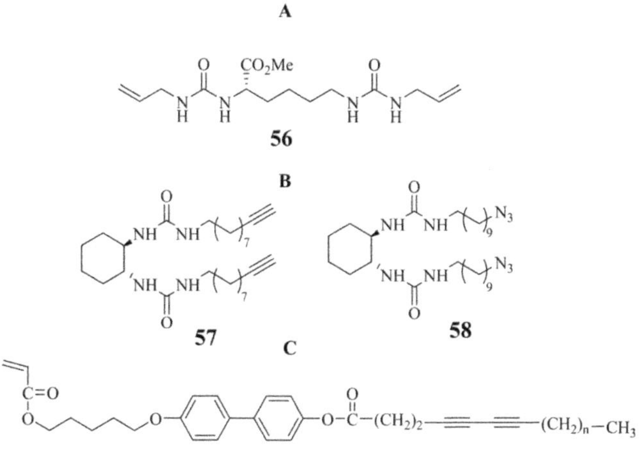

Scheme 5.16 Structures of gelators designed for covalent capture by (A) alkene metathesis, (B) azide-alkyne [3 + 2] cycloaddition ("click" reaction) and (C) combined covalent capture and polymer imprinting.

5.5   Catalytic Gels

5.5.1   Introduction: Supramolecular Catalysis

Catalysis is one of the main goals of Supramolecular Chemistry that started 30-40 years ago with the pioneering work of Pedersen, Cram and Lehn in that field. Indeed, it appeared as one of the most important applications of supramolecular complexes or supramolecules in their seminal publications and books.[68] Since then the supramolecular approach to the field of Catalysis has attracted the attention of many researchers in the most prestigious groups.[69]

Supramolecular approaches to catalysis have been classified into three categories: 1) molecular receptors that place a binding site close to a catalytic centre, 2) molecular receptors that simultaneously bind two reactants and promote their reaction, and 3) systems in which supramolecular interactions are used to construct a catalytic centre. In the case of catalytic supramolecular gels, noncovalent interactions are used for the construction of a multitopic catalyst. In this sense, they represent an extended and more complex version of the third family in that classification. For instance, gelators can be designed in a modular approach by combining an assembling fragment with a functional fragment (catalytic or precatalytic) (Figure 5.9).[6]

After self-assembly, an extended supramolecular object with built-in catalytic sites will be obtained. The functional fragment could be either a known catalyst or a precatalyst that will be activated after aggregation. Besides, the organisation of multiple catalytic sites in the fibre surface could generate additional catalytic features such as multivalent interactions, neighbouring effects and cooperativity. On the other hand, differences in polarity between solvent pools and inner regions of the fibres may lead to substrate selectivity. Additionally, the high degree of molecular order may ultimately induce regio- and stereoselective transformations. Overall, catalytic molecular gels may be considered as artificial enzymes.

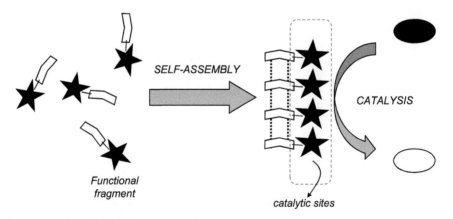

Figure 5.9   Design of a catalytic molecular gel.

Catalytic centres can be metals coordinated to the gel fibres (metallocatalysts) and catalytic organic fragments (organocatalysts). Further discussion will follow this fundamental distinction.

5.5.2   Metallocatalysis

Coordination of metals has been used quite often to obtain metallogels with interesting physical properties and applications. However, its use for the construction of gel-supported catalytic systems is still in its infancy and it is a promising field of development in gel materials research.[70]

Among the so-far limited examples found in literature three different types of catalytic metallogels could be considered (Figure 5.10): a) metal-supported gels, b) gels as supramolecular multitopic ligands for metals and c) gels made of organometallic gelators.

a) Metal-supported gels: In these systems metal coordination is responsible for the formation of a crosslinked coordination polymer that entraps the solvent and in consequence forms a gel. Additionally, this supramolecular material may present catalytic activity. One of the first examples of this class was described by Xu and coworkers in which pyridine-based ligands such as **60** were used to

A) Metal-supported gels

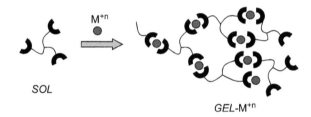

B) Gels as supramolecular multitopic ligands

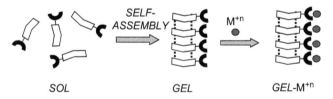

C) Organometallic gels

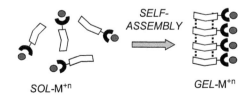

Figure 5.10   Classification of metallogels.

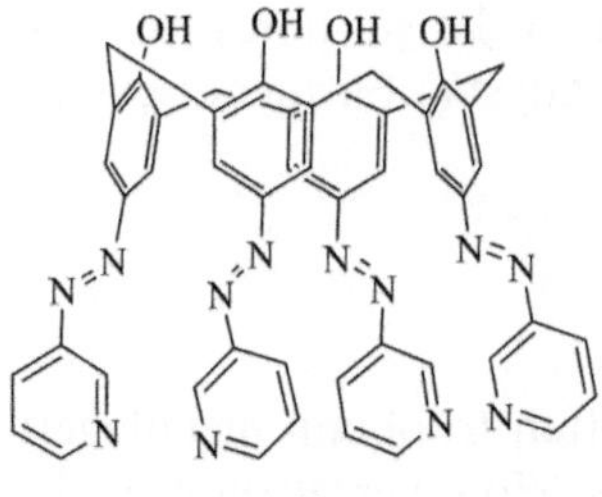

60

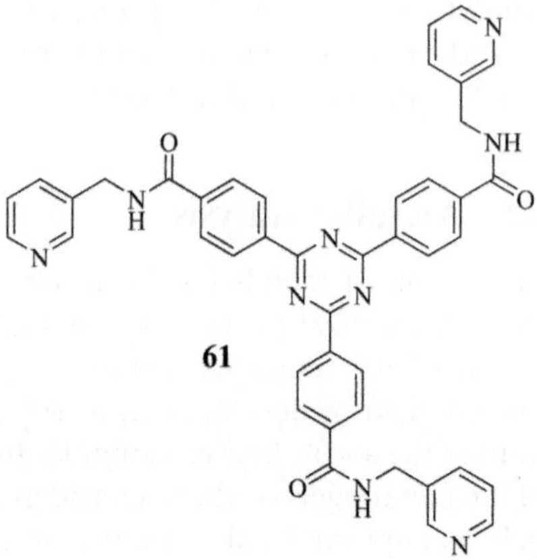

61

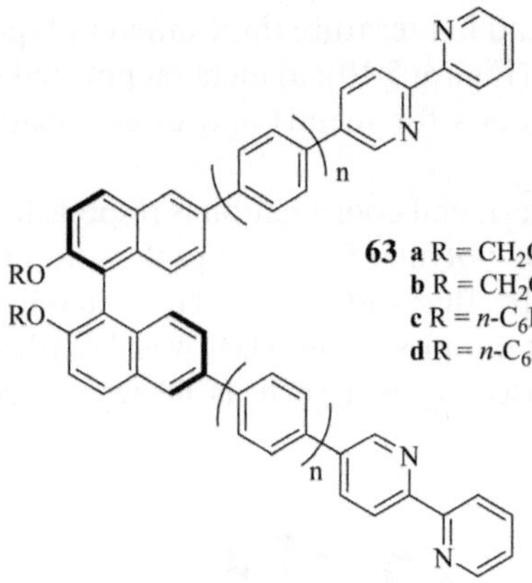

63 **a** R = CH_2OCH_3, n = 1
b R = CH_2CH_3, n = 1
c R = *n*-C_6H_{13}, n = 1
d R = *n*-C_6H_{13}, n = 2

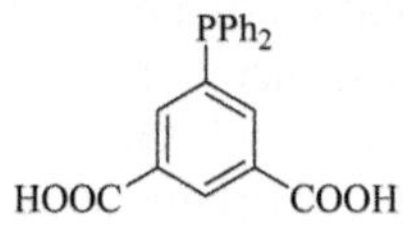

62

Scheme 5.17 Ligands for metal-supported gels.

construct Pd-metallogel networks in DMSO that were active catalysts for the aerobic oxidation of benzyl alcohol to benzaldehyde (Scheme 5.17).[71]

Catalytic reactions were analysed after 2 h with turnover number values twice those obtained using $Pd(OAc)_2$ as the catalyst, as a consequence of the superior stability of the catalytic gels.

More recently, Zhang and coworkers have prepared pyridine-based tripodal ligands for Pd(II) (**61**) that are able to form metallogels in methanol–chloroform mixtures and can be used as catalysts for the Suzuki–Miyaura C–C crosscoupling between 4-bromopyridine and phenylboronic acid in high yield after few hours (Scheme 5.17).[72] This gel-phase catalyst was also used for the cross coupling of other aryl halides such as iodobenzene being reused up to three times, with progressive inactivation. However, when the metallogel was dried and used directly as a xerogel, catalytic activity was comparable to the gel in the first run and maintained for at least five consecutive runs. More recently, the same authors have reported the study of Fe-metallogels based on 1,3,5-substituted benzene as ligand (**62**), their postmodification by diffusion of a $Pd(COD)Cl_2$ solution (COD: cyclooctadienyl), and their use as catalysts for the same organic reactions.[73] In that case, the catalyst was packed in a paper bag to avoid mechanical degradation and easy separation from reaction medium. The products were obtained in high yield and recycled for several runs.

Gao and coworkers reported recently the application of chiral binaphthylbispyridine-based copper(I) metallogels as catalysts for the Huisgen 1,3-dipolar cycloaddition ("click" reaction) (Scheme 5.17).[74] Compounds **63a–d** complexed with equimolar amounts of $Cu(CH_3CN)_4BF_4$ formed gels in CH_3CN-CH_2Cl_2 mixtures and their xerogels were tested for the "click" reaction between benzyl azide and several terminal alkynes. The best results were obtained when the xerogels were suspended in water as solvent, and were excellent for xerogel Cu(I) · **63c** bearing hexyl aliphatic chains. The authors suggested that these lipophilic chains could provide an apolar microenvironment that could facilitate the access of the apolar substrates to the catalytic sites in the fibres.

b) Gels as supramolecular multitopic ligands for metals: In this case the fibrillar network can be constructed either before the introduction of the metal or in the presence of the metal. In the first case a gelator molecule with free ligand sites is self-assembled and the metal is added *a posteriori* by diffusion. As an example, we have reported toluene gels of compound **7** armed with arrays of pyridyl fragments that could be loaded with Pd(II) and tested for the aerobic oxidation of benzyl alcohol.[75] The first results indicated that the catalysts were active in that reaction with low turnover numbers, however, it could be observed later that leaching of Pd species was most likely responsible for the observed catalytic behaviour. Other examples have been reported in which the role of the gel network is mainly to act as a support for the generation of nanoparticles with catalytic activity.[76]

Recently, Liu and coworkers have reported an example in which a hydrogelator (**64**) known to self-assemble into helical nanotubes is assembled in the presence of Cu(II) leading to a chiral catalyst active for a Diels–Alder reaction between cyclopentadiene and an aza-chalcone (Figure 5.11).[77] The presence of

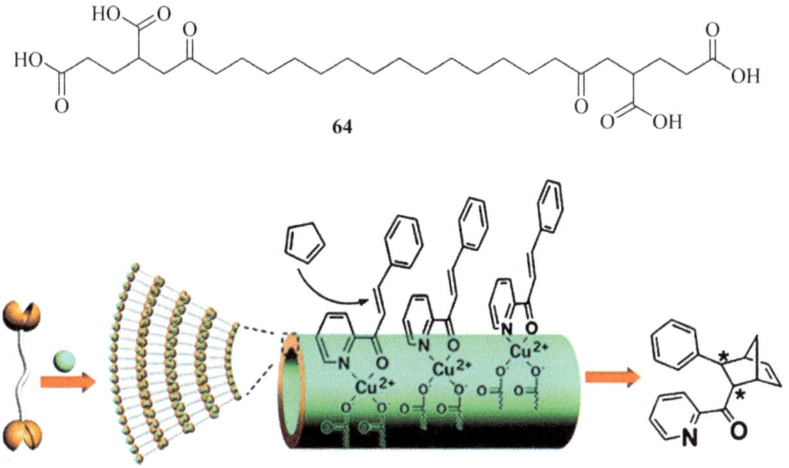

64

Figure 5.11 Structure of chiral gelator **64** and scheme of its self-assembly into nanotubes and catalytic activity.
Adapted with permission from ref. 77. Copyright 2011 American Chemical Society.

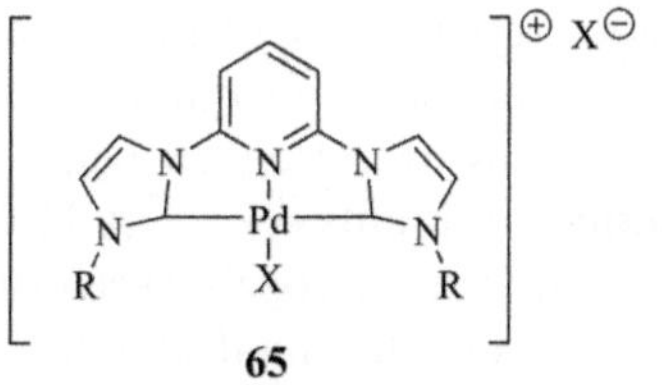

65

R = *n*-C$_{16}$H$_{33}$, X = I

Scheme 5.18   Organometallic gelator **65**.

the helical Cu(II)-**64** helical nanotubes was shown to accelerate the reaction as well as to enhance the stereoselectivity compared with other nanostructures formed by related compounds. Additionally, the enantiomeric excess of the product was related to the molecular and supramolecular chirality of the catalyst. L-**64** and D-**64** formed nanotubes of opposite handedness and the products of the Diels-Alder reactions were obtained with an ee of 47% and –51%, respectively. This is a nice example of the transmission of a molecular feature into the supramolecular and nanoscopic level.

c) Gels made of organometallic gelators: Metallogels can be built up from low-molecular-mass gelators that already contain a metal in their structure. This approach has been particularly explored by Dötz and coworkers who described the first example of a chromium metal-carbene based low-molecular-mass organometallic gelator.[78] Later they reported the use of palladium-CNC pincer biscarbene **65** as an air-stable organometallic LMWG able to catalyse a double Michael addition of α-cyanoacetate to methyl vinyl ketone in the gel phase (Scheme 5.18).[79]

Problems usually associated with heterogeneous metallocatalysis are also present in metallogels (leaching, catalyst inactivation, toxicity of metals for some applications, *etc.*) together with additional problems relative to the supramolecular nature of gel materials (disassembly of catalytic fragments in solution, slow diffusion, *etc.*). Despite these drawbacks research in this field is challenging due to the superior activity of metals in catalysis.

5.5.3   Organocatalysis

In the last decade organocatalysis, and in particular aminocatalysis, has experienced a "gold rush" as termed by Melchiorre *et al.*[80] Structurally diverse examples of homogeneous phase organocatalysis have been reported and widely reviewed.[81] Moreover, supramolecular organocatalysis has become a potent approach. In this sense different organised systems formed by low-molecular weight compounds have been studied in catalysis such as discrete self-assemblies,[82] micelles,[4] vesicles[83] or emulsions.[84] However, the study of supramolecular gels in organocatalysis is barely reported.

The first mention of the presence of a supramolecular gel in an organo-catalytic system was made by Inoue and coworkers by 1990 in their studies on the asymmetric addition of hydrogen cyanide to *m*-phenoxybenzaldehyde catalysed by cyclodipeptide (66) (Scheme 5.19).[85] They observed that low temperatures (and formation of a gel) were accompanied with an increase on stereoselectivity. Following this work, Danda reported that depending on the purification procedure of the catalyst either transparent or opaque gels or even solid suspensions were obtained in toluene.[86] It was observed that those gels were thixotropic – namely, viscosity decreased on increasing stirring rate – and that faster stirring rates gave increasing enantioselectivities. Other groups dedicated some attention to the study of the mechanism of catalysis employing kinetic, NMR and computational studies.[87] It was proposed that the process was second order on catalyst, with two imidazole residues necessary for the catalytic action.[88] However, not much attention was paid to the supramolecular structure of the gel and the heterogeneous nature of the system was regarded as "*A significant obstacle...which makes the interpretation and comparison of experimental data difficult*". Indeed, prevention of catalyst aggregation has been a major issue in many cases in which the catalyst presents groups suitable for intermolecular interactions.

It has been only recently that Guler and Stupp have described His-containing peptide 67 that forms a hydrogel at pH above 6.5 and at concentrations above 0.1 wt% capable of catalysing the hydrolysis of 2,4-dinitrophenyl acetate.[89] They showed that the nanofibres were better catalysts than other aggregates due to the high density of catalytic sites on their surface (Scheme 5.20).

Despite these seminal examples the field of catalytic molecular gels is still in its infancy compared with other applications of those materials.

Our research group has focused on this topic in recent years with the main interest on the design of molecular gelators provided with catalytic groups, the

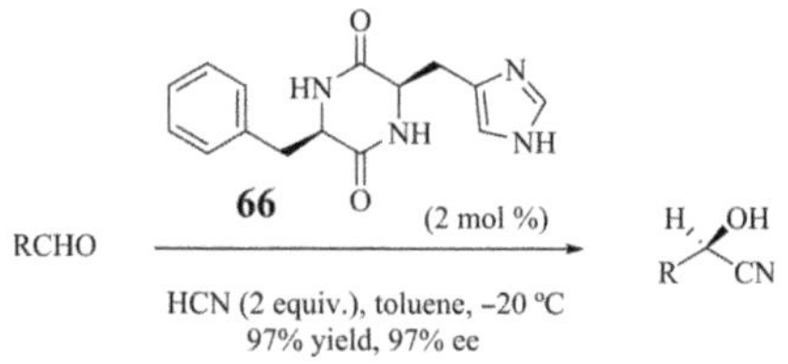

Scheme 5.19   Scheme of the asymmetric addition of hydrogen cyanide to aldehydes catalysed by a toluene gel of compound 66.

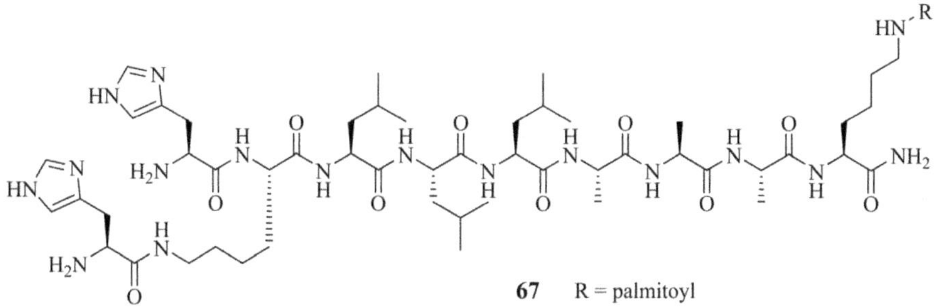

Scheme 5.20   Histidine-based catalytic peptide hydrogelator 67.

study of their catalytic activity and the analysis of the role played by the gel matrix in catalysis. Solvent, temperature, polymorphism, chirality, stereoelectronic effects and any parameter that may affect the catalytic process has to be considered in detail with the main goal of controlling the reaction performance by tuning the molecular and supramolecular gel structure. Moreover, the heterogeneous nature of the gel should not be considered as a drawback but as an additional advantageous feature that allows, for instance, easy recovery and recycling of the catalyst at a low synthetic cost. Moreover, the dynamic nature of supramolecular gels allows for an adaptable substrate–catalyst interaction reminiscent of the enzyme-substrate induced-fit mechanism.

We chose *L*-proline as the catalytically active fragment in the design of the gelator (Scheme 5.21). This amino acid has been extensively used as an active and stereoselective catalyst for C–C bond forming reactions such as aldol reaction or Michael addition reactions and its mechanism of action is well known. It is well accepted that these reactions follow mechanisms resembling that of natural Aldolase I enzymes in which an amine – L-Lysine residue in enzymes, L-proline in the current case – forms an enamine intermediate with a ketone substrate that further attacks the electrophylic reagent (aldehyde, α,β-unsaturated carbonyl, nitroalkene, *etc.*).[90,91] These catalytic groups have been connected to a well-known peptidic assembling fragment.[92]

Compounds **68a–c** have been shown to aggregate in organic solvents such as acetonitrile, ethyl acetate or toluene forming fibrillar networks. They have been shown to form gels under different conditions mainly by the formation of a 1D array of H-bonds between amide NH donors and carbonyl acceptors. Structural analysis has demonstrated that compounds **68a–c** have a tendency to fold in solution (in both acetonitrile and toluene) by forming several intramolecular H-bonds (Figure 5.12A). However, above the minimum gel concentration and after heating for dissolution and spontaneous cooling to room temperature, unfolded conformations are captured into metastable gels. In some cases such as compound **68a** in acetonitrile, a slow correction process converted the initial kinetically trapped gel into a thermodynamically more stable polymorphic structure.[93]

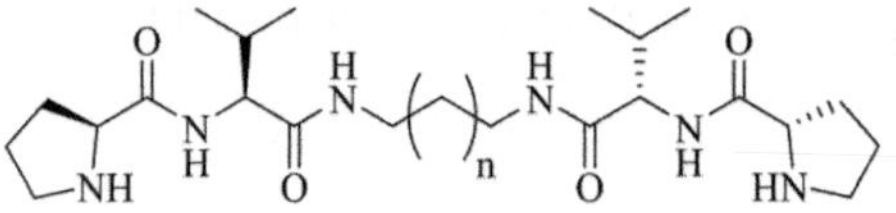

68a–c

n = 1, 4, 6

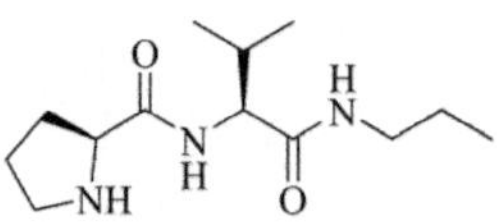

69

Scheme 5.21   *L*-Proline-based catalytic bolaamphiphilic gelators **68a–c**.

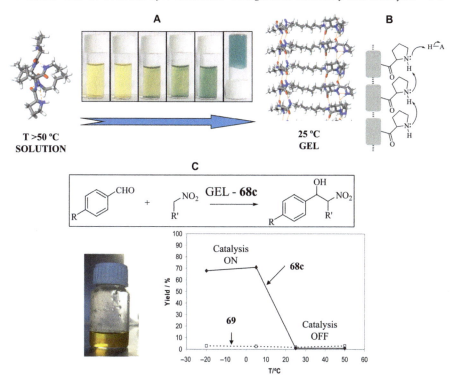

Figure 5.12 A) Molecular models of compound **68c** in solution and in the gel phase and dye colour changes associated to aggregation induced change in basicity. (B) Proposed proton relay system accounting for the increase of basicity. (C) Application of the catalytic gel for the Henry reaction (R = NO$_2$, R' = CH$_3$).

These compounds were initially tested for the aldol reaction between acetone and 4-nitrobenzaldehyde.[94] Reactions were performed at –20 °C in acetonitrile gels and in blank solutions in which the concentration was similar to that of catalyst that remains in the sol phase in equilibrium with the gel. Unfortunately, it was observed that the catalytic activity was similar in both systems, indicating that the reaction was not taking place in the gel phase but only in solution. Apparently, catalytic activity was inhibited by aggregation. A closer analysis of data brought our attention to the fact that if the gel reactions were left for several days under reaction conditions the product, initially obtained with a 1 : 5 enantiomer ratio, was racemised, whereas in pure diluted solutions remained unaffected. Several tests were performed and it was concluded that the gels were acting as a base, as confirmed by the use of a pH indicator, bromothymol blue, that turned to its basic colour (blue) only in the presence of the gels (Figure 5.12A). Additionally, the basicity of the gels was compared to the same compounds under nonassembling conditions (diluted or at high temperature) and it could be assigned a difference of 3 units in the pKa values in acetonitrile of the respective conjugated acids. Remarkably, self-assembly produced an enhancement of basicity most probably due to a cooperation of neighbouring proline moieties organised as a proton relay system (Figure 5.12B).

This fully new property that emerged after self-assembly of the catalyst into the nanofibres could be exploited for a base-catalysed transformation in which the gel phase was participating for the first time as the active phase.[95] The selected benchmark reaction was the Henry nitroaldol reaction in which a basic catalyst has to deprotonate a nitroalkane that afterwards attacks the carbonyl of an aldehyde (Figure 5.12C). Basic catalytic gels could be formed in a very convenient way in neat nitroalkanes (nitromethane and nitroethane) and gave excellent yields of the nitroaldol product after addition of aromatic aldehydes in contrast with nonaggregated samples of analogue compound **69** shown in Figure 5.12C. Moreover, the system showed a switchable catalytic performance controlled by temperature. Thus, cooling below T_{gel} yielded an active basic system, whereas heating above that temperature switched-off the catalytic activity after disassembly into inactive free molecules.

Why were gels formed by compounds **68a–c** in acetonitrile active as basic catalysts but inactive in the direct aldol reaction? It has been shown in the application to the Henry reaction that catalytic sites in the gel phase were accessible for nitroalkane substrates. On the contrary, it seems that enamine intermediates cannot be formed within the gel phase.

The solvent has been shown to play a relevant role in the catalytic performance. First, the solvent has an important influence on the structure of the catalytic centres. In particular, working with molecules such as those presented here with a relative flexibility, the solvent is the determinant for the crystalline packing of gelator. We have reported that closely related molecules bearing the same assembling fragment showed a high degree of polymorphism depending on solvent polarity.[96] Secondly, solvent polarity and solvophobic interactions will determine the solubility of gelator, namely concentration of free gelator molecules in solution in equilibrium with the gel phase. And finally, solvophobic interactions may be very important for the distribution of substrates between solution and the gel phase and consequently for the access of them to the catalytic site within the gel network. For instance, replacing acetonitrile by toluene as solvent has an enormous effect on the accessibility of the proline moiety allowing the formation of enamine intermediates. Indeed, toluene gels of compound **68b** have been shown to be catalytically active for the 1,4-conjugate addition of cyclohexanone to *trans*-β-nitrostyrene (Scheme 5.22). NMR and X-ray studies of these gels and nongelating analogues have revealed

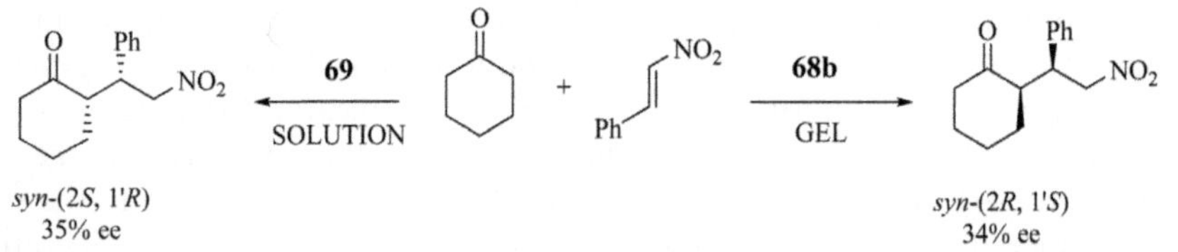

syn-(2*S*, 1′*R*)
35% ee

syn-(2*R*, 1′*S*)
34% ee

Scheme 5.22   1,4-conjugate addition of cyclohexanone to *trans*-β-nitrostyrene in a toluene gel of compound **68b** and a diluted solution of **69**.

that the local conformation of the *L*-Pro residue suffers a dramatic change going from nonaggregated states in which this residue is participating in a strong intramolecular H-bond to aggregates and gels where the proline lone pairs are oriented outwards available and accessible to form the enamines. Remarkably, the conformational change associated to the self-assembly is paralleled with an inversion of the stereoselectivity of the reaction compared to a solution of analogue **69**.[97]

The replacement of organic solvents by water is highly convenient both from an environmental as well as an economic point of view. Remarkably important is also the fact that water is the biological solvent being the hydrophobic effect the most important noncovalent interaction in biocatalysis. In this sense, the study of L-proline and related catalysts in water is receiving increasing attention.[98–101]

Recently, we designed an amphiphilic hydrogelator derived from *L*-proline (**44**).[102] This compound self-assembled in water forming hydrogels that were tested for the direct aldol reaction between cyclohexanone and 4-nitrobenzaldehyde (Figure 5.13). Reagents were easily added on top of the gel dissolved in toluene and the reaction was quantitatively completed after 24 h at 5 °C with high stereoselectivity (*anti*:*syn* 92:8, 88% ee). Moreover, the catalytic hydrogel could be reused after decanting of the toluene phase for at least three times with the same efficiency and stereoselectivity. Remarkably, in this system the hydrophobic effect plays a dual role: first as the driving force for gelator self-assembly and, secondly, conducting reagents to the hydrophobic catalytic sites.

In summary, molecular gels offer great possibilities in the field of catalysis. Several of the features of these materials are very convenient from a practical point of view: reversibility and sensitivity to external stimuli that allow control

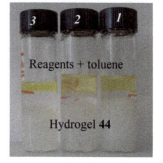

	Yield (%)	*anti*:*syn*	ee (%)
Run 1	98	92:8	88
Run 2	> 99	93:7	87
Run 3	> 99	92:8	90

Hydrogel: 0.026 mmol of gelator (0.2 eq) in water (4 mL);
Reagents: 4-nitrobenzaldehyde (1 eq), cyclohexanone (20 eq)
in toluene (1 mL). Time: 24 h. T = 5 °C.

Figure 5.13   Application of hydrogels of compound **44** as catalysts for the direct aldol reaction.

of catalytic events, straightforward molecular synthesis and possibilities for the recovery and regeneration of valuable catalysts, *etc*. After selecting a convenient self-assembling fragment, whose aggregation behaviour should be known, any catalytic moiety could be easily attached. On the other hand, different from conventional polymer-supported heterogeneous catalysts, in supramolecular gels the structure of the catalytic sites can be controlled at molecular level and, once the structure of the gel phase is well understood a fine tuning of catalytic performance can be achieved by subtle changes (solvent, structure modification, temperature, *etc*.).

As it has been shown that both metal and organic catalysis is feasible although the use of metals introduces additional variables such as complex stability, leaching, precipitation of metallic aggregates, *etc*. Regarding organocatalysis, molecular gels can be explored as enzyme mimics. Combination of catalytic sites with selective binding pockets should be one of the future goals in that field as it may render efficient and selective catalysis.

Furthermore, self-assembly may have different levels of action: i) self-construction of the catalytic system after simple input from a small-molecule design and ii) emergence of new catalytic properties associated with multi-molecular aggregates.

5.6 Conclusions and Outlook

In summary, molecular gels are dynamic and adaptable vessels for different physical and chemical processes in which molecules in the gel phase may be passive spectators or have a crucial effect on those processes. They constitute an example of the relevance of self-assembly, not only by the expression of molecular features into the supramolecular level but also, and more interestingly, by the emergence of new properties. Moreover, the combination of such enhanced properties with the responsiveness to external stimuli leads to complex smart systems reminiscent of natural counterparts. In this sense, the control of complex systems will be a target for the near future of the field. Furthermore, the paradigm of a complex dynamic multiresponsive system is the cell. Thus, an ambitious goal that has to be undertaken in the next years is the development of smart gellosomes mimicking the cell architecture and also the dynamic biological processes happening within.[103] Those systems, simple analogues of cells, could be of great value in the field of Prebiotic Chemistry where self-assembly and responsiveness are thought to play a relevant role.[104]

It has been proposed that small organic molecules, under appropriate concentration conditions in water, with the help of some sources of chemical energy and maybe mineral interfaces, could form membrane-bounded compartments that encapsulate subsets of components capable of react and produce new molecules, *i.e.* polymers. Posterior development of additional self-catalytic properties could lead to the beginning of replication and cellular life. In this context, molecular gels can be regarded as compartmentalised self-supported materials that, as has been shown in the last section, can develop new catalytic features after self-assembly, *i.e.* basic catalysis. It can be envisaged that the formation of hydrogels capable to develop self-catalysis could represent a suitable model for the understanding of emergence of life.

References

1. G. M. Whitesides and B. Grzybowski, *Science*, 2002, **295**, 2418.
2. J.-M. Lehn, *Proc. Natl. Acad. Sci. USA*, 2002, **99**, 4763.
3. J. Texter (ed.) *Reactions and Synthesis in Surfactant Systems in Surfactant Science Series*, vol. 100, Marcel-Decker, New York, 2001.
4. T. Dwars, E. Paetzold and G. Oehme, *Angew. Chem. Int. Ed.*, 2005, **44**, 7174.
5. Z. Dong, Q. Luo and J. Liu, *Chem. Soc. Rev.*, 2012, **41**, 7890.
6. B. Escuder, F. Rodríguez-Llansola and J. F. Miravet, *New J. Chem.*, 2010, **34**, 1044.
7. D. Díaz Díaz, D. Kuhbeck and R. J. Koopmans, *Chem. Soc. Rev.*, 2011, **40**, 427.
8. A. R. Hirst, I. A. Coates, T. R. Boucheteau, J. F. Miravet, B. Escuder, V. Castelletto, I. W. Hamley and D. K. Smith, *J. Am. Chem. Soc.*, 2008, **130**, 9113.
9. T. F. A. De Greef, M. M. J. Smulders, M. Wolffs, A. P. H. J. Schenning, R. P. Sijbesma and E. W. Meijer, *Chem. Rev.*, 2009, **109**, 5687.
10. C. Tang, A. M. Smith, R. F. Collins, R. V. Ulijn and A. Saiani, *Langmuir*, 2009, **25**, 9447.
11. L. Chen, S. Revel, K. Morris, L. C. Serpell and D. J. Adams, *Langmuir*, 2010, **26**, 13466.
12. T. K. Harris and G. J. Turner, *IUBMB Life*, 2002, **53**, 85.
13. A. Masuda, K. Ushida, H. Koshino, K. Yamashita and T. Kluge, *J. Am. Chem. Soc.*, 2001, **123**, 11468.
14. F. Galindo, M. I. Burguete, R. Gavara and S. V. Luis, *J. Photochem. Photobiol. A: Chem.*, 2006, **178**, 57.
15. K. Hanabusa, K. Hiratsuka, M. Kimura and H. Shirai, *Chem. Mater.*, 1999, **11**, 649.
16. D. C. Duncan and D. G. Whitten, *Langmuir*, 2000, **16**, 6445.
17. B. Escuder, M. Llusar and J. F. Miravet, *J. Org. Chem.*, 2006, **71**, 7747.
18. J. R. Moffat and D. K. Smith, *Chem. Commun.*, 2009, 316.
19. A. Brizard, M. Stuart, K. van Bommel, A. Friggeri, M. de Jong and J. van Esch, *Angew. Chem. Int. Ed.*, 2008, **47**, 2063.
20. P. Mukhopadhyay, Y. Iwashita, M. Shirakawa, S. Kawano, N. Fujita and S. Shinkai, *Angew. Chem. Int. Ed.*, 2006, **45**, 1592.
21. P. Mukhopadhyay, N. Fujita, A. Takada, T. Kishida, M. Shirakawa and S. Shinkai, *Angew. Chem. Int. Ed.*, 2010, **49**, 6338.
22. F. Rodríguez-Llansola, B. Escuder, J. F. Miravet, D. Hermida-Merino, I. W. Hamley, C. J. Cardin and W. Hayes, *Chem. Commun.*, 2010, **46**, 7960.
23. A. Barnard and D. K. Smith, *Angew. Chem. Int. Ed.*, 2012, **51**, 6572.

24. B. Escuder, J. F. Miravet and J. A. Sáez, *Org. Biomol. Chem.*, 2008, **6**, 4378.
25. J. A. Sáez, B. Escuder and J. F. Miravet, *Chem. Commun.*, 2010, **46**, 7996.
26. D. K. Smith, *Chem. Soc. Rev.*, 2009, **38**, 684.
27. H. Jintoku, M. Takafuji, R. Oda and H. Ihara, *Chem. Commun.*, 2012, **48**, 4881.
28. X. Chen, Z. Huang, S.-Y. Chen, K. Li, X.-Q. Yu and L. Pu, *J. Am. Chem. Soc.*, 2010, **132**, 7297.
29. T. Tu, W. Fang, X. Bao, X. Li and K. H. Dötz, *Angew. Chem. Int. Ed.*, 2011, **50**, 6601.
30. J. A. Foster, M.-O. Piepenbrock, M. G. O. Lloyd, N. Clarke, J. Howard, A. K. and J. W. Steed, *Nature Chem*, 2010, **2**, 1037.
31. A. Shumburo and M. C. Biewer, *Chem. Mater.*, 2002, **14**, 3745.
32. S. Kiyonaka, K. Sada, I. Yoshimura, S. Shinkai, N. Kato and I. Hamachi, *Nature Mater*, 2004, **3**, 58.
33. Y. Gao, F. Zhao, Q. Wang, Y. Zhang and B. Xu, *Chem. Soc. Rev.*, 2010, **39**, 3425.
34. Q. Wang, L. Li and B. Xu, *Chem. Eur. J.*, 2009, **15**, 3168.
35. Q. Wang, Z. Yang, X. Zhang, X. Xiao, C. K. Chang and B. Xu, *Angew. Chem. Int. Ed.*, 2007, **46**, 4285.
36. J. A. Zorn, H. Wille, D. W. Wolan and J. A. Wells, *J. Am. Chem. Soc.*, 2011, **133**, 19630.
37. W. Gu, L. Lu, G. B. Chapman and R. G. Weiss, *Chem. Commun.*, 1997, 543.
38. R. J. H. Hafkamp, B. P. A. Kokke, I. M. Danke, H. P. M. Geurts, A. E. Rowan, M. C. Feiters and R. J. M. Nolte, *Chem. Commun.*, 1997, 545.
39. F.-X. Simon, N. S. Khelfallah, M. Schmutz, N. Diaz and P. J. Mésini, *J. Am. Chem. Soc.*, 2007, **129**, 3788.
40. T.-T.-T. Nguyen, F.-X. Simon, N. S. Khelfallah, M. Schmutz and P. J. Mésini, *J. Mater. Chem.*, 2010, **20**, 3831.
41. J. R. Moffat and D. K. Smith, *Chem. Commun.*, 2011, **47**, 11864.
42. H. Koshima, W. Matsusaka and H. Yu, *J. Photochem. Photobiol., A*, 2003, **156**, 83.
43. J. J. D. de Jong, L. N. Lucas, R. M. Kellogg, J. H. van Esch and B. L. Feringa, *Science*, 2004, **304**, 278.
44. J. F. Miravet and B. Escuder, *Org. Lett.*, 2005, **7**, 4791.
45. J. F. Miravet and B. Escuder, *Tetrahedron*, 2007, **63**, 7321.
46. T.-T.-T. Nguyen, F.-X. Simon, M. Schmutz and P. J. Mesini, *Chem. Commun.*, 2009, 3457.
47. A. Dawn, N. Fujita, S. Haraguchi, K. Sada and S. Shinkai, *Chem. Commun.*, 2009, 2100.
48. A. Dawn, T. Shiraki, S. Haraguchi, H. Sato, K. Sada and S. Shinkai, *Chem. Eur. J.*, 2010, **16**, 3676.
49. Q. Chen, D. Zhang, G. Zhang and D. Zhu, *Langmuir*, 2009, **25**, 11436.
50. Q. Chen, Y. Lv, D. Zhang, G. Zhang, C. Liu and D. Zhu, *Langmuir*, 2010, **26**, 3165.

51. F. Rodríguez-Llansola, J. F. Miravet and B. Escuder, *Chem. Commun.*, 2011, **47**, 4706.
52. N. Sreenivasachary and J.-M. Lehn, *Proc. Natl. Acad. Sci. USA*, 2005, **102**, 5938.
53. B. Buchs, W. Fieber, F. Vigouroux-Elie, N. Sreenivasachary, J.-M. Lehn and A. Herrmann, *Org. Biomol. Chem.*, 2011, **9**, 2906.
54. M. M. Smith, W. Edwards and D. K. Smith, *Chem. Sci.*, 2013, **4**, 671.
55. J. Boekhoven, A. M. Brizard, K. N. K. Kowlgi, G. J. M. Koper, R. Eelkema and J. H. van Esch, *Angew. Chem. Int. Ed.*, 2010, **49**, 4825.
56. M. Masuda, T. Hanada, K. Yase and T. Shimizu, *Macromolecules*, 1998, **31**, 9403.
57. M. de Loos, J. van Esch, I. Stokroos, R. M. Kellogg and B. L. Feringa, *J. Am. Chem. Soc.*, 1997, **119**, 12675.
58. W. Guijun and D. H. Andrew, *Chem. Eur. J.*, 2002, **8**, 1954.
59. M. George and R. G. Weiss, *Chem. Mater.*, 2003, **15**, 2879.
60. K. Aoki, M. Kudo and N. Tamaoki, *Org. Lett.*, 2004, **6**, 4009.
61. M. Shirakawa, N. Fujita and S. Shinkai, *J. Am. Chem. Soc.*, 2005, **127**, 4164.
62. L. Hsu, G. L. Cvetanovich and S. I. Stupp, *J. Am. Chem. Soc.*, 2008, **130**, 3892.
63. J. R. Moffat, I. A. Coates, F. J. Leng and D. K. Smith, *Langmuir*, 2009, **25**, 8786.
64. D. K. Smith and I. A. Coates, *Chem. Eur. J.*, 2009, **15**, 6340.
65. D. Díaz Díaz, K. Rajagopal, E. Strable, J. Schneider and M. G. Finn, *J. Am. Chem. Soc.*, 2006, **128**, 6056.
66. D. Díaz Díaz, J. J. Cid, P. Vázquez and T. Torres, *Chem. Eur. J.*, 2008, **14**, 9261.
67. S. H. Kang, B. M. Jung, W. J. Kim and J. Y. Chang, *Chem. Mater.*, 2008, **20**, 5532.
68. J. –M. Lehn, *Supramolecular Chemistry: Concepts and Perspectives*, Wiley-VCH, Weinheim, 1995.
69. P. W. N. M. van Leeuwen, Ed., *Supramolecular Catalysis*, Wiley-VCH, Weinheim, 2008.
70. F. Fages, *Angew. Chem. Int. Ed.*, 2006, **45**, 1680.
71. B. Xing, M.-F. Choi and B. Xu, *Chem. Eur. J.*, 2002, **8**, 5028.
72. Y.-R. Liu, L. He, J. Zhang, X. Wang and C.-Y. Su, *Chem. Mater.*, 2009, **21**, 557.
73. J. Y. Zhang, X. B. Wang, L. S. He, L. P. Chen, C. Y. Su and S. L. James, *New J. Chem.*, 2009, **33**, 1070.
74. Y. He, Z. Bian, C. Kang, Y. Cheng and L. Gao, *Chem. Commun.*, 2010, **46**, 3532.
75. J. F. Miravet and B. Escuder, *Chem. Commun.*, 2005, 5796.
76. J. H. Lee, S. Kang, J. Y. Lee and J. H. Jung, *Soft Matter*, 2012, **8**, 2557.
77. Q. Jin, L. Zhang, H. Cao, T. Wang, X. Zhu, J. Jiang and M. Liu, *Langmuir*, 2011, **27**, 13847.

78. G. Bühler, M. C. Feiters, R. J. M. Nolte and K. H. Dötz, *Angew. Chem. Int. Ed.*, 2003, **42**, 2494.

79. T. Tu, W. Assenmacher, H. Peterlik, R. Weisbarth, M. Nieger and K. H. Dotz, *Angew. Chem. Int. Ed.*, 2007, **46**, 6368.

80. P. Melchiorre, M. Marigo, A. Carlone and G. Bartoli, *Angew. Chem. Int. Ed.*, 2008, **47**, 6138.

81. For instance, see a thematic issue on organocatalysis: *Chem. Rev.*, 2007, **107**, 5413.

82. For instance, M. L. Clarke and J. A. Fuentes, *Angew. Chem. Int. Ed.*, 2007, **46**, 930.

83. J. E. Klijn and J. B. F. N. Engberts, *J. Am. Chem. Soc.*, 2003, **125**, 1825.

84. K. Holmberg, *Eur. J. Org. Chem.*, 2007, 731.

85. K. Tanaka, A. Mori and S. Inoue, *J. Org. Chem.*, 1990, **55**, 181.

86. H. Danda, *Synlett*, 1991, 263.

87. E. A. C. Davie, S. M. Mennen, Y. Xu and S. J. Miller, *Chem. Rev.*, 2007, **107**, 5759.

88. F. Schoenebeck and K. N. Houk, *J. Org. Chem.*, 2009, **74**, 1464.

89. M. O. Guler and S. I. Stupp, *J. Am. Chem. Soc.*, 2007, **129**, 12082.

90. B. List in *Modern Aldol Reactions*, Vol. 1 (ed. R. Mahrwald), Wiley-VCH, Weinheim, 2004, pp. 161–200.

91. D. Enders, M. R. M. Hüttl, C. Grondal and G. Raabe, *Nature*, 2006, **441**, 861.

92. B. Escuder, S. Martí and J. F. Miravet, *Langmuir*, 2005, **21**, 6776.

93. F. Rodríguez-Llansola, J. F. Miravet and B. Escuder, *Chem. Commun.*, 2009, 209.

94. F. Rodríguez-Llansola, J. F. Miravet and B. Escuder, *Org. Biomol. Chem.*, 2009, **7**, 3091.

95. F. Rodríguez-Llansola, B. Escuder and J. F. Miravet, *J. Am. Chem. Soc.*, 2009, **131**, 11478.

96. D. S. Tsekova, J. A. Sáez, B. Escuder and J. F. Miravet, *Soft Matter*, 2009, **5**, 3727.

97. F. Rodríguez-Llansola, J. F. Miravet and B. Escuder, *Chem. Eur. J.*, 2010, **16**, 8480.

98. A. P. Brogan, T. J. Dickerson and K. D. Janda, *Angew. Chem. Int. Ed.* 2006, 45, 8100.

99. Y. Hayashi, *Angew. Chem. Int. Ed.*, 2006, **45**, 8103.

100. D. G. Blackmond, A. Armstrong, V. Coombe and A. Wells, *Angew. Chem. Int. Ed.*, 2007, **46**, 3798.

101. J. Paradowska, M. Stodulski and J. Mlynarski, *Angew. Chem. Int. Ed.*, 2009, **48**, 4288.

102. F. Rodríguez-Llansola, J. F. Miravet and B. Escuder, *Chem. Commun.*, 2009, 7303.

103. A. M. Brizard and J. H. van Esch, *Soft Matter*, 2009, **5**, 1320.

104. D. Deamer, S. Singaram, S. Rajamani, V. Kompanichenko and S. Guggenheim, *Philos. Trans. R. Soc. B.*, 2006, **361**, 1809.

CHAPTER 6

Biomedical Applications of Molecular Gels

WARREN TY TRUONG, LEV LEWIS AND
PALL THORDARSON*

School of Chemistry and the Australian Centre for Nanomedicine,
The University of New South Wales, Sydney, NSW 2052, Australia
*Email: p.thordarson@unsw.edu.au

6.1   Introduction

Since antiquity, humanity has used naturally available materials like wood to augment the body due to injuries sustained over time (Figure 6.1A). With advances in technology, these materials were superseded by man-made synthetic polymers, ceramics and metal alloys, which provided better performance, increased functionality and enhanced reproducibility (in terms of their properties) than their naturally derived counterparts.[1]

The field of biomedical materials has seen great progress from the crude prosthetics carved from wood,[2] to the nylon-based sutures used today for wound closure and the advent of contact lenses and drug-infused wafers[3–5] (Figure 6.1B)[5] that are implanted within the body.

These natural and synthetic-derived materials have had a profound impact on the progression in therapeutic treatments. However, as the field is developing, there is a pressing need for better biomimetic materials. One solution that featured heavily in the development of these biomaterials has been molecular gels.

RSC Soft Matter No. 1
Functional Molecular Gels
Edited by Beatriu Escuder and Juan F. Miravet
© The Royal Society of Chemistry 2014
Published by the Royal Society of Chemistry, www.rsc.org

A   B

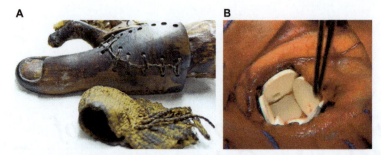

Figure 6.1   Biomedical materials through time. (A) A false toe discovered in 2000 AD
near Luxor, Egypt in the necropolis of Thebes, this wood-and-leather
prosthetic toe belonged to Tabaketenmut, a high priest's daughter who
lived between 950 and 710 BC.[2] (Reprinted from The Lancet from ref. 2
Copyright © (2011), with permission from Elsevier) (B) Gliadel® drug-
infused wafers are inserted into excised tumour sites in a patient's brain
and release carmustine temporally.[4] (Reprinted from ref. 5 Copyright ©
(2003), with permission from Elsevier).

In this chapter we aim to provide a glimpse into the breadth of work dedicated to the development of molecular gels for biomedical applications. An
emphasis is placed on the ability of these materials to mimic the natural biological environment whilst imparting localised therapeutic activity. The discussion will begin by outlining the general need to develop biomaterials to
mimic the extracellular matrix and construct the framework within which these
materials exhibit bioactivity. The ability of molecular gels to reproducibly self-
assemble into well-defined structures has warranted their use in biomedical
applications.

In the remainder of this chapter an exploration into the use of molecular gels
in a wide range of biomedical applications; including drug delivery, three-
dimensional (3D) cell culture, tissue engineering and regenerative medicine will
be conducted.

The chapter will conclude by highlighting the key design considerations to be
addressed in the development in molecular gels for biomedical applications and
the future directions of this field.

6.2   Biomimetic Materials

6.2.1   The Extracellular Matrix (ECM)

Currently, there is a pressing need for biomedical materials that better mimic
biological environments. On the nanoscale level, this environment would be the
extracellular matrix (ECM) found in living systems, as it has been shown that
the interactions at the interface are critical in determining the fate of cells. One
of the key reasons that molecular gels are considered such promising materials
for medical applications is the nanostructure (Figure 6.2) closely mimics that of
ECM.[6]

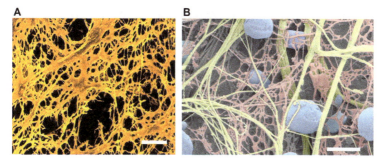

Figure 6.2 Scanning electron micrographs (SEM) of synthetic *versus* biological molecular gels. (A) A SEM of a typical small-peptide-based molecular gel from the authors. (B) The extracellular matrix (ECM – pink) intertwined with nerves and nerve bundles (yellow) and ganglion cells (blue) in front of the retina of the eye.[7]
(Reproduced with permission from ref. 7 Copyright © Photo courtesy of University of Rochester/URnano). Scale bars represent 5 μm and 2 μm, respectively.

The ECM is a complex intertwined mesh of fibrous proteins (such as collagen) and polysaccharides (such as glycosaminoglycans) secreted by cells (Figure 6.2B).[7] It serves as a structural scaffold in tissue by providing anchorage to cells, regulating cellular behaviour and communication *via* structural and chemical signals; playing a major role in their development, function and physiology.[8]

Cells receive cues (mechanical, chemical) from the ECM, whilst simultaneously interacting with, and constantly remodelling their environment, thus reinforcing their phenotype. By excreting proteases, such as matrix metalloproteinases (MMPs), cells can degrade the matrix around them and replace it with newly synthesised ECM proteins. This process in turn affects cellular behaviour, such as proliferation, migration, apoptosis, and differentiation during development, tissue remodelling, and angiogenesis.[9] One major driving force in mimicking the ECM in synthetic systems arises from (besides inherent biocompatibility) the critical role the ECM has in influencing the spreading and differentiation of stem cells;[10] this could be applied to particular advantage in the areas of regenerative medicine/tissue engineering and three-dimensional cell culture.

Over the years, natural materials such as collagen, chitosan[11] and other natural polymer-based gels, have featured heavily in the development of new biomaterials for wound healing and other tissue engineering.[12] Based on the success of these naturally based polymeric gels, numerous chemically engineered polymeric gels have been designed and tailored for the similar purposes in recent times.[13] However, this chapter's focus will not be on polymeric gels as this area has already been covered in several excellent reviews.[14,15]

The following section will aim to introduce the motivation for their utilisation in biomedical applications.

6.2.2 Molecular Gels as Extracellular Matrix Mimics

Molecular gels are good candidates as biomimetic materials, as they possess the
following characteristics:

- Structurally, they are similar to the ECM, as shown in Figure 6.2. As the
  majority of them are assembled with peptide building blocks and have
  high H_2O (90-99% w/w) content, they are inherently biocompatible.
- On the macroscale, a material will need to mimic the rigidity of tissues for
  mechanical functionality.[16] Molecular gels have been shown to be able to
  be fine tuned to mimic the viscoelasticity of tissues.[6]
- Easily administered as they can reversibly interchange from liquid to gel
  states with:
  ○ Physical and chemical stimuli (see Chapter 3).
  ○ Enzymes (see Chapter 4).
- They are chemically well defined as they are usually made from two
  simple low molecular weight components; the gelator and water or other
  solvent. This is in contrast to materials based on covalent polymers where
  polydispersity and problems with batch-to-batch reproducibility can:
  ○ Make it harder to interpret biological assays / studies based on these
    materials.
  ○ Complicate matters when seeking regulatory approvals (e.g. from the
    FDA) for the use of these materials in food or medicine.
- Easily modified – modular structures, rationally designed, easy to synthesise, biodegradability, ability to decorate with biologically active
  motifs. A change at the molecular structure (*e.g.* peptide sequence),
  directly translates to a change in bulk material property.[17]
- Porous; nutrients, cell signalling factors and oxygen (through haemo-
  globin) can diffuse through their fibrous network structures – this is
  critical to the survival of cells if they are to be encapsulated within molecular gels for regenerative medicine or tissue engineering.

6.2.2.1 Chemical and Physical Definitions

Structurally, gels consist of two components, a solvent that is encapsulated
within a three-dimensional network of entangled fibres (Figure 6.3).[18] It can be
noted that this is reminiscent of the numerous protein fibrils present in the
extracellular matrix, such as collagen, fibronectin and elastin.[15]

In most gels, it is the solvent that makes up the majority (by weight) of the
gel. Molecular gels typically consist of 0.1–10% w/w of the gelator, whereas in
polymeric gels, the weight percentage tends to be higher. If the solvent is organic, then we can classify the gel as an organogel (Figure 6.3, right), whereas if
the solvent is water we can classify it as a hydrogel (Figure 6.3, left).
A distinguishing feature of molecular gels is the supramolecular interactions
connecting the gelator molecules to form the fibres (see Chapter 1), in contrast

Molecular gels

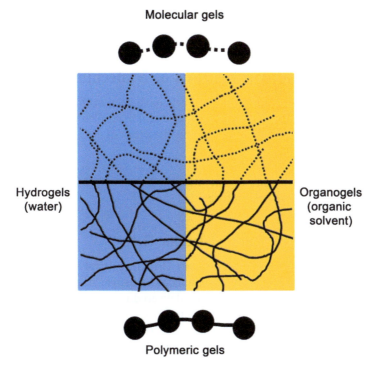

Hydrogels
(water)

Organogels
(organic
solvent)

Polymeric gels

Figure 6.3 Classifications of gels based on fibres and solvent. Gels can be either made up of polymeric (below the horizontal line) or molecular gel fibres (above the horizontal line). The solvent that makes up the majority of the gels (by weight) can either be water (left: hydrogels) or organic solvent (right: organogels). Molecular gels are also known as self-assembled or physical gels.[18]

to the covalent bonds of polymeric gel fibres (*e.g.* poly(2-hydroxyethyl methacrylate) PHEMA).

6.2.2.2 *Molecular Gels in Nature*

It is worth emphasising that self-assembling fibrillar structures are found in nature, and although they do not form gels, their functional properties do arise from their fibrillar nature rather than their building blocks *per se*. The most recognisable examples of this kind are the β-amyloids that are associated with diseases such as Alzheimer's disease.[19] A few natural peptide and protein hormones self-assemble into nondisease-related aggregates near or within their storage sites.[20] Finally, it is important to note that molecular gels are dynamic in nature[21] akin to naturally occurring self-assembling systems, such as actin filaments and the above-mentioned β-amyloids. The dynamic nature of molecular gels allows them to adapt better to their environments and changes in their surroundings, including inside living tissue. It is quite probable that the dynamic nature of molecular gels is the underlying explanation for the apparent advantage that they seem to have over conventional polymer-based gels in applications such as a tissue engineering and drug delivery.

6.3   Biomedical Applications of Molecular Gels

Numerous applications for molecular gels have been outlined in the literature; this chapter's main focus will be on successful examples of biomedical applications of water-based molecular gels in the areas of:

- drug delivery;
- tissue engineering and regenerative medicine;
- three-dimensional cell culture.

6.3.1   Drug Delivery and Molecular Gels

Conventional methods of drug delivery (Figure 6.4) rely on the body's own systemic and cellular transport mechanisms to deliver drug molecules to their target destination. Drugs are generally delivered into the body through oral or intravenous routes. The disadvantage of these methods is that, in certain situations, such as in chemotherapy, toxic drug molecules can come into contact with healthy tissues, thereby causing major side effects that prohibit treatment. This situation is often the cause of chemotherapy failure when bone marrow cell death prevents the patient from undergoing a complete treatment.[22] Localised drug delivery, on the other hand, offers numerous advantages compared to conventional delivery methods including improved efficacy, reduced toxicity, and improved patient compliance and convenience.[22]

Currently, localised therapeutic delivery systems rely heavily on synthetic polymers to carry the drugs[22] (Figure 6.4).[23–28] A good example of this is the FDA approved Gliadel® polymer inserts for the treatment of glioblastoma

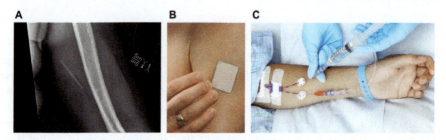

Figure 6.4   Examples of current delivery methods. (A) Implanon® is a subdermal implant (4 cm × 2 mm) that contains 68 mg of progestogen, etonogestrel. It releases approximately 40 μg of etonogestrel/day and is a very effective birth control method. (Reproduced from ref. 23 Copyright © 2010 with permission from BMJ Publishing group Ltd) (B) Reprinted by permission from Macmillan Publishers Ltd from ref. 24 Copyright © 2004) (C) Intravenous drip. (Photo by Calleamanecer, "Neutron Patency Device" 1 May, 2013 *via* Wikipedia, Creative Commons Attribution from ref. 27).

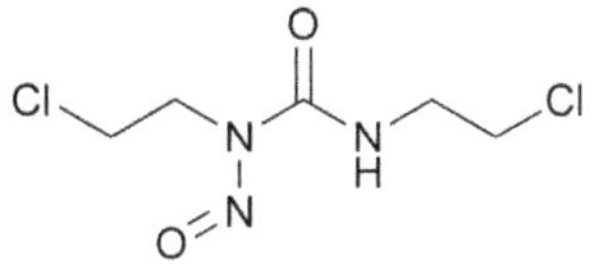

6.1

Chart 6.1

multiforme, an aggressive form of brain cancer, as an adjunct to surgery and radiation. The Gliadel® polymer comes in the form of wafers that are implanted in resected tumour sites.

A chemotherapeutic, carmustine (BCNU) (Chart 6.1) is then released from within the wafers. The aim of this localised form of delivery is to prevent any cancerous cells that were not removed during resection from metastasising.

There are a few approaches for localised delivery with gels, polymeric or molecular. The first is to dope a therapeutic within the voids of a gel.[29] This can usually be achieved by dissolving the therapeutic in the solvent the gel is to be comprised of before formation of the gel (Figure 6.5A). This method can be categorised as a passive delivery method. Since the pioneering work of Friggeri *et al.* on the *in vitro* release of quinoline derivatives from a *N,N'*-dibenzoyl-cystine gelator,[30] numerous papers have looked at the passive *in vitro* release of drugs and drug mimics from molecular gels, however, in this chapter the focus will instead be on the medically relevant *in vivo* studies on passive drug delivery from molecular gels.

Formation of the gel is usually triggered through a change in pH,[28] by heating,[25] or enzymatically triggered.[26] The gel can then be topically applied or injected and it is then degradation of the gel and/or diffusion of the therapeutic that allows the drug to act at the target site (Figure 6.5).[22]

The second approach to the delivery of drugs is *via* stimuli-responsive, molecular gel-therapeutic delivery systems. This approach relies on the modulation of therapeutic release *via* various stimuli such as a matrix metalloproteinase-2 (MMP-2) sensitive peptide sequence or a labile hydrazone linkage. As of writing this review, it can be noted a search of the literature reveals this is not so common in polymer-based gels. Unlike the passive delivery mentioned earlier, these systems dynamically release their cargo.

A third approach is to design a gelator motif that incorporates a therapeutically active sequence into its structure (Figure 6.5B).[31] We define this approach as *therapeutic molecular gels*. This method does not rely on the passive release or triggered release of therapeutics from within another ideally inert gel; the gel itself rather is the therapeutic.

Molecular gels are now showing significant potential in the field of localised drug delivery. Inherently, gels can easily mould into any shape that is required; this is necessary for easy application and efficacy, especially in the field of localised drug delivery. As it has been shown that molecular gels can be triggered to gelate by means of various stimuli,[18] they offer specific advantages for

A

Inert gel loaded with a drug   Diffusion of drug

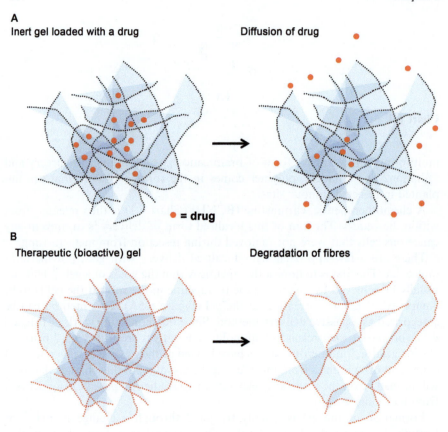

● = drug

B

Therapeutic (bioactive) gel   Degradation of fibres

Figure 6.5   Drug-delivery approaches with molecular gels. (A) Doping a therapeutic
within a molecular gel. (B) Therapeutic molecular gel.

localised drug delivery relative to other forms of drug-delivery methods as they
can be tuned/designed to respond in specific biological environments.

6.3.1.1   Passive In Vivo Drug Delivery using Molecular Gels

The simplest methodology for the delivery of drugs using molecular gels is
passively, *i.e.* diffusion and/or degradation of the gel system. The following
examples provide a glimpse into the research that has been done with passive
molecular gel therapeutic delivery systems.

Nitric oxide (NO) has been shown to inhibit neointimal hyperplasia
(the narrowing of arteries), after arterial interventions in several animal models.
Kapadia and coworkers[32] created a molecular gel-based delivery system for NO
for perivascular application.

Gels were formed by mixing equal volumes of the peptide amphiphile **6.2** in
ultrapure water, diazeniumdiolate in phosphate-buffered saline (PBS) and
heparin in PBS (Chart 6.2).

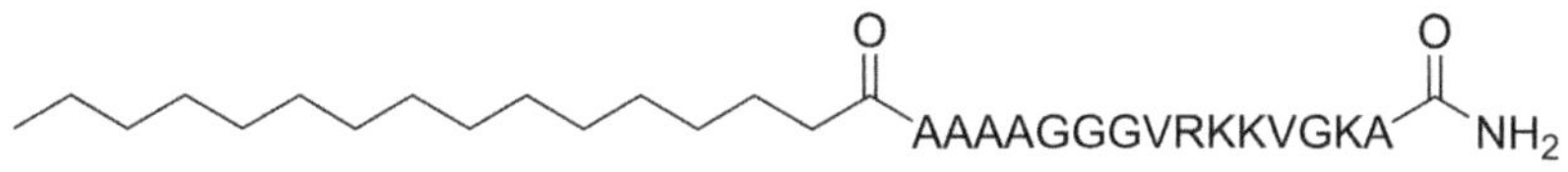

$$\underset{\text{AAAAGGGVRKKVGKA}}{\overset{\displaystyle O}{\|}} \overset{\displaystyle O}{\|} NH_2$$

6.2

Chart 6.2

It was noted that the mixing of the nitric oxide within the gel extended the release of the nitric oxide significantly to four days *in vitro*. This mixture was then applied directly to the exterior of an injured blood vessel (rat model) after angioplasty (the mechanical widening of narrowed/obstructed arteries). As an example of this first approach to localised drug delivery, the system showed clinically promising results in the limiting of abnormal narrowing of arteries or valves after surgery (neointimal hyperplasia) by up to 77% compared with the controls and also limited inflammation in the injury site.[32]

Topical application of gel–drug formulations at the site of injury/disease undoubtedly offers additional advantages of delivering the active drug compound to the specific site. For instance, in a study by Xu's group, it was shown that the topical application of a molecular hydrogel based on the mixture of Fmoc-L **6.3**, the uranyl nitrate binding ligand pamidronate **6.4**, and ε-Fmoc-K **6.5** (Figure 6.6) could be used to treat wounds on the skin of mice that had been contaminated with uranyl nitrate. The treated mice recovered, whereas untreated mice weighed 35% less or died, presumably from radiation exposure caused by the uranyl-nitrate-contaminated skin wounds.[33]

In localised drug delivery, the rate of drug release is an important aspect. The controlled release kinetics of molecular hydrogels have been demonstrated and investigated. Liang and coworkers did the first *in vivo* imaging of a molecular hydrogel formed from a naphthalene-ᴰFᴰF dipeptide **6.6** and studied the controlled release of radiolabeled iodine ^{125}I from a NaI-loaded gel formed from **6.6**.[34]

6.3.1.2 Enzymatically Induced Drug Delivery from Molecular Gels

Stimuli-induced molecular gel-therapeutic delivery systems can be designed to modulate the release rates of therapeutics over prolonged periods, through responding to enzymes in the body for example.

In work by Kim and coworkers,[35] a novel method for the targeted delivery of cisplatin **6.7** was achieved through a bioresponsive molecular-gel delivery system. Their peptide amphiphile (PA) system consisted of a peptide sequence sensitive to matrix metalloproteinase-2 (MMP-2) (GTAGLIGQRGDS) **6.8**; MMP-2 is known to be overexpressed in different kinds of invasive tumours, playing a critical role in tumour progression, angiogenesis, and metastasis.[36,37] Upon mixing **6.8** with **6.7** (Figure 6.7), complexation of the carboxylic acids in the peptide sequence with cisplatin would form a gel–cisplatin complex.[38–40]

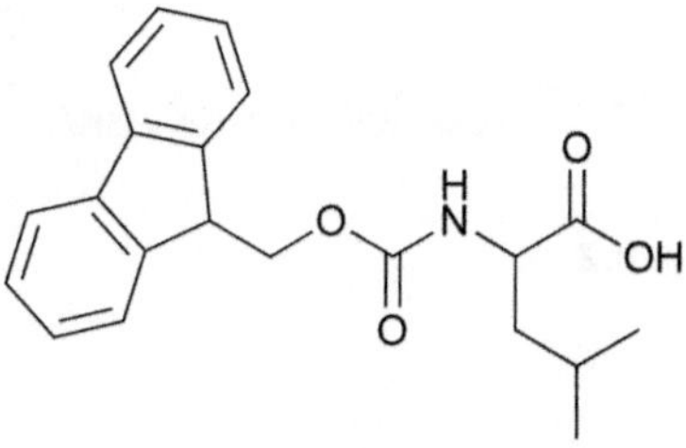

6.3

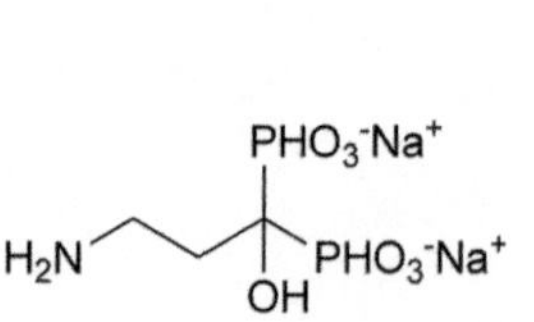

6.4

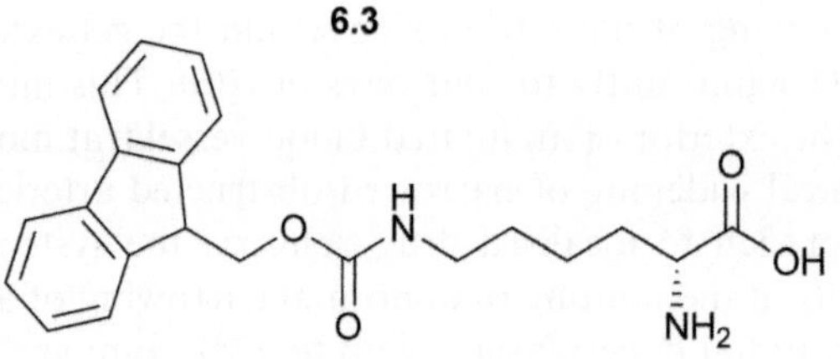

6.5

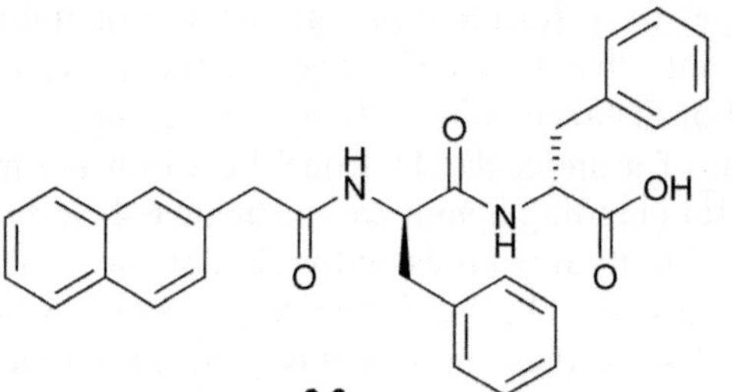

6.6

Figure 6.6 Passive drug release from molecular gels. The molecular gelators Fmoc-Leu **6.3** and ε-Fmoc-Lys **6.5** that were used with the ligand pamidronate **6.4** to treat skin wounds from uranyl oxide contamination.[33] The naphthalene-ᴰFᴰF **6.6** gelator was used to demonstrate the release of radiolabelled iodine from NaI-loaded gels.[34]

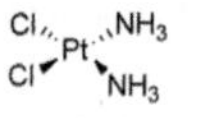

cisplatin

6.7

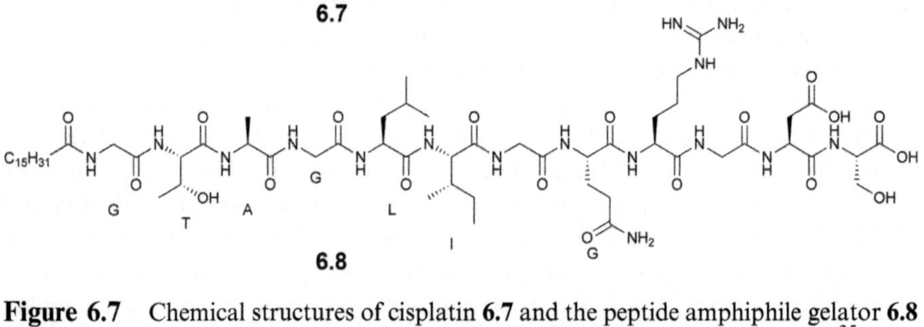

6.8

Figure 6.7 Chemical structures of cisplatin **6.7** and the peptide amphiphile gelator **6.8** used by Kim and coworkers for targeted delivery of cisplatin **6.7**.[35]

Cisplatin **6.7** is one of the most extensively used chemotherapeutics for the treatment of various cancers such as testicular cancer and glioma.[41] However, severe side effects such as acute nephrotoxicity and neural toxicity have limited the clinical use of cisplatin.[42,43]

Hence, in an effort to mitigate these adverse effects and enhance its anticancer activity, the tumour-specific accumulation and controlled release of cisplatin at the site was investigated.

As expected, cisplatin **6.7** release from the peptide amphiphile–gel complex was triggered by the cleavage of the MMP-2-sensitive sequence in the peptide amphiphile and was found to be dependent on the concentration of the enzyme (Figure 6.8). Also, the amounts of cisplatin **6.7** loaded in the gel were found to be approximately 2.5–3-fold greater than its aqueous solubility. Although this study was preclinical, it demonstrates the potential of stimuli-responsive molecular gels for targeted drug delivery (Chart 6.3).

Another chemotherapeutic delivery system was investigated by Gao and coworkers, who modified Taxol® (paclitaxel) through the 2′-position with a linker, self-assembling motif and enzyme-cleavable group to yield **6.10**.[44] Upon addition of alkaline phosphatase, an enzyme present in all tissues throughout the body, dephosphorylation of **6.10** gives **6.11**, which readily forms molecular gels in water. Paclitaxel is a notoriously highly insoluble hydrophobic anticancer drug, and the group established a new, facile method to convert this drug

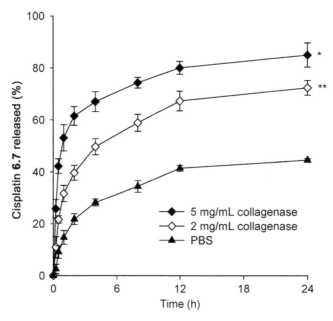

Figure 6.8 Release profiles of cisplatin from cisplatin **6.7**-PA **6.8** gels showing dependency on type-IV collagenase (MMP-2) concentration. The release study was performed with Franz diffusion cells at 37 °C.[35]
(Reproduced with permission from ref. 35 Copyright © 2009 American Chemical Society).

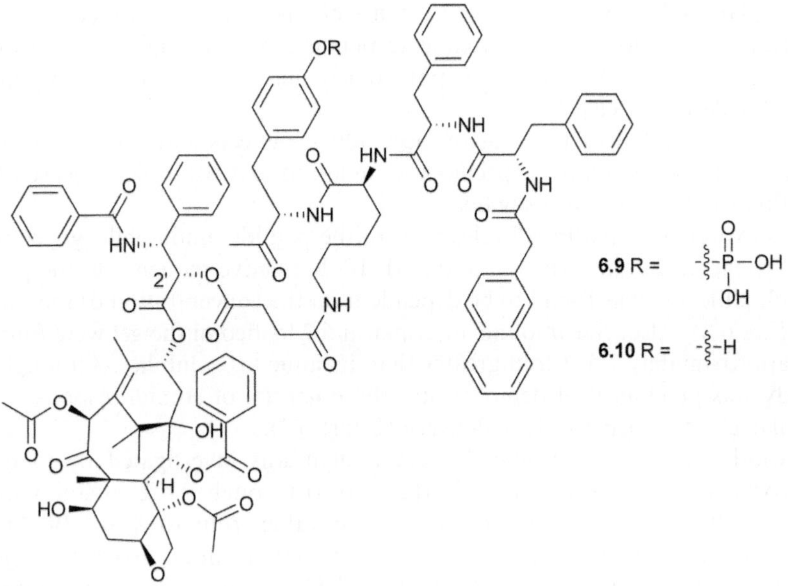

6.9 R = $-\xi-\overset{\displaystyle O}{\underset{\displaystyle OH}{\overset{\displaystyle \|}{P}}}-OH$

6.10 R = $-\xi-H$

Chart 6.3

into a gel without compromising its biological activity. This work demonstrates the versatility of molecular gels in being the first enzyme-instructed, self-assembly and hydrogelation of a complex, bioactive small molecule. It is a further proof of principle that a molecular gel can act as a prodrug.

Bremmer *et al.* have developed a modular approach to the design of enzymatically active gelators.[25] The system consists of a sequence to promote solubility that is covalently linked to an enzymatic recognition sequence that cleaves a gelator **6.11** from the system. The cleaved gelator can then self-assemble into a gel triggered by enzymatic activity (Figure 6.9) (see Chapter 4).

The enzymatically active system exhibits high selectivity due to the protease-specific recognition sequence at the cleavage site. The modularity of this system allows for the facile interchanging of recognition sequences, allowing Bremmer *et al.*[25] to engineer a system that has the potential to have induced gelation through a wide variety of enzymes.

Three systems were fabricated to contain protease-specific recognition sequences for thrombin (LTPR), chymotrypsin (AAPF), and Glu-C (DAFE). All three enzymes triggered gelation at physiologically relevant concentrations; Thrombin (400 pM), Chymotrypsin (50 nM), Glu-C (50 nM); and were unable to induce gelation on either of the other systems highlighting the specificity of the protease recognition sequence.

The thrombin-triggered system has particular importance in the application of regenerative medicine or drug delivery through its role in the formation of blood clots by converting soluble fibrinogen into insoluble strands of fibrin. Although the time to induce gelation was dependent of enzyme concentration and took ~2 h at physiologically relevant concentrations, the use of this

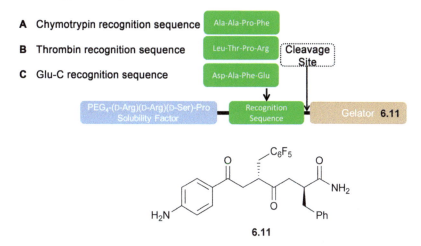

A Chymotrypin recognition sequence   Ala-Ala-Pro-Phe

B Thrombin recognition sequence   Leu-Thr-Pro-Arg   Cleavage Site

C Glu-C recognition sequence   Asp-Ala-Phe-Glu

PEG₄-(D-Arg)(D-Arg)(D-Ser)-Pro Solubility Factor — Recognition Sequence — Gelator **6.11**

6.11

Figure 6.9 Schematic of an enzymatically active gelator system with facile interchangeable recognition sequences that are cleaved by the following proteases: (A) Chymotrypsin (B) Thrombin (C) Glu-C. Upon recognition of specific proteases, gelation occurs as gelator **6.11** is liberated.[25]

enzyme induced gelator is envisioned to be incorporated into surface wound garments that would stem bleeding and promote blood clotting for injured patients. Furthermore, this system has applications in the detection and diagnosis of selective protease activity. Bremmer *et al.* have highlighted the modularity of their system to elicit high specificity to enzymatic activity.[25] The utilisation of modular recognition sequences for varying protease activity adds significant versatility in the design and function of this system.

6.3.1.3 Therapeutic Molecular Gels for Drug Delivery

Therapeutic molecular gels are here defined as medicinal compounds such as antibiotics or chemotherapeutics that have been modified with a bulky aromatic group or β-sheet forming peptide sequence to induce gelation under the appropriate conditions, circumventing the need for drug encapsulation. These systems have been shown to display enhanced release kinetics for drug delivery relative to diffusion-based release from encapsulating gel networks.[45]

One early and prominent example of the therapeutic molecular gel approach is in the modification of vancomycin with a pyrene group by Xing and coworkers, which enabled the formation of a vancomycin molecular gel **6.12** (Chart 6.4). Vancomycin is an antibiotic used in the prevention and treatment of infections caused by Gram-positive bacteria. In levels of antibiotic activity, the pyrene-modified vancomycin **6.12** showed an 11-fold increase, relative to plain vancomycin.[31]

Undoubtedly, gel–drug complexes can contribute hugely to the treatment of various cancers through localised and sustained chemotherapy. As an adjunct to surgery, local delivery using molecular hydrogels is especially well suited to

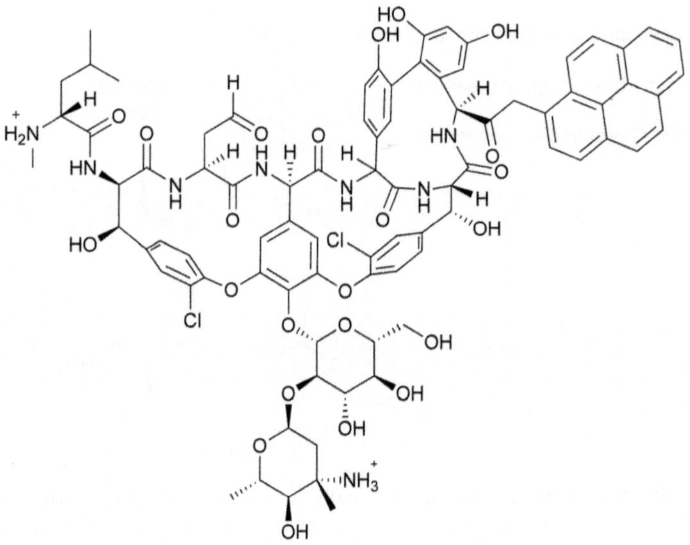

6.12

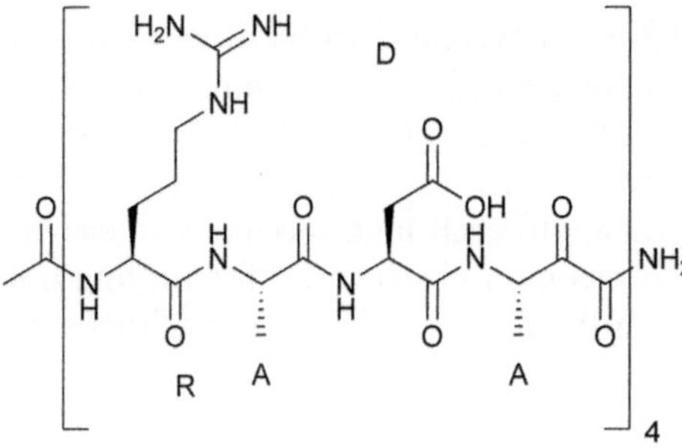

6.13

Chart 6.4

deliver chemotherapeutics to the site of a recently resected tumour. This concept has been successfully applied in Gliadel® polymer wafers.

Another demonstrated application for therapeutic molecular gels is in stemming bleeding during surgery (haemostasis). Ellis-Behnke and coworkers demonstrated[46] that the known self-assembling peptide Ac-(RADA)$_4$-NH$_2$ **6.13**,[47] forms a molecular gel on contact with bodily fluids such as blood. They reported that the self-assembling peptide **6.13** establishes a nanofibre barrier (<15 s) to achieve complete haemostasis when applied directly to a wound in the brain, spinal cord, femoral artery, liver or skin of mammals.[46] In theory, this could be topically applied after tumour resection, as an adjunct to surgery. On contact, the gel would mould into the shape of the cavity, having an increased surface area of contact for diffusion of encapsulated therapeutics, thereby being an efficacious therapy relative to Gliadel® (polymer) wafers for example in addition to achieving haemostasis.[46] The use of Gliadel® wafers, however, requires invasive surgery while the results above suggest that gel–drug mixtures from **6.7** could form *in vivo* after a simple injection to the site of interest.

A self-assembling peptide amphiphile (PA **6.15**) designed to deliver carbon monoxide (CO) developed by Matson *et al.*[48] (Figure 6.10) represents another advance in bioactive gelators for drug delivery. CO is known to play a protective role through its anti-inflammatory, antiapoptotic, and antiproliferative functions in tissue.[49,50] However, systemic overexposure and CO poisoning pose inherent risks for inhaled delivery techniques. Therefore, synthetic CO-releasing molecules (CORMs) circumvent these risks by providing greater precision over concentration, dose, kinetics, and localisation of therapeutic CO delivery.

Although CORM-based therapies provide greater control over delivery than inhaled CO gas, these small molecules are expected to diffuse quickly after administration, thus limiting their ability to localise within a specific tissue. Furthermore, other recently reported CORM systems[51,52] require UV activation for CO release, which limits their biological applications where UV light may harm tissues and cells. Therefore, greater control over localisation is

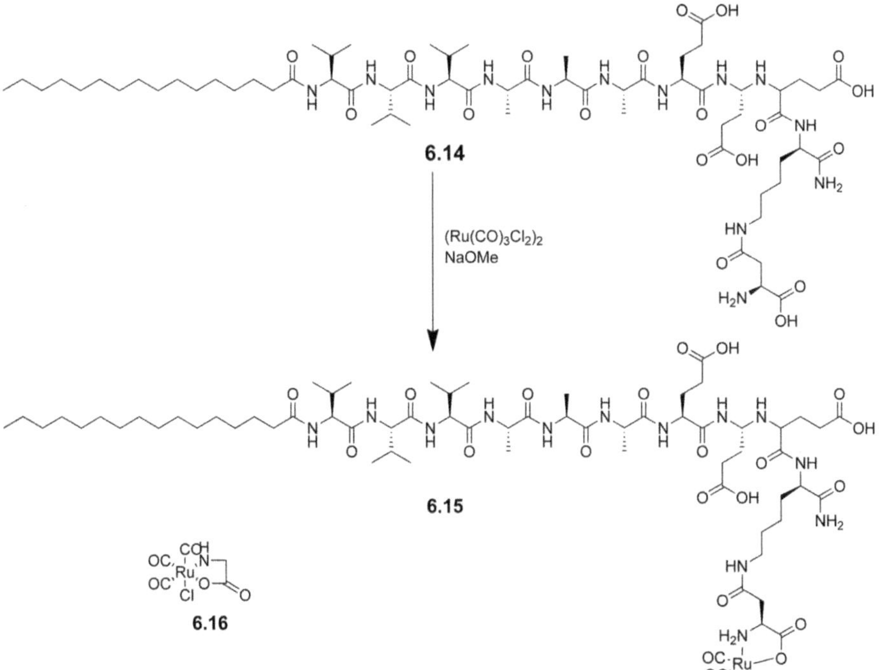

Figure 6.10   The synthesis of the CO-releasing molecular (CORM) gelator **6.15**. Complexation of peptide amphiphile **6.14** with the water-soluble ruthenium tricarbonyl (CORM-3) **6.16** yields the PA-CORM-3 complex **6.15**.[48]

required and spontaneous sustained CO release is preferred over UV activation for effective therapy.

As a result, PA-CORM-3 **6.15** was designed to address these issues through self-assembling peptide-based materials. The peptide amphiphile contains an alkyl tail covalently attached to a short peptide sequence that is able to self-assemble into supramolecular nanofibre gel networks. The peptide sequence was functionalised with a $Ru(CO)_3Cl$(glycinate) motif inspired by CORM-3 **6.16**, the most extensively evaluated CORM to date.[53]

The water-soluble ruthenium tricarbonyl **6.16** exhibits spontaneous release of one equivalent of CO under physiological conditions with a half-life of ~ 2 min in solution.[48] The self-assembling nature of PA-CORM-3 **6.15** into a supramolecular gel network offers therapeutic control as delivery is easily achieved *via* sol-state injection and localisation is attained upon gelation in the tissue of interest. Gelation was found to dramatically prolong the CO release half-life by 8 times (~ 18 min), addressing the problem of rapid depletion of the gas in most other structures. The Ru-carbonyl-containing amphiphile is the first known example of a gelating CO-releasing material. The self-assembling gel framework offers potential for therapeutic localisation within a target tissue, limiting nontargeted effects of CO therapy and the spontaneous and sustained release kinetics minimises total dose requirements, making PA-CORM-3 **6.15** an effective CO therapeutic.

More recent work by Stupp and coworkers has further developed the potential of peptide amphiphiles (PA) as therapeutic gelators, in this case to suppress localised acute inflammatory response without systemic immune suppression. PA **6.17** was conjugated with the potent anti-inflammatory drug dexamethasone[54] (Dex) **6.19** *via* a labile hydrazone linkage (Figure 6.11).[45] The bioactive Dex-PA **6.18** was able to self-assemble into a gel network, reminiscent of PA **6.17**. The controlled release of Dex from the molecular hydrogel network was modulated *via* the hydrazone linkage.[55]

The ability of hydrazone linkage to be hydrolysed under physiological conditions enabled the sustained release of Dex over 32 days with $\sim 40\%$ of the total Dex released from the gel (minimal burst release, zero-order release kinetics). However, noncovalently bound Dex encapsulated within the gel network displayed a greater burst release and faster release profile over the 32 days. Furthermore, over 50% of the total Dex had been released from the gel within the first five days, highlighting the burst release characteristics of encapsulated Dex relative to the prolonged release kinetics achieved using a hydrazone linkage (Figure 6.12).

Ultimately, this system is envisioned to be used in tandem with the anti-inflammatory CO-releasing PA-CORM-3 **6.15** to regulate the innate inflammatory response for therapeutic delivery and transplantation.

Wang *et al.* further pushed the notion of molecular gelators exhibiting bioactivity in the form of Taxol® covalently attached to short peptides and amino acids **6.20–6.25** (Chart 6.5).[56] Taxol® acts as a chemotherapeutic, in the treatment of a wide range of cancers, by disrupting cell division through the stabilisation of microtubules.[57]

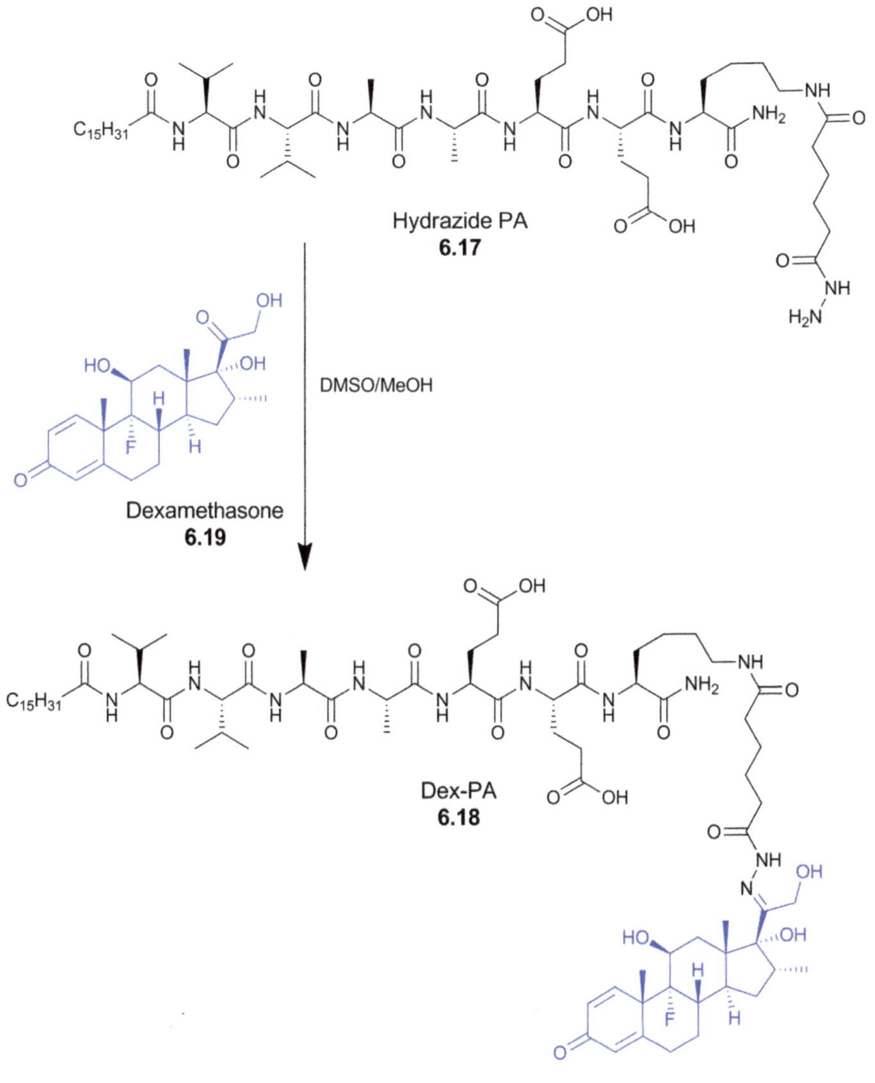

Figure 6.11 The synthesis of a gelator **6.18** from the anti-inflammatory immuno-suppressant drug Dexamethasone **6.19** and the PA hydrazide gelator **6.17**.[45]

The ability of the systems, developed by Wang *et al.* (Figure 6.13),[56] to form molecular gels, allows for the localised delivery of hydrophobic therapeutics, such as Taxol®, without the need for carriers. Moreover, the molecules were engineered to gel in response to a reducing agent. The reduction of the disulfide bond by glutathione induced self-assembly of the bioactive gelator. The response of the molecular gels to the ubiquitous reducing action of glutathione in biological systems allows for *in vivo* gelation upon injection to the target site.[58]

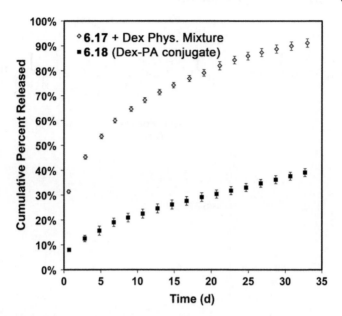

Figure 6.12 Cumulative release of dexamethasone conjugated to Dex-PA **6.18** nano-fibre gel of this same composition over 32 days as well as release of dexamethasone that is physically mixed in a gel of the control PA **6.17** over the same timeframe. Error bars denote the standard deviation of the cumulative release.[45]
(Reproduced with permission from ref. 45 Copyright © 2012 Elsevier).

Another interesting example of a molecular gel with potential biomedical applications was developed by Luo's group. They were able to successfully fabricate molecular hydrogels using DNA.[59] Although this system is based on a biopolymer, DNA, its properties are based on the single-strand noncovalent pairing of DNA and hence the gels are reversible just as is the case with molecular gels from low molecular weight gelators (LMWG).

The DNA hydrogel exhibits remarkable mechanical properties, exhibiting hysteresis. In the absence of water the DNA hydrogel forms a viscous liquid, however, upon the addition of water the hydrogel will conform to the shape it was initially moulded into. Figure 6.14 highlights these transitions in state and clearly depicts the ability of the DNA hydrogel to exhibit shape memory. Upon the removal of water, the hydrogel structure breaks down and the material exhibits liquid-like properties by conforming to the shape of the vessel. However, the addition of water causes the DNA hydrogel to assume its original self-supporting structure; in the form of D-, N-, and A-shaped hydrogels.

In addition to the water-sensitive structure of the DNA hydrogel, the material is capable of remoulding into different shapes by heating beyond its denaturing temperature. Figure 6.15 highlights the ability of the DNA hydrogel to adopt and retain new shapes when heated to 90 °C and then cooled to room temperature in a particular mould. The annealing process causes the hydrogel to assume the new shape with the same water-sensitive structure exhibited

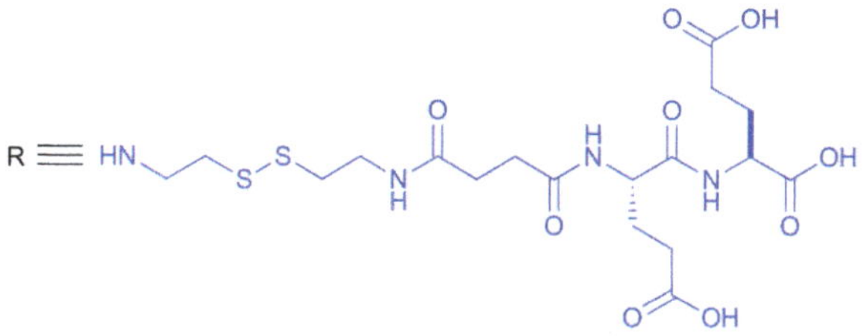

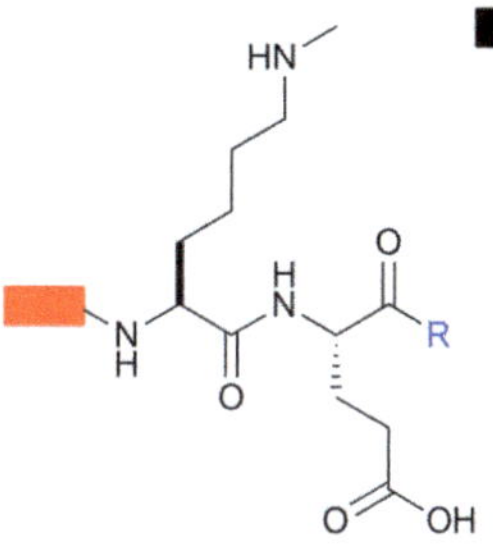

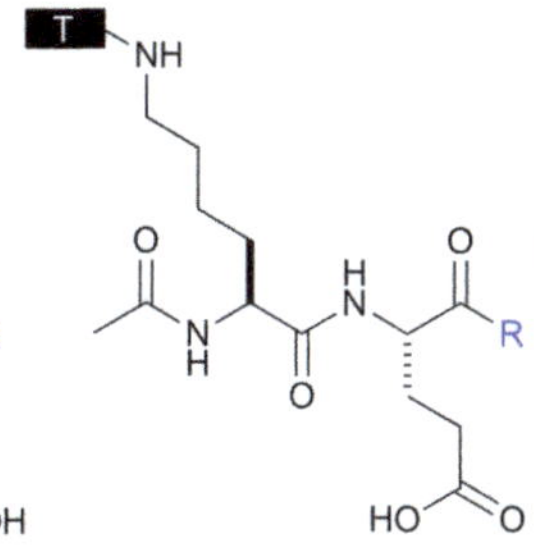

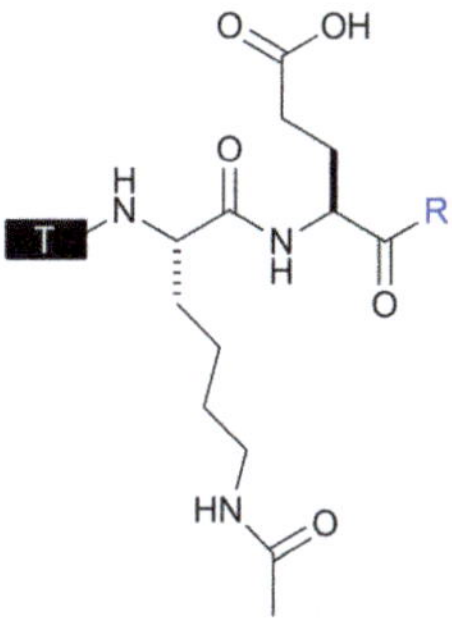

Dex-K(Taxol)E-ss-EE
6.20

Ac-K(Taxol)E-ss-EE
6.21

Taxol-K(Ac)E-ss-EE
6.22

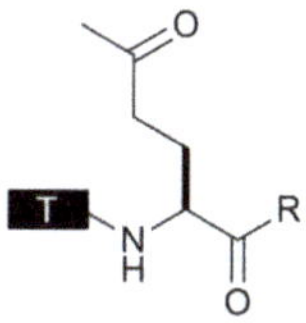

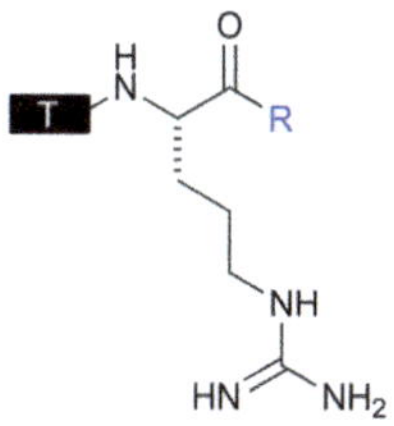

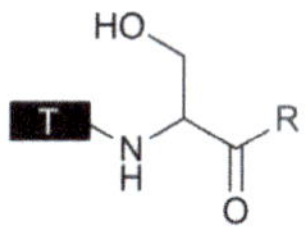

Taxol-E-ss-EE
6.23

Taxol-R-ss-EE
6.24

Taxol-S-ss-EE
6.25

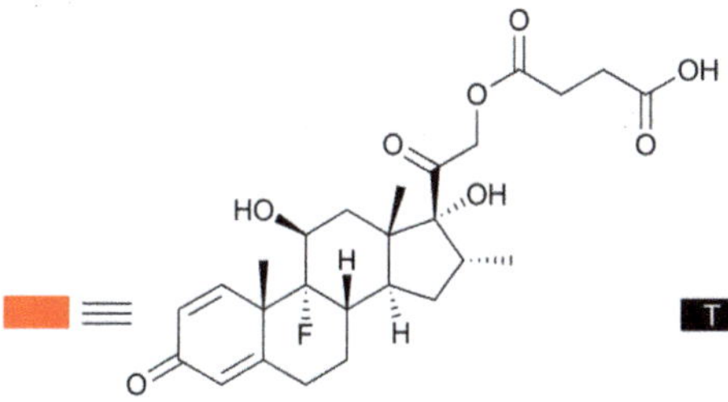

Dex-SA

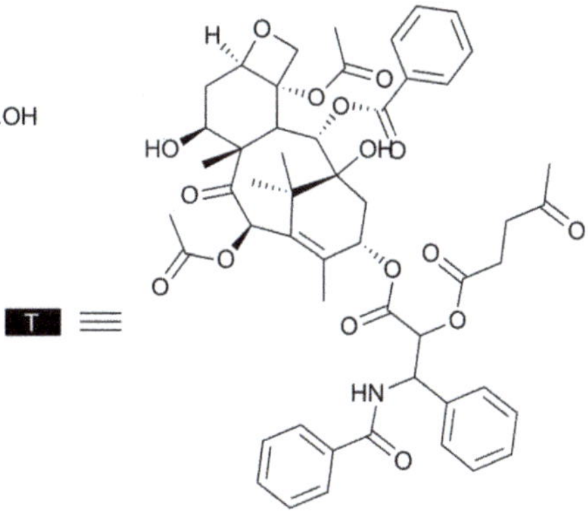

Taxol-SA

Chart 6.5

Figure 6.13   Optical images of hydrogels formed by treating PBS solutions containing 1.0% (w/w) of different precursors with 4 equiv. of glutathione. (A) Dex-K(Taxol)E-gel **6.20** (B) AcK(Taxol)E-gel **6.21** (C) Taxol-K(Ac)E-gel **6.22** D) Taxol-E-gel **6.23** (E) Taxol-R-gel **6.24** and (F) Taxol-S-gel **6.25**. (Reproduced with permission from ref. 56).

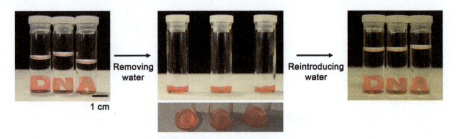

Figure 6.14   Shape transitions of DNA hydrogel showing the shape memory (gels were premoulded prior to immersion in water).[59]
(Reprinted by permission from Macmillan Publishers Ltd: *Nature Nanotechnology* from ref. 59 Copyright © 2012).

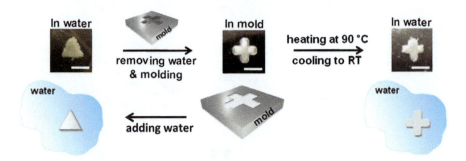

Figure 6.15   Schematic of remoulding of DNA hydrogel shape through heating/cooling and hydrating/dehydrating.[59]
(Reprinted by permission from Macmillan Publishers Ltd: *Nature Nanotechnology* from ref. 59 Copyright © 2012).

previously. Therefore, the DNA hydrogel can be repeatedly returned to its original structure upon the addition of water and is capable of being re-moulding into any shape through the annealing process.

The reversible sol–gel transitions and shape memory gives the DNA hydrogel unique properties, which may allow for its use in localised drug delivery. There is the potential to encode therapeutic benefits directly into the DNA hydrogel sequence to elicit specific functionality. Although, it must be noted that being able to predict the final structure may become difficult. Luo's group explored the potential of the DNA hydrogel to passively delivery encapsulated therapeutics to the surrounding environment. The unique factor of this system is that the DNA building blocks themselves can be used as the drug carrier. This property was demonstrated by examining the controlled release of two drugs, doxorubicin and insulin. Doxorubicin is a chemotherapeutic that intercalates between two base pairs in DNA, hence the cavities in the hydrogel and the DNA itself both act as conduits for drug delivery.

The intrinsic interaction between the DNA hydrogel and doxorubicin was clearly demonstrated by the release profile exhibiting sustained release over 15 days compared to the burst release profile of insulin due to enhanced retention through the intercalation in the DNA structure (Figure 6.16).

6.3.2 Molecular Gels in Tissue Engineering and Regenerative Medicine

Tissue engineering can be defined as an interdisciplinary field applying the principles of life sciences and engineering to the development of functional biological substitutes that restore, maintain or improve tissue function.[15]

Molecular gels are suited to tissue engineering and regenerative purposes because synthetic biomaterials minimise the risk of carrying biological

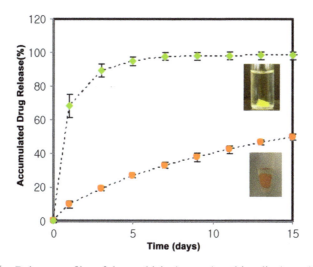

Figure 6.16 Release profiles of doxorubicin (orange) and insulin (green) from within DNA-based hydrogels.[59]
(Reprinted by permission from Macmillan Publishers Ltd: *Nature Nanotechnology* from ref. 59 Copyright © 2012).

pathogens and contaminants compared to naturally derived polymer gels such as collagen and alginate.[60] Moreover, they can be easily decorated with bio-logical motifs, tailoring its properties for specific cell response. A key strategy in the development of molecular gels for these purposes is to combat the innate host immune response; this is a vital step towards the successful implantation of fragile cells or tissues in regenerative medicine. Hence, the design and delivery of biomedical gelators incorporating anti-inflammatory therapeutics prove to address fundamental challenges in the biomedical applications of molecular gels. The fields of tissue engineering and regenerative medicine are set to benefit from the ongoing research in molecular gels, for many of the reasons already previously discussed.

In pioneering work by Stupp and coworkers,[61] rodent neural progenitor cells (NPCs) were encompassed within a 3D network of nanofibres formed *via* self-assembly of peptide-amphiphile **6.26** molecules. The self-assembly process was triggered by mixing cell suspensions in media with (0.5–1% w/w H$_2$O) aqueous solutions of the molecules, with the cells surviving the growth of the fibres around them during this process. NPCs were chosen owing to their potential advantages in replacing lost central nervous system cells after degenerative or traumatic injuries.[62]

The peptide-amphiphile gelator **6.26** incorporated the peptide sequence IKVAV, an epitope found in laminin, and is known to promote neurite sprouting and direct neurite growth. As a control for bioactivity, gelator **6.27** with the nonbioactive epitope EQS was also synthesised; forming a physically similar gel (Figure 6.17).

The ability to have a dense population of biologically active factors (IKVAV) incorporated in the nanofibres presenting themselves to the NPCs was

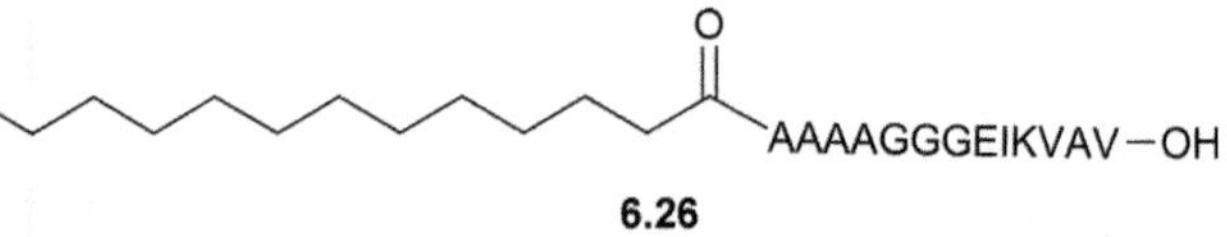

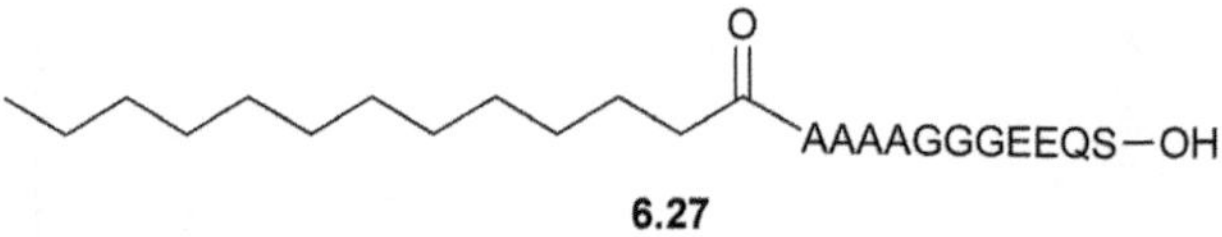

Figure 6.17 Examples of peptide amphiphiles (PA) used in tissue engineering. The PA **6.26** and control PA **6.27** used by Stupp and coworkers in the study on the stimulation of neural progenitor cells (NPC).[61] The KDL12 peptide **6.28** reported by Kisiday and coworkers is intended for cartilage repair by supporting chondrocyte growth.[64]

determined to be the critical factor in the observed rapid and selective differentiation of cells into neurons compared to the peptide amphiphile control **6.27**.[61]

In the 99–99.5% (w/w H_2O) gels, it was observed that cells survived the growth of the nanofibres around them, remaining viable for the length of observation (22 days), whilst at the same time being able to migrate throughout the network. In contrast, gels formed in denser, more rigid networks (98% w/w H_2O) did not survive.

Relative to laminin or soluble peptide, the gel–cell suspension induced very rapid differentiation of cells into neurons, whilst discouraging the development of astrocytes. This rapid, selective differentiation is linked to the high density of epitopes on the fibres presented to the cells. The artificial scaffolds formed by the self-assembling gelators (99.5% w/w H_2O) allowed for a mechanically supportive matrix to form at low concentrations of the peptide amphiphiles.

Self-assembly of the scaffold was also demonstrated within tissue; 10 to 80 µL of 1% (99% w/w H_2O) peptide amphiphile solutions **6.26** were injected into freshly enucleated rat eye preparations and *in vivo* into rat spinal cords following a laminectomy to expose the cord. This process localises the network in tissue and prevents passive diffusion of the molecules away from the epicentre of an injection site. Furthermore, it was demonstrated that animals survive for prolonged periods after injections of the peptide amphiphile solutions into the spinal cord, a finding of relevance to the present study.

Following on from their initial work, Stupp and coworkers in 2008[63] used the same peptide amphiphile **6.26** without exogenous proteins or cells, as a therapy in a mouse model of spinal cord injury (SCI). When a liquid solution of the peptide amphiphile was injected, changes in ionic strength of the *in vivo* environment triggered self-assembly within the extracellular spaces of the spinal cord, thereby resulting in nanoscale gel-like structures. In this work, *in vivo* treatment with the gel after SCI reduced astrogliosis (an inhibitor of axon regeneration), reduced cell death, and increased the number of oligodendroglia (cells that support and insulate axons) at the site of injury. Furthermore, the nanofibres promoted regeneration of both descending motor fibres and ascending sensory fibres through the lesion site. Treatment with the peptide amphiphile **6.26** also resulted in significant behavioural improvement; at nine weeks, the control groups demonstrated no hind limb movement, whereas the peptide amphiphile IKVAV epitope group had hind limb movement.[63]

Another example from the field of tissue engineering comes from the group of Kisiday *et al.*,[64] which designed a peptide **6.28** (KLD-12) (Figure 6.17) related to the aforementioned RADA peptide **6.13**. The KLD-12 peptide **6.28** formed a molecular hydrogel that was used as a scaffold to support chondrocyte growth and development for cartilage repair. During one month of culture *in vitro*, chondrocytes seeded within the hydrogel retained their morphology and developed a cartilage-like extracellular matrix rich in proteoglycans and type-II collagen, which is indicative of a stable chondrocyte phenotype. As time progressed, the stiffness of the material increased, thereby indicating that new, mechanically functional cartilage was formed. The outcome of this experiment established the potential of a self-assembling peptide molecular gel as a tool for the synthesis and accumulation of a cartilage-like extracellular matrix for tissue regeneration. The versatility of molecular gels in medical applications is neatly demonstrated in their combination with a traditional prosthetic material for regenerative medicine. The regeneration or replacement of hard tissue in the body has proven to be a challenge due to its mechanical properties.

One solution is the use of metal prostheses to replace the hard tissue. As an example of how molecular gels could serve to assist the regeneration of hard tissue, Sargeant and coworkers took a biologically inert Ti-6Al-4V prosthesis and incorporated a peptide amphiphile **6.29** (Chart 6.6) based molecular gel into the bone implant. This hybrid material was shown to be able to mineralise with calcium phosphate over time and cells could be encapsulated in these hybrids in a controlled manner. *In vivo* experiments showed that *de novo* bone is formed adjacent to and inside the PA **6.29**:Ti hybrid by four weeks, thus offering strong evidence of osteoconduction, the growth of bone on the surface and into the pores of the implant.[65]

In order for the successful application of hydrogels in tissue engineering there is clearly a need to be able to fabricate scaffolds with controlled and intricate precision. This will allow scientists to engineer tissues with the architectures needed for their application.

Molecular gels have been used to control the 3D structure of polymeric gels in tissue engineering. Komatsu *et al.* have demonstrated this using a molecular hydrogel **6.30** with the ability to degrade in response to pH and UV irradiation, which allows for the controlled fabrication of intricate 3D structures.[66] The technique utilises precise photofabrication, using focused UV laser light (266 nm), which causes the hydrogel to transition to a liquid state and results in the formation of fine channels in the hydrogel scaffold. Subsequent casting of alternative gelators within the hydrogel mould results in the formation of the desired structure. The cast material, in this example collagen, can be infused with various cultures to create interweaving structures with differing cell lines.

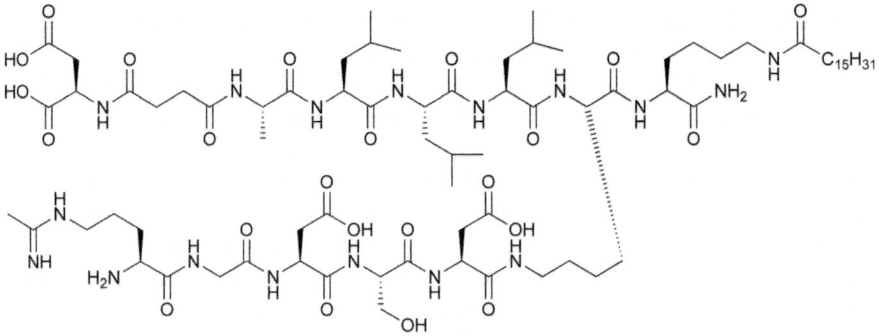

6.29

Chart 6.6

The remaining hydrogel scaffold can be easily degraded in mild basic or acidic conditions leaving behind the cast collagen matrix. In this manner, Komatsu *et al.* fabricated weaved grid patterns in the hydrogel using UV light.[66] Two cell lines, rhodamine-stained HeLa cells and stably expressing EGFP-Chinese hamster ovary (CHO) cells, were then cultured and poured with collagen solution into the various channels resulting in an interlocking pattern of two isolated cell lines (Figure 6.18).

This technique represents a step towards the controlled fabrication of intricate 3D structures for applications in tissue engineering and regenerative medicine. The key consideration needs to be how to maintain cell viability and incorporate the required cell function throughout the 3D structure. This system provides a platform from which this can be achieved by allowing controlled design of the hydrogel mould for precise tissue architectures.

The ability to recover cells postculture for further analysis is a fundamental feature in traditional *in vitro* biological assays. The design of 3D tissues for tissue-engineering purposes has relied on the gel scaffold to support cell growth and exhibit appropriate functionality for implantation. However, the ability of the gel to suppress the host immune response has largely been overlooked and

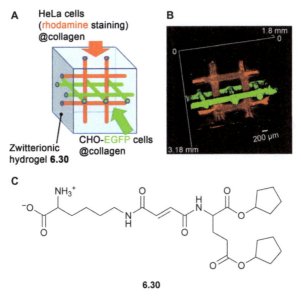

Figure 6.18 Schematic highlighting the 3D patterning of two different cell lines in a collagen gel matrix moulded from microchannels in hydrogel **6.30**. (A) HeLa and CHO cells in collagen solution were poured into separate microchannels of hydrogel **6.30** to create an interweaving mesh framework. (Reproduced with permission from ref. 66 Copyright © 2011 Wiley-VCH). (B) A 3D render of a confocal laser scanning microscope (CLSM) z-stack image showing the spatial patterning of CHO-EGFP cells (green) and rhodamine stained HeLa cells (red) embedded in a collagen gel matrix.[66] (Reproduced with permission from ref. 66 Copyright © 2011 Wiley-VCH). C) Structure of hydrogel **6.30**.

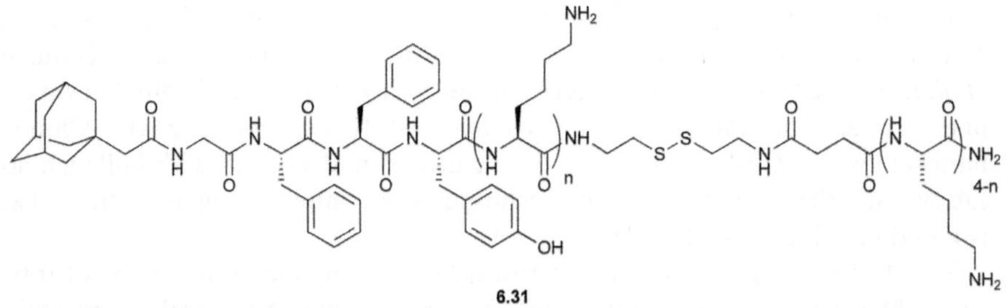

6.31

Figure 6.19 Chemical structures of Adamantane-peptide precursors **6.31** (Ada-GFFYK$_n$-ss-K$_{(4-n)}$-CONH$_2$, $n = 0-3$).

may hinder the biocompatibility of the engineered tissue. One strategy to address this is with biodegradable gels that can switch to a sol state with the appropriate triggers.

Yang and coworkers have developed an adamantane-peptide-based self-assembling **6.31** molecular gel (Figure 6.19) that can be switched easily to a sol state upon complexation with methyl-β-cyclodextrin (M-β-CD).[67] The hydrogel is capable of gelation through disulfide bond cleavage by glutathione and was shown to effectively foster the growth of mouse fibroblast 3T3 cells. The hydrogel promoted cell adhesion and 3T3 cells were cultured on top of the gels for 3 days. The cells were then collected postculture using M-β-CD.

The β-CD derivative forms a tight complex with the adamantane (Ada) portion of the molecule and disrupts the hydrophobic interactions of the supramolecular structure, resulting in the dramatic increase in solubility of the gelator. The strong interaction between M-β-CD and Ada results in the breakdown of the hydrogel structure and induces a sol transition of the gelator, which can then facilitate the postculture recovery of 3T3 cells using centrifugation. These mild conditions for gelation and recovery improve the biocompatibility for cell encapsulation and suggest prospective applications for postculture cell analysis. This system highlights the potential of fabricating 3D gel scaffolds with postculture cell recovery for regenerative medicine and tissue engineering, allowing for *ex vivo* engineering of viable cells in recoverable hydrogel scaffolds for tissue transplantations. These scaffolds should be biodegradable to maximise their ability to foster the replacement of tissues.[68]

An elegant example in the application of molecular gels for tissue engineering (regenerative endodontics) was reported by Galler and coworkers.[69] They hypothesised that the incorporation of dental pulp-derived stem cells in an enzyme-cleavable molecular gel **6.32** (Figure 6.20), with the addition of growth factors, would serve as a better alternative filling to the inert fillings currently used in root canal treatment. The idea in transplanting cells within a molecular gel (besides imparting easy delivery during dental surgery) is to support them structurally whilst they proliferate and differentiate, regenerating the lost tissue. With an enzyme-cleavable gel, the cells would ideally spread throughout the gel scaffold and degrade it, remodelling it over time, mimicking *in vivo* systems. Their molecular gel toolkit is based on customisable "multidomain peptides"

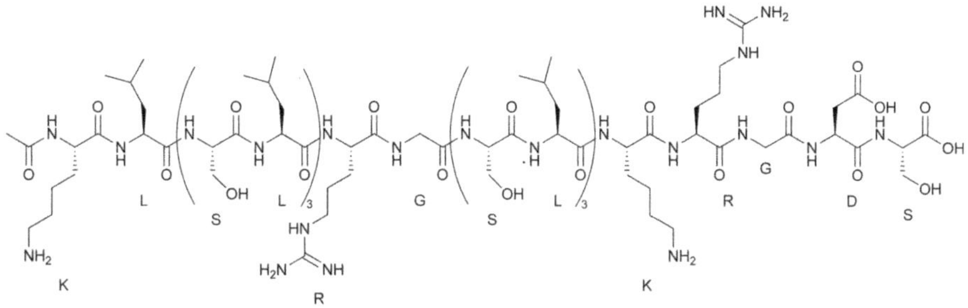

6.32

Figure 6.20 Chemical structure of "multidomain peptide" gelator KL(SL)₃RG(SL)₃KRGDS **6.32**. The SLRG sequence is a cleavage motif that is sensitive to MMP-2, an enzyme expressed by fibroblasts and odontoblasts in healthy dental pulp to remodel their environment.[69] The RGD motif is important for cell adhesion.

(MDPs), developed previously in the Hartgerink laboratory.[70] Each "domain" is designed to have differing functions that are able to be changed and optimised independently of each other, thus allowing the tailoring of self-assembled materials. The domains are arranged in ABA block motifs with an amphiphilic core (the "B" block) flanked by charged regions (the "A" block). The B block is responsible for the supramolecular assembly of the peptides into a sandwich-like β-sheet nanofibre in which hydrogen bonding occurs parallel to the growing fibre axis. The charged amino acid residues in the A block (*e.g.* glutamate (–ve charged) or lysine (+ve charged)) provide water solubility and counteract fibre assembly *via* electrostatic repulsion.

MDP gelators based on positively charged lysine residues self-assemble upon complexation with negatively charged heparin, inducing gelation. To promote cell proliferation, differentiation and angiogenesis, growth factors, vascular endothelial growth factor (VEGF), transforming growth factor β1 (TGFβ1) and fibroblast growth factor basic (FGF2) were incorporated *via* heparin binding. The incorporation of TGFβ1 and FGF2 have previously been reported to stimulate proliferation, cytodifferentiation and mineralisation of dental pulp derived stem cells.[71] Their artificial 3D scaffold and cell suspension was shown to be able to support the formation of a vascularised soft connective tissue similar to dental pulp after transplantation *in vivo* into immunocompromised mice.

6.3.3 Molecular Gels as Three-Dimensional Cell Cultures

It is well known that cells behave structurally and functionally different when seeded on thin two-dimensional (2D) surface-coated substrates *versus* a thick layer of polymeric three-dimensional (3D) molecules, which more closely mimics their natural environment.[72,73] In their seminal paper in 2006, Engler and coworkers demonstrated this clearly by reporting on the previously undocumented influence of microenvironment stiffness on stem-cell specification, establishing that matrix elasticity directed mesenchymal stem cell (MSC) differentiation. MSCs grown on soft matrices that mimic the stiffness of brain were shown to become neurogenic, stiffer matrices that mimic muscle are myogenic and matrices mimicking collagenous bone prove osteogenic.[6] As a results, biomedical researchers have become increasingly aware of the limitations of the long-established 2D cell culture and over the years, much attention has been paid to artificial scaffolds that mimic the natural ECM.[74,75]

There is also an increasing awareness that evaluating drug efficacy in pre-clinical high-throughput studies should be ideally performed in 3D *in vitro* cell culture models instead of 2D on glass.[76] 3D cell cultures can offer a much better approximation of the natural micro- and local environment compared to the 2D cultures.[77] Moreover, the functional properties of cells can be observed and manipulated in ways that are not possible in animal models. Surprisingly, the mentioned differences in cell behaviour between 2D and 3D cell cultures are independent of whether the culture substrates are derived from purified extracellular matrix components or components obtained synthetically. Therefore, hydrogels (polymeric or molecular) hold great promise as an alternative 3D cellular microenvironment for tissue studies.[78] This is not only because their elastic moduli can be adjusted to resemble that of natural tissues, but also because their composition can be tailored to bear the appropriate chemical, physical and biological cues that facilitate the development of tissue-like, and hence organoid-type cultures *in vitro*.[79] The successful utilisation of 3D cell cultures would have profound implications for biomedical engineering practise.

Of particular interest are the discoveries by Assaf Mahler and colleagues,[80] who demonstrated that simple 9-fluorenylmethoxycarbonyl (Fmoc) dipeptides such as Fmoc-FF **6.33** (Chart 6.7) can form hydrogels upon a change in pH. It was reported that this system could serve as a 3D cell-culture scaffold, successfully culturing CHO cells. Overall, the cells grown on the gel showed the same viability (>90%), morphology and rate of proliferation as the control cells that were grown on nonhydrogel substrates.

This is significant not only because 3D gels mimic tissue, but also because the composition and matrix density of the gel can be fully controlled by chemical tailoring approaches. This is a huge advantage when compared to the commercially available cell-culture matrices, whether naturally derived such as Matrigel™ or purified/synthesised by means of large-scale biotechnological methods such as alginate. For instance, Matrigel™ is prepared by reconstituting proteins extracted from Engelbreth–Holm–Swarm sarcomas, a tumour grown in mice. The potential tumorigenicity and immunogenicity, as well as poorly defined composition and potential batch-to-batch variation of this material, however, limit its use for *in vivo* therapeutic applications and for *in vitro* drug toxicity studies.

The flexibility of the approach of using these small self-assembling molecules is demonstrated by Zhou and coworkers.[81] One key to success to the practical application of hydrogels is the functionalisation of the highly hydrated gel complex with amino acids that code for the adhesion peptide sequence RGD or

RGDS. This has proven be a highly valuable molecular alteration to increase cell–gel interactions.[82] These short amino acid sequences are well known to play a key role in many recognition systems involved in cell-to-cell and cell-to-matrix adhesion.[83]

Zhou *et al.* mixed Fmoc-FF **6.33** and Fmoc-RGD **6.34** (Chart 6.7). This bioactive mixture resulted in a three-dimensional biomimetic nanoscaffold, which successfully allowed the adhesion, spreading, and proliferation of human adult dermal fibroblasts. This specific mixture was shown to form a gel at physiologically relevant 37 °C and a pH of 7.0.[78]

Similarly, Jayawarna *et al.*[84,85] investigated whether chemical functionality introduced through mixtures of Fmoc protected amino acids and Fmoc-FF **6.33** enhances compatibility with different cell types. A mixture of Fmoc-S-OH **6.35** (Chart 6.7) and Fmoc-FF **6.33** in a 1 : 1 ratio, (introducing –OH groups *via* the CH$_2$OH side chain in S) was found to enhance the compatibility of the gel

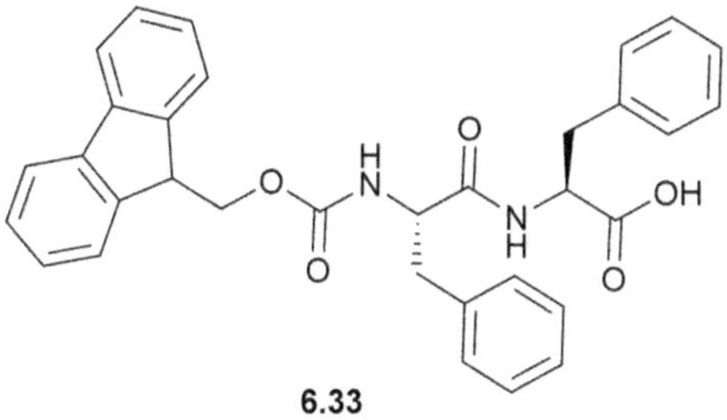

6.33

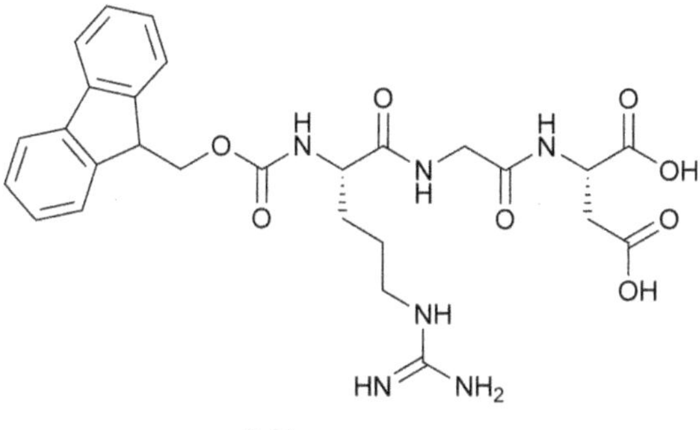

6.34

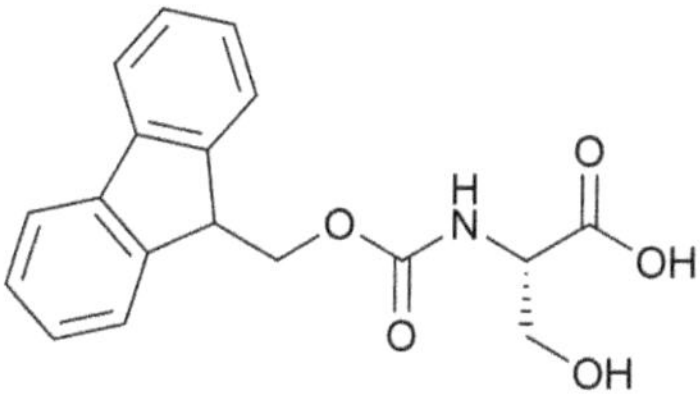

6.35

Chart 6.7

with various mesenchymal cell lines and a superior hydrogel scaffold was achieved for cell culture.[85]

Interestingly, when reviewing the literature, a key factor to successfully maintaining the viability of cells exposed to molecular gels in culture is largely dependent on both the cell type and the introduction of chemical functionality groups. Fibroblasts, chondrocytes or other cell types of mesenchymal origin seem to be a key to success when it comes to *in vitro* biocompatibility studies. This is not surprising as those cells have the cell-membrane receptors to interact with the surrounding connective tissue (*e.g.* ECM). Some of these cell types secrete natural ECM components by themselves. Evidently, the addition of smart functional groups to the gel, such as a multitude of -NH_2 or -OH groups, creates the ideal microenvironment for tight interactions of cell-surface receptors (*e.g.* integrins) with the gel components.

The previously mentioned Ac-(RADA)$_4$-NH_2 **6.13** peptide gelator was originally reported by Yokoi and coworkers[47] to form a molecular hydrogel (0.5–1.0% w/w H_2O) and later patented under the trademark PuraMatrix™. The peptide sequence enables cell attachment and migration, thus enabling the rapid formation of cell–cell contacts. The gelator **6.13** is soluble at low pH and osmolarity; when changed to physiological pH and osmolarity, it quickly forms fibres on the order of 5–10 nm and assembles into interwoven 3D scaffolds. The high volume fraction ($\sim 99\%$) of water within these hydrogels and the structural resemblance of these peptides to natural collagen along with the ability to customise the peptide backbone have yielded successful culturing of osteocytes,[86] neural cells[87] and chondrocytes.[64]

Relative to Matrigel™ which is extracted from reconstituted proteins from mice, the synthetically produced PuraMatrix™ yielded cells with comparable specific liver function. Given the well-defined and synthetic nature of PuraMatrix™, these results suggest that PuraMatrix™ is a biomaterial with a good potential for a long-term hepatocyte (liver cell) culture in several applications, including clinical artificial liver development and high-throughput *in vitro* drug metabolism and toxicity evaluation.

An elegant demonstration of the rational addition of functional groups to a molecular gel and their effects for cell compatibility was shown in the Hartgerink lab. They synthesised a library of molecular gels based on the previously mentioned MDP gelators. Through customisation of each domain, functionality was rationally incorporated into the polypeptide sequence. They noted that the fibre assembly and β-sheet structure of the MDP aggregates was reminiscent of amyloid aggregates observed in Alzheimer's disease. Due to these similarities, programmed susceptibility to proteolytic degradation was incorporated into the MDP with the addition of a MMP-2 sensitive sequence, to address potential nonbiodegradability with the added benefit of helping cells remodel their environment, just as is accomplished in the natural extracellular matrix.[8]

Three analogues of **6.36** were synthesised; **6.37** includes the cell adhesion motif RGD, **6.38** incorporated an MMP-2 cleavage site, and **6.39** combines both (Figure 6.21).[88] The combination of both the RGD and MMP-2 motifs in

6.36 = KK-SLSLSLSLSLSL---KK

6.37 = KK-SLSLSLSLSLSL---KKGRGDS

6.38 = K-SLSL$\underset{\downarrow}{\underline{\text{SL}}}$RGSLSLSL-K

6.39 = K-SLSL$\underset{\downarrow}{\underline{\text{SL}}}$RGSLSLSL-KGRGDS

Figure 6.21   The sequences of the multidomain peptides **6.36**–**6.39** developed by the Hartgerink group that form molecular gels for cell-culture studies.[88] The parent ABA-block peptide **6.36** forms molecular gels in water. Peptides **6.37** and **6.39** incorporate the RGD cell adhesion motif (italics). Peptides **6.38** and **6.39** include a MMP-2 cleavage motif (underlined) with the cleavage site shown by an arrow.[88]

A   **MAX1:** VKVKVKVK-V^{D}PPT-KVKVKVKV-NH$_2$
                        6.40
      MAX8: VKVKVKVK-V^{D}PPT-KVEVKVKV-NH$_2$
                        6.41

B

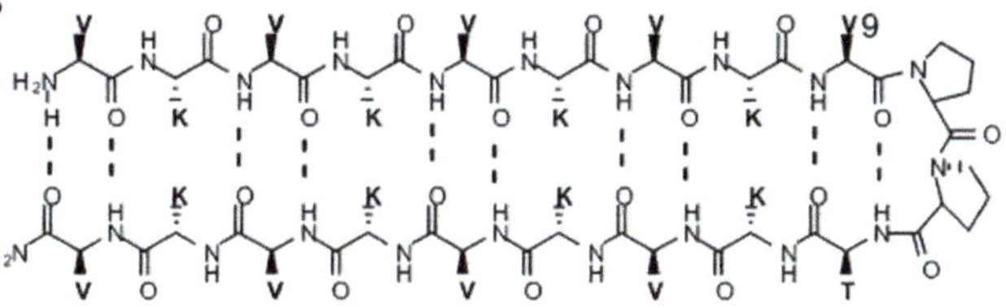

6.40

Figure 6.22   Shear-thinning MAX-peptides. (A) The structure of MAX1 **6.40**[86] and MAX8 **6.41**.[88] (B) The proposed β-hairpin of MAX1 **6.40**.[89,91] (Reproduced with permission from ref. 89 Copyright © 2002 American Chemical Society).

6.39 resulted in the largest improvements in cell viability as well as marked differences in cell spreading and morphology. This could be attributed to the MMP-2 sequence allowing cells to migrate into the gel network as they do naturally in the ECM.

In summary, the peptide functionalisation data discussed above highlights once more the importance of functional tailoring of molecular hydrogels to allow optimal cell and tissue function. The incorporation of the MMP-2 cleavage motif allowed the degradation of the gel network by cells – allowing the cells to penetrate into the matrix better.

Schneider and coworkers have developed a family of peptides capable of shear-thinning behaviour. This allows for the ability to deliver these self-assembling systems *via* traditional syringe injection techniques, simplifying the implementation of these molecular gelators in therapeutic applications. MAX1 **6.40**, one of the first peptides to be developed in this family, consists of 8 alternating valine (V) and lysine (K) amino acid residues followed by a four residue type II′, β-hairpin turn (-V^{D}PPT-) (Figure 6.22B) and capped with another 8 alternating V and K residues.[89]

Peptide Name	Sequence	
VK$_9$	VKVKVKVKV-NH$_2$	**6.42**
VK$_{10}$	VKVKVKVKVK-NH$_2$	**6.43**
VK$_{11}$	VKVKVKVKVKV-NH$_2$	**6.44**
VK$_{12}$	VKVKVKVKVKVK-NH$_2$	**6.45**
VK$_{13}$	VKVKVKVKVKVKV-NH$_2$	**6.46**
AK$_{13}$	AKAKAKAKAKAKA-NH$_2$	**6.47**
IK$_{13}$	IKIKIKIKIKIKI-NH$_2$	**6.48**
LK13	LKLKLKLKLKLKL-NH$_2$	**6.49**
VR$_{13}$	VRVRVRVRVRVRV-NH$_2$	**6.50**

Figure 6.23   The peptide library **6.42–6.50** reported by Geisler and Schneider. The *N*-terminus is the free amine and the *C*-terminus the carboxamide.[92]

The self-assembling peptide was capable of traditional syringe injection due to its shear-thinning behaviour; however, cell sedimentation occurred when 3D encapsulation was attempted. Continuing on with this work, MAX8 **6.41** was developed, exhibiting successful homogenous encapsulation (with no cell sedimentation) and syringe delivery of C3H/10T1/2 mesenchymal stem cells within a 3D network of hydrogel.[90,91] This strategy introduces greater flexibility for the use of molecular gels in regenerative medicine as the hydrogel can be delivered using standard delivery techniques whilst maintaining cell viability.

Both the linear and β-stranded peptides were capable of shear thinning for traditional syringe delivery and exhibited self-healing properties to reform self-supporting hydrogels. This family of peptides highlights a key mechanical property required from self-assembling systems for their ease of translation into the clinic.

Moreover, Schneider and Geisler have prepared a family of injectable short linear amphiphilic peptides with varying hydrophobic and hydrophilic amino acid ratios with the ability to form gels upon addition of a buffered saline solution (Figure 6.23), in addition to displaying shear-thin recovery behaviour – allowing syringe delivery like their previous systems.[92] They reported that sequence length and hydrophobic character are a critical factor in determining rigidity of their gel systems. From their evolution based design, the optimal sequence of LK13 **6.49** was found to be able to form cytocompatible hydrogels with cell culture media as the triggering solution allowing the direct encapsulation of cells in 3D. Upon injection from syringes, their gelator systems showed shear-thinning behaviour with self-healing properties to reform self-supporting hydrogels upon cessation of injection. It is envisaged LK13 **6.49** gels could be utilised within microfluidic devices for high throughput screening of drug compounds against 3D cultures of cells due to the advantageous shear-thinning property exhibited by these systems.

6.4   Emerging Trends and Applications

Although we are able to design and synthesise new and more elaborate molecular gels, the question as to whether we have gained a real understanding of

PEP6R: VKVRVRVRV$^\mathrm{D}$PPTRVRVRVKV-NH$_2$

6.51

Chart 6.8

the gelation process still beckons.[93] This may ultimately prove to be the "black box" halting the progression of molecular gels in biomedical applications. As this is still a young field, more study directed toward the structure-activity relationship is needed and from this, a development of predictive models based on their physicochemical properties. Schneider and Veiga *et al.* demonstrated an elegant method of achieving structure-activity relationships through their development of a library of β-hairpin self-assembling peptide gelators rich in arginine with inherent antibacterial activity. The gels formed were found to be extremely effective at killing gram-positive and gram-negative bacteria and the multidrug resistant *P. aeruginosa.*[16] Antimicrobial peptides (AMPs) have long been investigated for their use in preventing bacterial infection, a particularly prevalent problem for hospitalised patients. The antibacterial properties of arginine-rich peptides have previously been shown to effectively disrupt the cell membrane of bacteria leading to cell death.[94] Through their use of peptide based gelators, changes in the monomer level can be facilely accomplished – translating directly into changes in the bulk material properties, allowing the investigation of structure-function relationship to be easily determined.

Their studies culminated in the optimised gelator PEP6R **6.51** (Chart 6.8). When dissolved in water, the peptide gelator **6.51** exists in a random coil conformation; upon addition of bis-tris propane buffer (100 mM, 300 mM NaCl, pH 7.4), it transitions into an amphiphilic β-hairpin conformation. PEP6R **6.51** contains two β-strands of alternating valine (V) and arginine (R) residues connected by a four residue type II′, β-turn sequence (-V$^\mathrm{D}$PPT-). This conformation allows for the pairing of hydrophobic and polycationic faces of the peptide to more effectively form β-sheets. The cationic face engages into the outer leaflet of the net negatively charged cell membrane *via* electrostatic interactions and hydrogen bonding. The hydrophobic face then disrupts the cell membrane by inserting itself into the hydrophobic pocket resulting in cell death. PEP6R **6.51** displayed potent activity against *E. coli, S. aureus* and multidrug resistant *P. aeruginosa*, whilst being cytocompatible to human red blood cells and mammalian mesenchymal stem cells. Again, like their previous systems, these gelators exhibit self-healing abilities, allowing for syringe delivery.

6.5 Conclusions

Molecular gels have demonstrated potential to be used as vehicles for therapeutic delivery, a substrate for cell culture and conduits for tissue regeneration. However, before fully capitalising on the potential of molecular gels (ECM mimicking materials), it requires a proper understanding of their structure–function relationship in order for us to understand how their bio-mimetic properties arise.[90]

In terms of the range of applications covered in this chapter, the future appears bright. Within the field of drug delivery from molecular gels, stimuli-triggered drug release and the design of better therapeutic molecular gels appear particularly active at the moment. There is also a strong overlap between therapeutic molecular gels and molecular gels for applications in tissue engin-eering. It is not surprising that the emphasis in the field is shifting towards these emerging areas as these are generally more synthetically and intellectually challenging than "simple" molecular gels for passive drug delivery. That said, much is still to be learned in drug release, both passive and stimuli-triggered (*e.g.* enzymatic), from molecular gels.

Another very interesting emerging area is the use of molecular gels to create 3D cell cultures. Medical research has largely been reliant on data obtained from 2D cell culture studies but this potentially creates a lot of misleading information as the target (usually human) tissue is in 3D and cell behave quite different in a 3D matrix than on the bottom of a glass Petri dish. Molecular gels are chemically well defined, yet it is easy to create structural and functional complexity from molecular gels making them particularly attractive for the creation of 3D cell culture models.

There are so far no clear trends or guidelines on how to *a priori* design molecular gels for a particular drug release at a particular rate. Questions such as what the correlations between the size, shape or hydrophobicity ($\log P$) of a drug molecule *vs.* the expected release rate have not yet been answered. To a certain extent this problem can be traced back to the lack of understanding of how molecular gels form and what controls their properties (see Chapter 1).[93]

Finally, it is worth noting that in nearly all cases, the molecular gels in this chapter are peptide based. On the one hand, this can be explained by the rich bioactive nature of peptides in combination with the relative ease that peptide derivatives can form molecular gels in water or even biological media. But on the other hand, the absence of molecular gels based on other structural motifs, such as urea compounds, carbohydrates or steroids when it comes to medical applications is surprising even though water-based molecular gels based on such motif are quite well known (see Chapter 1). Clearly, there appears to be an opportunity to explore these motifs further in terms of medical applications, with carbohydrate-based molecular gels as potential ECM mimicking looking particularly intriguing in this context.

The future for molecular gels in terms of medical applications appears very bright and there are still plenty of opportunity for creative and clever minds to push the boundaries of what is known about molecular gels while potentially also solving pressing medical challenges for the benefit of society as a whole. This includes the making of better therapeutic molecular gels, the combination of molecular gels with other cutting-edge research such as stem-cell therapy and the utilisation of 3D cell cultures based on molecular gels to deepen our understanding of how drugs interact with cells, of cell–cell interactions and supramolecular molecular biology in general, leading to now unforeseen advances in medicine.

References

1. N. Huebsch and D. J. Mooney, *Nature*, 2009, **462**, 426.
2. J. Finch, *Lancet*, 2011, **377**, 548.
3. M. S. Lesniak and H. Brem, *Nat. Rev. Drug Discovery*, 2004, **3**, 499.
4. E. P. Sipos, B. Tyler, S. Piantadosi, P. C. Burger and H. Brem, *Cancer Chemother. Pharmacol.*, 1997, **39**, 383.
5. M. A. Moses, H. Brem and R. Langer, *Cancer Cell*, 2003, **4**, 337.
6. A. J. Engler, S. Sen, H. L. Sweeney and D. E. Discher, *Cell*, 2006, **126**, 677.
7. http://www.optics.rochester.edu/workgroups/cml/opt307/spr06/joe (accessed 2/5/2013).
8. B. Alberts, A. Johnson, J. Lewis, M. Raff, K. Roberts and P. Walter, in *Molecular Biology of the Cell*, Garland Science, New York, 4th edn., 2002.
9. I. Stamenkovic, *J. Pathol.*, 2003, **200**, 448.
10. O. Chaudhuri and D. J. Mooney, *Nature Mater.*, 2012, **11**, 568.
11. P. Roughley, C. Hoemann, E. DesRosiers, F. Mwale, J. Antoniou and M. Alini, *Biomaterials*, 2006, **27**, 388.
12. D. G. Wallace and J. Rosenblatt, *Adv. Drug Delivery Rev.*, 2003, **55**, 1631.
13. J. P. Jung, J. Z. Gasiorowski and J. H. Collier, *Biopolymers*, 2010, **94**, 49.
14. N. A. Peppas, J. Z. Hilt, A. Khademhosseini and R. Langer, *Adv. Mater.*, 2006, **18**, 1–345.
15. J. Kopecek, *J. Polym. Sci. Part A*, 2009, **47**, 5929.
16. R. Langer and J. P. Vacanti, *Science*, 1993, **260**, 920.
17. A. S. Veiga, C. Sinthuvanich, D. Gaspar, H. G. Franquelim, M. A. R. B. Castanho and J. P. Schneider, *Biomaterials*, 2012, **33**, 8907.
18. W. T. Truong, Y. Su, J. T. Meijer, P. Thordarson and F. Braet, *Chem.-Asian J.*, 2011, **6**, 30.
19. I. W. Hamley, *Angew. Chem. Int. Ed.*, 2007, **46**, 8128.
20. C. Keeler, M. E. Hodsdon and P. S. Dannies, *J. Mol. Neurosci.*, 2004, **22**, 43.
21. N.-V. Buchete, R. Tycko and G. Hummer, *J. Mol. Biol.*, 2005, **353**, 804.
22. K. E. Uhrich, S. M. Cannizzaro, R. S. Langer and K. M. Shakesheff, *Chem. Rev.*, 1999, **99**, 3181.
23. D. Mansour, *J. Fam. Plann. Reprod. Health Care*, 2010, **36**, 187.
24. H. Pearson, *Nature News Online*, 2004, doi: 10.1038/news040419-2.
25. S. C. Bremmer, J. Chen, A. J. McNeil and M. B. Soellner, *Chem. Commun.*, 2012, **48**, 5482.
26. S. W. Toledano, R. J. Williams, V. Jayawarna and R. V. Ulijn, *J. Am. Chem. Soc.*, 2006, **128**, 1070.
27. http://en.wikipedia.org/wiki/ICU_Medical (accessed 2/5/2013).
28. D. J. Adams, M. F. Butler, W. J. Frith, M. Kirkland, L. Mullen and P. Sanderson, *Soft Matter*, 2009, **5**, 1856.

29. Z. M. Yang, H. W. Gu, L. Zhang, L. Wang and B. Xu, *Chem. Commun.*, 2004, 208.
30. A. Friggeri, B. L. Feringa and J. v. Esch, *J. Control. Release*, 2004, **97**, 241.
31. B. Xing, C.-W. Yu, K.-H. Chow, P.-L. Ho, D. Fu and B. Xu, *J. Am. Chem. Soc.*, 2002, **124**, 14846.
32. M. R. Kapadia, L. W. Chow, N. D. Tsihlis, S. S. Ahanchi, J. W. Eng, J. Murar, J. Martinez, D. A. Popowich, Q. Jiang, J. A. Hrabie, J. E. Saavedra, L. K. Keefer, J. F. Hulvat, S. I. Stupp and M. R. Kibbe, *J. Vasc. Surg.*, 2008, **47**, 173.
33. Z. M. Yang, K. M. Xu, L. Wang, H. W. Gu, H. Wei, M. J. Zhang and B. Xu, *Chem. Commun.*, 2005, 4414.
34. G. L. Liang, Z. M. Yang, R. J. Zhang, L. H. Li, Y. J. Fan, Y. Kuang, Y. Gao, T. Wang, W. W. Lu and B. Xu, *Langmuir*, 2009, **25**, 8419.
35. J. K. Kim, J. Anderson, H. W. Jun, M. A. Repka and S. Jo, *Mol. Pharmaceutics*, 2009, **6**, 978.
36. J. D. Mott and Z. Werb, *Curr. Opin. Cell Biol.*, 2004, **16**, 558.
37. G. Klein, E. Vellenga, M. W. Fraiije, W. A. Kamps and E. S. d. Bont, *Crit. Rev. Oncol. Hematol.*, 2004, **50**, 87.
38. X. Yan and R. A. Gemeinhart, *J. Control. Release*, 2005, **106**, 198.
39. J. R. Tauro and R. A. Gemeinhart, *Bioconjugate Chem.*, 2005, **16**, 1133.
40. H. Ye, L. Jin, R. Hu, Z. Yi, J. Li, Y. Wu, X. Xi and Z. Wu, *Biomaterials*, 2006, **27**, 5958.
41. J. W. Ho, *Recent Pat. Anti-Cancer Drug Discovery*, 2006, **1**, 129.
42. V. Pinzani, F. Bressolle, I. J. Haug, M. Galtier, J. P. Blayac and P. Balmes, *Cancer Chemother. Pharmacol.*, 1994, **35**, 1.
43. D. Screnci and M. J. McKeage, *J. Inorg. Biochem.*, 1999, **77**, 105.
44. Y. Gao, Y. Kuang, Z.-F. Guo, Z. Guo, I. J. Krauss and B. Xu, *J. Am. Chem. Soc.*, 2009, **131**, 13576.
45. M. J. Webber, J. B. Matson, V. K. Tamboli and S. I. Stupp, *Biomaterials*, 2012, **33**, 6823.
46. R. G. Ellis-Behnke, Y.-X. Liang, D. K. C. Tay, P. W. F. Kau, G. E. Schneider, S. Zhang, W. Wu and K.-F. So, *Nanomed. Nanotech. Biol. Med.*, 2006, **2**, 207.
47. H. Yokoi, T. Kinoshita and S. Zhang, *Proc. Natl. Acad. Sci. USA*, 2005, **102**, 8414.
48. J. B. Matson, M. J. Webber, V. K. Tamboli, B. Weber and S. I. Stupp, *Soft Matter*, 2012, **8**, 6689.
49. L. E. Otterbein, F. H. Bach, J. Alam, M. Soares, H. T. Lu, M. Wysk, R. J. Davis, R. A. Flavell and A. M. K. Choi, *Nature Med.*, 2000, **6**, 422.
50. R. Tenhunen, H. S. Marver and R. Schmid, *Proc. Natl. Acad. Sci. USA*, 1968, **61**, 748.
51. C. S. Jackson, S. Schmitt, Q. P. Dou and J. J. Kodanko, *Inorg. Chem.*, 2011, **50**, 5336.
52. H. Pfeiffer, A. Rojas, J. Niesel and U. Schatzschneider, *Dalton Trans.*, 2009, 4292.

53. J. E. Clark, P. Naughton, S. Shurey, C. J. Green, T. R. Johnson, B. E. Mann, R. Foresti and R. Motterlini, *Circ. Res.*, 2003, **93**, e2.
54. R. I. Scheinman, P. C. Cogswell, A. K. Lofquist and A. S. Baldwin, Jr., *Science*, 1995, **270**, 283.
55. J. B. Matson, C. J. Newcomb, R. Bitton and S. I. Stupp, *Soft Matter*, 2012, **8**, 3586.
56. H. Wang, L. Lv, G. Xu, C. Yang, J. Sun and Z. Yang, *J. Mater. Chem.*, 2012, **22**, 16933.
57. M. A. Jordan and L. Wilson, *Nature Rev. Cancer*, 2004, **4**, 253.
58. A. Meister and M. E. Anderson, *Annu. Rev. Biochem.*, 1983, **52**, 711.
59. J. B. Lee, S. Peng, D. Yang, Y. H. Roh, H. Funabashi, N. Park, E. J. Rice, L. Chen, R. Long, M. Wu and D. Luo, *Nature Nanotech.*, 2012, **7**, 816.
60. T. C. Holmes, *Trends Biotechnol.*, 2002, **20**, 16.
61. G. A. Silva, C. Czeisler, K. L. Niece, E. Beniash, D. A. Harrington, J. A. Kessler and S. I. Stupp, *Science*, 2004, **303**, 1352.
62. H. Okano, *J. Neurosci. Res.*, 2002, **69**, 698.
63. V. M. Tysseling-Mattiace, V. Sahni, K. L. Niece, D. Birch, C. Czeisler, M. G. Fehlings, S. I. Stupp and J. A. Kessler, *J. Neurosci.*, 2008, **28**, 3814.
64. J. Kisiday, M. Jin, B. Kurz, H. Hung, C. Semino, S. Zhang and A. J. Grodzinsky, *Proc. Natl. Acad. Sci. USA*, 2002, **99**, 9996.
65. T. D. Sargeant, M. O. Guler, S. M. Oppernheimer, A. Mata, R. L. Satcher, D. C. Dunand and S. I. Stupp, *Biomaterials*, 2008, **29**, 161.
66. H. Komatsu, S. Tsukiji, M. Ikeda and I. Hamachi, *Chem.–Asian J.*, 2011, **6**, 2368.
67. C. Yang, D. Li, Z. Liu, G. Hong, J. Zhang, D. Kong and Z. Yang, *J. Phys. Chem. B*, 2012, **116**, 633.
68. B. V. Slaughter, S. S. Khurshid, O. Z. Fisher, A. Khademhosseini and N. A. Peppas, *Adv. Mater.*, 2009, **21**, 3307.
69. K. M. Galler, J. D. Hartgerink, A. C. Cavender, G. Schmalz and R. N. D'Souza, *Tissue Eng.: Part A*, 2012, **18**, 176.
70. L. Aulisa, H. Dong and J. D. Hartgerink, *Biomacromolecules*, 2009, **10**, 2694.
71. H. He, J. Yu, Y. Liu, S. Lu, H. Liu, J. Shi and Y. Jin, *Cell Biol. Int.*, 2008, **32**, 827.
72. W. Mueller-Klieser, *Am. J. Physiol.*, 1997, **273**, C1109.
73. R. Z. Lin and H. Y. Chang, *Biotechnol. J.*, 2008, **9–10**, 1172.
74. J. L. Ifkovits, H. G. Sundararaghavan and J. A. Burdick, *J. Vis. Exp.*, 2009, **32**, 1589.
75. M. H. Kim, M. Kino-oka and M. Taya, *Biotechnol. Adv.*, 2010, **28**, 7.
76. G. D. Prestwich, *Acc. Chem. Res.*, 2008, **41**, 139.
77. F. Zhao, M. L. Lung and B. Xu, *Chem. Soc. Rev.*, 2009, **38**, 883.
78. M. W. Tibbitt and K. S. Anseth, *Biotechnol. Bioeng.*, 2009, **10**, 655.
79. M. C. Cushing and K. S. Anseth, *Science*, 2007, **316**, 1133.
80. A. Mahler, M. Reches, M. Rechter, S. Cohen and E. Gazit, *Adv. Mater.*, 2006, **18**, 1365.

81. M. Zhou, A. M. Smith, A. K. Das, N. W. Hodson, R. F. Collins, R. V. Ulijn and J. E. Gough, *Biomaterials*, 2009, **30**, 2523.
82. M. Colombo and A. Bianchi, *Molecules*, 2010, **15**, 178.
83. C. A. Buck and A. F. Horwitz, *Annu. Rev. Cell Biol.*, 1987, **3**, 179.
84. V. Jayawarna, M. Ali, T. A. Jowitt, A. F. Miller, A. Saiani, J. E. Gough and R. V. Ulijn, *Adv Mater*, 2006, **18**, 611.
85. V. Jayawarna, S. M. Richardson, A. R. Hirst, N. W. Hodson, A. Saiani, J. E. Gough and R. V. Ulijn, *Acta Biomater.*, 2009, **5**, 934.
86. A. Horii, X. Wang, F. Gelain and S. Zhang, *PLoS ONE*, 2007, **2**, e190.
87. F. Gelain, D. Bottai and A. Vescovi, *PloS ONE*, 2006, **1**, e119.
88. K. M. Galler, L. Aulisa, K. R. Regan, R. N. D'Souza and J. D. Hartgerink, *J. Am. Chem. Soc.*, 2010, **132**, 3217.
89. J. P. Schneider, D. J. Pochan, B. Ozbas, K. Rajagopal, L. Pakstis and J. Kretsinger, *J. Am. Chem. Soc.*, 2002, **124**, 15030.
90. C. Yan, M. E. Mackay, K. Czymmek, R. P. Nagarkar, J. P. Schneider and D. J. Pochan, *Langmuir*, 2012, **28**, 6076.
91. L. Haines-Butterick, K. Rajagopal, M. Branco, D. Salick, R. Rughani, M. Pilarz, M. S. Lamm, D. J. Pochan and J. P. Schneider, *Proc. Natl. Acad. Sci. USA*, 2007, **104**, 7791.
92. I. M. Geisler and J. P. Schneider, *Adv. Funct. Mater.*, 2012, **22**, 529.
93. J. H. van Esch, *Langmuir*, 2009, **25**, 8392.
94. D. I. Chan, E. J. Prenner and H. J. Vogel, *Biochim. Biophys. Acta, Biomembr.*, 2006, **1758**, 1184.

CHAPTER 7

Optic and Electronic Applications of Molecular Gels

JOSEP PUIGMARTÍ-LUIS AND DAVID B. AMABILINO*

Institut de Ciència de Materials de Barcelona (CSIC), Campus Universitari, 08193 Bellaterra, Spain, Email: jpuigmarti@icmab.es
*Email: amabilino@icmab.es

7.1 Introduction: Why Gels for Functional Materials with Optical and Electronic Properties?

Functional molecular materials are perhaps most usually formed through processing techniques that involve vapour deposition, crystallisation from solution or upon cooling from the melt, thin film casting methods and so on; a plethora of possible routes to give materials with novel optical, electrical and magnetic properties along with combinations of them.[1] The many methods available to process materials each have their own advantages and drawbacks, and perhaps one of the challenges in all the routes used to any of these materials is the preparation of fibres of molecules that have widths from the nanometre to micrometer scale (Figure 7.1). Gels, of course, have this feature as their almost unifying characteristic.[2] The fibres that immobilise the solvent in which they are formed comprise a material that can contain well-ordered chains of molecules with the suitable packing to provide a unique property. In addition, the intrinsic porosity of the gels affords opportunities for applications not available to bulk-processed materials.[3]

In particular, many molecular systems find their use on surfaces or at interfaces of other materials.[4] When depositing molecules from homogeneous

RSC Soft Matter No. 1
Functional Molecular Gels
Edited by Beatriu Escuder and Juan F. Miravet
© The Royal Society of Chemistry 2014
Published by the Royal Society of Chemistry, www.rsc.org

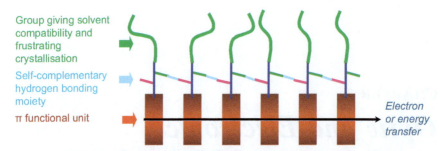

Group giving solvent compatibility and frustrating crystallisation

Self-complementary hydrogen bonding moiety

π functional unit

Electron or energy transfer

Figure 7.1   A general representation of fibres formed by a functional molecule in a gel with the roles performed by each component of the molecular structure for the case of organogels showing charge or energy transport, which has parallels in other material properties.

solution onto surfaces, the combination of interactions at the interface[5] – between surface and molecule, surface and solvent, solvent and molecule and between the molecules themselves – in combination with any drying effects of the drop can lead to complex, undesired, and/or inhomogeneous layers. The surface also plays a dramatic role in determining the type of nanostructure formed at a surface, which in turn affects properties of the molecular layer.[6,7] In the preparation of a gel, the fibrous network is preformed prior to any casting technique, and therefore is collapsed, but in principle changed little structurally at the molecular level upon deposition on a surface. On the other hand, the many interconnections present in the gel network are advantageous for transfer of a physical property through the material.

Molecular systems forming nanomaterials that exhibit bulk electrical conductivity is of tremendous current interest for nanoscience and technology in general, and particularly molecular electronics.[8] The possibility of connecting electronic components through an organic material is an important goal in this area, and the self-assembly of conducting supramolecular wires is a viable and potentially efficient pathway to it. Fibres for molecular electronics have been prepared from solution that go some way to meeting the challenge,[9] but deposition on surfaces can influence greatly the nature of the material in terms of morphology and supramolecular structure. The bottom-up gel route to form these wires is especially attractive for this goal because the formation of the fibres can be followed by efficient doping of the material without concern about surface effects on the self-assembly process.[10]

In a similar way, molecular materials with optical properties of interest can find unique arrangements in the gel and xerogel states.[11] While photoinduced charge separation may benefit from the fibrous nature of gels in bulk material once solvent is removed, the gel state itself can be of utility for samples that exhibit luminescence. In particular, the gelling of nonvolatile fluids that contribute to the mechanical or structural properties of the immobilised state can lead to materials with unique behaviour.[12]

The gelling of solvents also facilitates the preparation of composite nanostructured materials.[13] Here, two poorly miscible materials can be trapped in an arrangement that would not be possible through conventional solvent processing methods when phase separation can be prevalent. The rapid formation of the gel network gives the molecule's metastable aggregated state and the second material can be incorporated in various ways depending on the solvation and supramolecular compatibility of the two components. Indeed, the use of gel-based systems for the formation of complex matter is extremely promising in the development of nanocomposites that can incorporate more than one property or whereby performance can be enhanced by synergic effects.

7.2   Gels, Aerogels and Xerogels: Consequences for Materials

The formation of a molecular gel is the starting point for the use of the network of fibres generated in the process as a material that can show interesting behaviour, in the case of the systems reviewed in this chapter electronic and optical properties. The material properties of the gel, aerogel or xerogel will be distinct from one another because of the solvent presence or absence, the interconnectivity and the porosity. In the following subsections we discuss generally the pros and cons of these different materials that can be made from the same source compounds.

7.2.1   The Gel-State Properties – Property Dilution and Mobility

The gel state usually incorporates a minor component (in general the functional part for our purposes) in a majority of fluid. Therefore, although the mechanical properties of the whole are dominated by the fibres of the gelator, any other property that depends on the sample – as a simple example the absorption – will be low on a weight for activity basis when compared with the pure gelator molecule: One can say that the property has been diluted. In general, the dilution of property in a material is not desirable unless some other feature is afforded in the process. The use of plasticisers, for instance, provides a desirable mechanical feature to certain polymers. However, when a particular system requires relatively dense material any additional filler will reduce the effectiveness per unit weight.

The dependences of magnetic or electrical properties of a given molecule on surroundings are prime examples of the dilution effect. In the case of magnetism, when a molecule carrying, say, an unpaired electron is made bigger through chemical substitution then the distance between the part of the molecules bearing the paramagnetic region is increased and therefore the coupling between these units in different molecules is decreased, as is the spin content per unit weight. Considering now an electrical conductor, very often the highest current density per unit mass is sought. Therefore, the increased material mass per unit of conductor goes against high current density. Furthermore, charge transport is favoured by a high number of contacts between fibres. The gel state has the fibres solvated and therefore with relatively few contacts, and the doping of the material also requires movement of charge. So, fabrication of effective conductive gels (with solvent present) is very difficult.

Opportunities for biological applications of gels, such as the use of optical response to certain analytes, would require that the solvent – probably water – be present so that the molecule being sensed is transported through the gel network. The mobility of this molecule is essential for the sensing to be efficient and reversible. Technological applications in displays also require the presence of the majority component whose property is aided by the fibrous web: As we shall see, liquid-crystal displays based on anisotropic gels of mesogens have promising properties.

7.2.2 Aerogels – Porous Materials for Multiple Applications

Aerogels in general are very attractive materials because of their high porosity, great thermal and acoustic insulation, transparency, and strength per unit weight. While most aerogels are based on silica sol–gel chemistry,[14] there are interesting cases where polymers have been used to make the materials[15,16] which can have unique chemical properties in the area of nanomaterials.[17,18] Also, carbon-based aerogels have remarkable properties, for example as so-called "aero-capacitors".[19] Though these systems do not fit the molecular remit of this volume, they are certainly inspiration for the development of systems based on molecules.

The preparation of aerogels of pure molecular materials is not feasible presently, as they cannot achieve the structural strength of purely covalent networks because the gel framework collapses to give the xerogel (*vide infra*). However, given the opportunities offered by this kind of porous material, it would not take a great leap of imagination to come up with a route for stabilising molecular gels such that functional aerogels could be created in the near future.

7.2.3 Xerogels – Collapse as an Opportunity!

The evaporation of solvent from a gel leaves only the gelator in a form that is most usually a tangled mesh of fibres, although some discotic molecules do not always have this feature present obviously.[20] The drying process could clearly play a role on both the density and form of the fibres, as shall be seen in the case of a gel-derived molecular conductor, and the use of cryoelectron microscopies is often the preferred route for analysis of the morphology of the colloidal component.

That said, the collapse of a gel to give a dense, tangled mesh can be advantageous. The compacting process under solvent evaporation leads to an increase in fibre–fibre contacts when compared with the gel state. The material that results is still porous because of the generally anisotropic nature of the gels which leads to a randomly criss-crossing mat. Provided that interfibre contacts are sufficiently close, the processes of excited or ground-state electron transport and magnetic coupling can be transferred across the boundary.

The great advantage of the gel state as a processing medium when compared with other supramolecular fibres is that the former generally creates longer yarns with a greater degree of interconnection. Also, the practically guaranteed porosity means that procedures such as oxidation or reduction (chemically or electrochemically) can be achieved throughout the material. Their surface area is much higher than the crystal state or the gel state (where the solvent can impede efficient diffusion) and on the other hand can contain sufficient supramolecular order to exhibit desirable properties.

7.3 Gel-Based Materials with Optical Properties

One of the major hurdles in nanotechnology is to precisely control the bottom-up self-assembly and the molecular orientation of building block moieties towards the construction of functional nanoscaled structures with relevant optical properties. To engineer nanoscaled structures with photoactive synthetic molecules has, however, proven difficult, even in conjugated oligomers where well-defined molecular systems can be finely achieved by chemical modification. Changes in the molecular structure can certainly facilitate variations in the intermolecular interactions of the self-assembled structure, which may lead to new appealing features including enhanced electro-optical properties. In previous sections we have seen that the gel state can potentially organise low-molecular-weight materials in one-dimensional aggregates with short intermolecular distances between moieties. Hence, to induce a supramolecular organisation of chromophores or π-conjugated molecules in fibrous gel networks will undoubtedly influence charge-carrier mobilities and the luminescent effects of the distinct single molecules. In addition, any approach enabling a fine tuning of the fibrous gel network design can potentially lead to optical and electronic properties needed for device fabrication. In this section, we shall highlight how this methodology is recently emerging as a powerful strategy to increase the number of practical applications of gel-based systems.

7.3.1 Isotropic and Anisotropic Gels, Birefringence, and Liquid-Crystalline Gels

The physical gelation of liquid crystals (LCs) by low molecular weight organogelators is of current interest in advanced technologies because of the unique and favourable properties accompanying the LCs, *i.e.* high molecular order and a dynamic state that can be influenced by physical fields.[21-26] Owing to the two independent thermally induced phase transitions present in LC physical gels – the sol–gel transition of the gelator and the isotropic–anisotropic transition of the liquid crystal – unusual anisotropic phase separations can be achieved. For example, when the self-assembly of the gel network occurs at a temperature higher than the isotropic–anisotropic liquid-crystal phase transition temperature of the LC ($T_{\text{iso–lc}}$), a randomly dispersed fibre network is formed and liquid-crystalline polydomains are feasible (Figure 7.2a). However, if the $T_{\text{iso–lc}}$

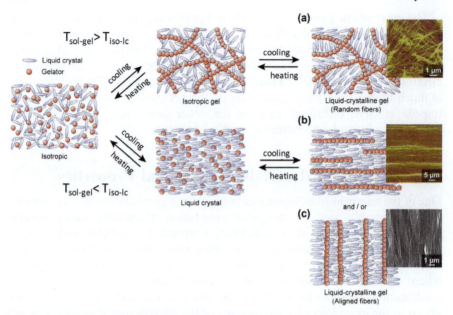

Figure 7.2   Schematic representations of the two types of structural changes of liquid-crystalline physical gels. The insets in (a) and (b) are AFM images of dispersed fibrous aggregates formed in an isotropic phase, and in an oriented LC phase, respectively. In (c) a SEM image of a solid gel network formed from a LC template growth.
(Reproduced with permission from RSC.)

is higher than the sol–gel transition temperature ($T_{\text{sol-gel}}$), the solid fibrous networks of the gelator are formed in the anisotropic states of liquid crystals (Figures 7.2b and c). In a sense, if $T_{\text{iso-lc}}$ is higher than $T_{\text{sol-gel}}$ the LC phase serves as an anisotropic functional fluid that templates and guides the self-assembly of the fibrous gel network. Indeed, this induction of anisotropic phase-separated structures leads to the creation of new functions and to the enhancement of properties (*vide infra*).

Kato's group has prepared and studied a vast number of liquid-crystalline physical gels.[27–30] In an early example, an octadecyl-benzyl amino acid based organogelator was self-organised in a smectic molecular structure thanks to the gelation of a commercially available ferroelectric liquid crystal, SCE8 (Hoechst)[31] (Figure 7.3a). In this case, the fibrous gel network was assembled in the direction perpendicular to the long axis of the LC, *i.e.* the direction of the smectic layer. However, one should consider that in smectic gelled liquid crystals the self-assembly of the organogelator network can occur in two different orientations, one being the direction of the long axis of the LC, and the other the direction of the smectic layer. For example, fibrous aggregates of the same gelator but aligned in the direction of the long axis of a liquid crystal was achieved with an LC containing a heptyl-oxy-cyanobiphenyl group[32] (Figure 7.3b). Furthermore, anisotropic control of an anthracene

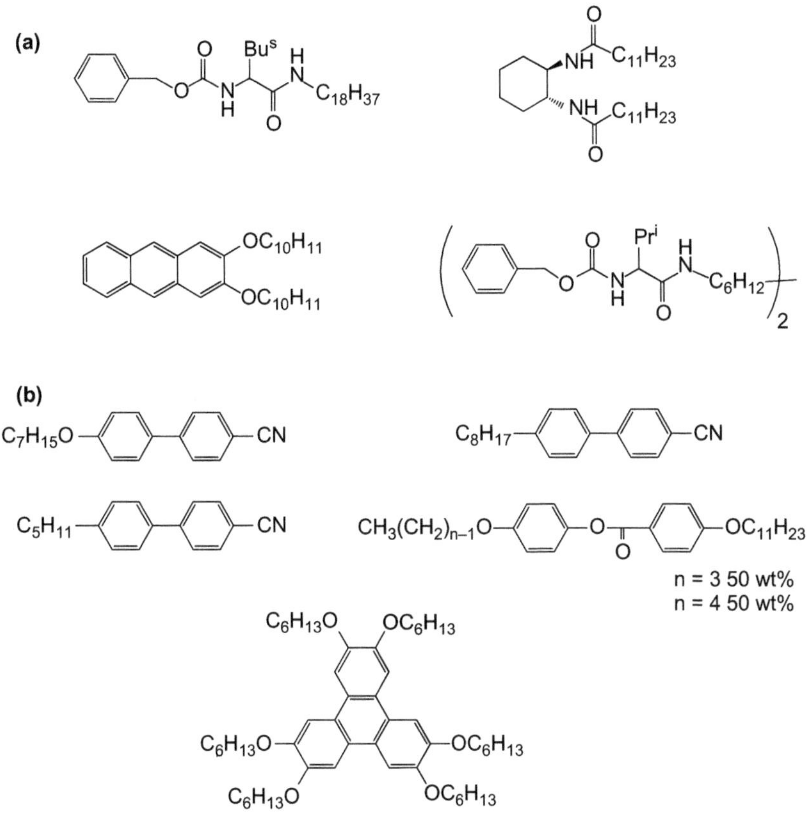

Figure 7.3 Molecular structures of: (a) organogelators and (b) LCs.

organogelator derivative was demonstrated with pentyl-cyanobiphenyl and octyl-cyanobiphenyl LCs[33] (Figure 7.3). In this example, a preferred growth direction of the self-assembled gel fibres in both anisotropic functional liquids was possible because T_{iso-lc} was higher than the $T_{sol-gel}$. Importantly, this work also proved the dependency on the organogelator and liquid crystal choice towards the alignment of the fibrous gel network. Here, the π–π interactions between the LC and the anthracene moiety favoured an oriented parallel alignment of the organogelator that self-assembled perpendicular to the long axis of the LC phase.

On the other hand, randomly self-assembled fibrous gel networks can be generated from isotropic phases, where the oriented nature of the fluid LCs can be preserved in crystalline polydomains (Figure 7.2a). For example, the nematic liquid crystal pentyl-cyanobiphenyl was successfully immobilised by a gelator containing a diaminocyclohexane group[34] (Figure 7.3). In this case, the xerogel showed no molecular alignment ($T_{sol-gel} > T_{iso-lc}$), however, the liquid-crystal phase was homogeneously aligned (unlike the situation in most gels). Even though randomly dispersed fibrous gel networks are generated in isotropic liquid states with $T_{\text{sol–gel}} > T_{\text{iso–lc}}$, enhanced functionalities can be displayed. For example, a discotic LC physical gel synthesised combining a hole-transporting triphenylene LC derivative and a biphenyl amino acid based gelator was proved to form randomly disperse fibrous gel networks with microphase-separated states of the triphenylene LC derivative ($T_{\text{sol–gel}} > T_{\text{iso–lc}}$)[35] (Figure 7.3). In this LC physical gel a hole mobility three times higher than that of the columnar LC structure alone was demonstrated by the introduction of micro-phase separated structures into the discotic LC. It was assumed that the fibrous gel network effectively suppressed the LC fluctuations, thereby leading to a faster hole transport.

Additionally, liquid-crystalline polydomains formed from liquid-crystalline physical gels with $T_{\text{sol–gel}}$ higher than $T_{\text{iso–lc}}$ have effectively been introduced as reversibly switchable films with light-scattering functionalities. For example, Smith and coworkers fabricated white-light-scattering films with a commercial nematic LCs (LC-E7) and a sorbitol organogelator derivative[36] (Figure 7.4). In this case, it was shown that electro-optically active cells formed by this LC physical gel were reversibly switchable between a transparent and an opaque state by applying alternating current (AC) electric fields (Figure 7.4). Upon

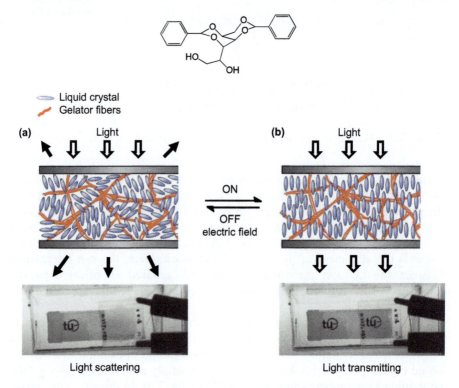

Figure 7.4   The molecular structure of a sorbitol organogelator derivative and sche-matic illustrations (a and b) showing the switching between an opaque state and the transparent state in a nematic LC gel, respectively. At the bottom, pictures of the OFF (left) and ON (right) cell states.

application of an electric field, the LC molecules oriented in the direction of the electric field, thereby inducing aligned domains which enabled the transmission of light (Figure 7.4b). Even though small amounts of gelator were necessary to fabricate these films, large voltages were required for this LC-based switching system.

Later, Kato and coworkers described a liquid-crystalline physical gel formed from a benzonitrile-based twisted-nematic liquid crystal and a L-lysine-based gelator.[37] (Figure 7.5) In this case, the twisted nematic phase of the benzonitrile-based liquid crystal was used to organise the self-assembly of the organogelator molecules with a twist orientation between different layered aggregates ($T_{iso-lc} > T_{sol-gel}$) (Figures 7.5a and b). Surprisingly, these structures showed fast and high-contrast electro-optical responses with low driving voltages, unlike other previous work with randomly dispersed networks and partial light scattering systems.[38,39] Strikingly, this observation showed that liquid-crystalline physical gels with T_{iso-lc} higher than $T_{sol-gel}$ could also be applied for the fabrication of reversibly switchable light-scattering electro-optical materials.

In an extension of this work, a rewritable light-scattering memory element was described by a thermoreversible formation of composite structures in nematic gels upon an electric field activation.[40] Here, the L-lysine-based gelator showed previously was used to gellify a commercial nematic liquid crystal, LC-E63 (Merck). Because T_{iso-lc} was higher than $T_{sol-gel}$, and because the external electric field could be used to manipulate the orientation of the

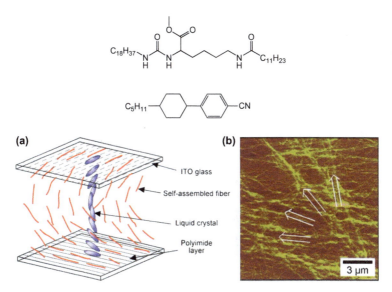

Figure 7.5 Molecular structures of the L-lysine-based gelator and of the benzonitrile-based liquid crystal. Schematic illustration of a twisted nematic cell (a) with oriented fibrous networks, and AFM image (b) showing the oriented fibrous aggregates of the L-lysine-based gelator.
(Reproduced with permission from Wiley-VCH.)

nematic liquid-crystal phase before gelation occurred, two bistable nematic phases were described through the fixation of the gel fibre network. On the one hand, a stable light-transmitting state was obtained by the formation of oriented fibrous aggregates of the gelator under an electrically aligned nematic phase, (Figure 7.6, (a) → (b) → (c)) and on the other hand, a stable turbid appearance was generated in absence of an electric field (Figure 7.6, (a) → (d) →(e)). In the latter nematic gel state it was proved that even though a dispersed random fibre network was formed, a reversible turbid to transparent transition was possible upon activation of an electric field (Figure 7.6f). In addition, both bistable structures – the stable light-transmitting and turbid nematic gel state – were fully recovered upon heating the system to the isotropic state, hence permitting a complete erasable and rewritable system.

Another very promising approach to construct rewritable gel-based systems consisted in the incorporation of photochromic or polymerisable parts into the molecular structure of the organogelator.[41] For example, a rewritable LC physical gel was achieved by mixing a chiral hydrogen-bond-forming gelator, incorporating photochromic azobenzene moieties, with the pentyl-cyanobiphenyl nematic liquid crystal[42] (Figure 7.7). When an isotropic liquid of the LC and the organogelator in its *trans* conformation was cooled to room

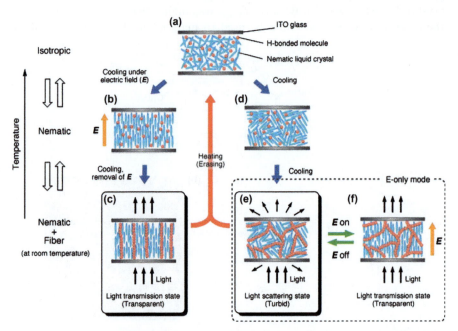

Figure 7.6   Schematic illustration of the fixation of the fibre gel network with an electric field applied, (b) and (c), and without an electric field, (d) and (e). (f) Schematic illustration of the alignment of the nematic LC phase in a dispersed fibre network upon an electric field is applied.
(Reproduced with permission from Wiley-VCH.)

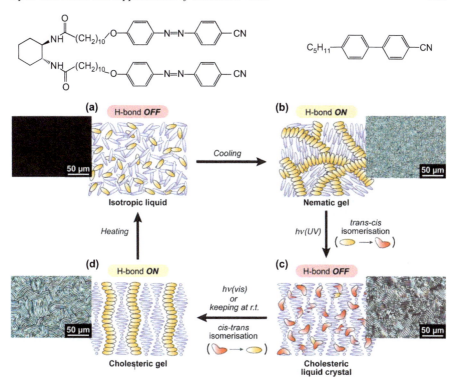

Figure 7.7 Molecular structures of the chiral gelator incorporating photochromic azobenzene moieties and of the pentyl-cyanobiphenyl-based liquid crystal. From (a) to (d) scheme including schematic illustrations and optical polarised micrographs of the structural changes accomplished with a LC physical gel prepared with 3 wt% of the azobenzene derivative gelator and the pentyl-cyanobiphenyl liquid crystal. (Reproduced with permission from RSC.)

temperature, a nematic LC gel was generated ($T_{sol-gel} > T_{iso-lc}$) (Figure 7.7, (a) → (b)). Upon UV-light irradiation, *trans–cis* photoisomerisation of the azobenzene moieties induced a phase transition to a cholesteric liquid-crystal phase, where small chiral domains were appreciable which arise from the cholesteric phase induced by the *cis*-azobenzene units (Figure 7.7, (b) → (c)). After visible-light irradiation, the reverse *cis–trans* photoisomerisation of the azobenzene was accomplished, and hence, the assembly of the gelator molecules was again induced, but now along with the fingerprint structure of the cholesteric liquid-crystal phase (Figure 7.7, (c) → (d)). In this case, two rewritable and bistable structures were accomplished, the nematic gel and the cholesteric gel. In addition, patterned arrays of the two rewritable structures were demonstrated by UV irradiation of the nematic gel phase through photomasks.

Apart from liquid-crystalline gels prepared by mixing a LC and an organogelator, examples exist in which a single molecular component can display

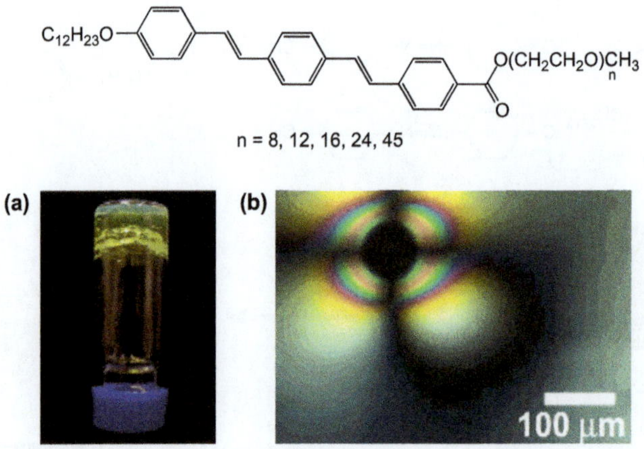

C₁₂H₂₃O— ... —O(CH₂CH₂O)ₙCH₃

n = 8, 12, 16, 24, 45

(a)

(b)

100 μm

Figure 7.8 Molecular structure of an OPV derivative incorporating a hydrophilic PEG segment. (a) Photograph of an aqueous OPV gel, and (b) polarised optical microscope image of a homeotropically aligned aqueous LC.

birefringent LC gel phases with enhanced electro-optical properties in dried films. For example, Stupp and coworkers reported the formation of very long and uniformly aligned domains of an OPV derivative by precisely controlling the length of the poly(ethylene glycol) (PEG) segment present into the structure[43] (Figure 7.8). The authors proved that longer PEG chains enabled the formation of lyotropic LC gels in water, which upon controlled thermal conditions could generate homeotropically aligned OPV structures on glass thanks to the strong interaction between the PEG moieties and the hydrophilic glass surface. This work shows how a single synthetic modification can lead to appreciable changes not just in the molecular aggregation, but also in the electro-optical properties of the assemblies.

On the contrary, optical and electronic properties of a single dye component can be tuned through the gel state without the need for any synthetic modification at all. For example, Kikkawa and coworkers have recently demonstrated the potential of complementary multiple hydrogen-bonding interactions for the complexation and gel formation of a merocyanine dye bearing a barbituric acid group with a barbituric acid receptor containing a bisamine-triazine derivative[44] (Figure 7.9). In this case, the authors observed that upon complete complexation of the dye with the bisamine-triazine derivative (1 : 1), the binary system formed a gel and displayed a threadlike birefringent texture characteristic of a nematic gel state (Figures 7.9a–c). This binary system showed a blue-shift in the absorption spectra as compared to the spectra of the free dye in solution, and further AFM analysis of the dried nematic gel indicated the formation of parallel fibrous gel networks on the surface (Figure 7.9d). Photocurrent integration analysis of the dried nematic gel showed enhanced charge-carrier mobility for the binary system attributed to the organised molecular arrangement and the efficient π-overlap of the conjugated chromophores within the structure.

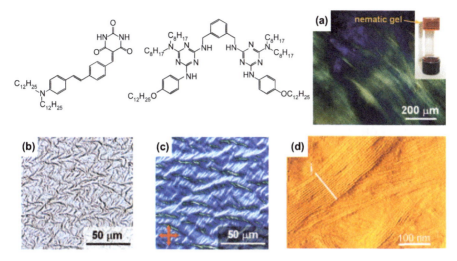

Figure 7.9 Molecular structure of the merocyanine dye derivative and of the dye acceptor. (a) and (c) polarised optical microscope images of the binary system (1 : 1), and in (b) without crossed polarisers. (d) AFM image of the dried nematic gel. Inset in (a): picture of the nematic gel in a vial.

7.3.2 Fluorescent Gels

Self-assembly of synthetic luminescent molecules is also a very suitable tool, not just for the preparation of new photofunctional materials, but because tuning the supramolecular organisation can markedly influence the optoelectronic properties of the constituent chromophores. It is noteworthy that to modulate luminescent properties of small synthetic molecular systems can represent a major step towards preparing photofunctional materials "*a-la-carte*", an ultimate goal for a sustainable development of future advanced nanoscale optoelectronic and photonic devices. In this sense, supramolecular alignment of chromophore-based molecules through gelation can provide essential insights towards this aimed goal. Therefore, a large variety of tailor-made molecules incorporating chromophores or dye moieties have been designed with the goal of self-organising into extended fibre-like networks capable of entrapping large volumes of solvent. Cooperative hydrogen bonding, π–π stacking, and van der Waals interactions are the most common noncovalent interactions used. From the immense number of π-conjugated systems employed, *e.g.* oligo(p-phenylene vinylene) (OPV), thiophene, phenylene, azobenzene, naphthopyran, stilbene, phthalocyanines, porphyrins, benzimidazole, [n]acenes, pyrene, fluorene, tetrathiafulvalene, phenanthroline, hexabenzocoronene, perylene bisimide, triazine derivatives, *etc.*,[45–68] this section will highlight and put into perspective a few recent advances in the field of fluorescent organogelators. A detailed look to excitation energy transfer and cascade transfer processes occurring in the gel state will be remarked upon in the following section.

By means of a controlled Wittig–Horner reaction, George and Ajayaghosh synthesised a variety of OPV derivatives that proved to be efficient gelators of

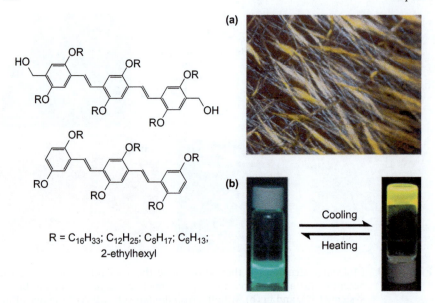

Figure 7.10 Molecular structures of OPV derivatives with and without hydroxyl end groups. (a) Micrograph image of a hydroxylated OPV derivative gel in decane under crossed polarisers. (b) Pictures of the sol–gel transition of a hydroxylated OPV derivative under UV-light illumination.

nonpolar hydrocarbon based solvents[69] (Figure 7.10). The authors explored the role of the hydrogen bond in the gelation process by designing OPV derivatives without hydroxyl end groups (Figure 7.10). It was proved that hydrogen-bond donor groups were necessary for localising the OPV moieties within the π-stacked assembly, thereby favouring the gel formation. Furthermore, polarised optical microscopy studies of the OPV gels indicated the formation of strongly birefringent domains, which authenticated the high alignment of the chromophore's moieties in the gel fibre network (Figure 7.10a). In addition, this extraordinarily ordered OPV self-assembled structures induced a remarkable change in the optical and photophysical properties as compared to the molecularly dissolved species. Illumination of the OPV solutions before gelation exhibited a strong greenish-blue fluorescence, whereas upon gelation a greenish-yellow emission was evident (Figure 7.10b). Importantly, the shifts in the absorption and emission bands were attributed to the self-assembled OPV molecules and not to the formation of excimers. This investigation clearly evidences the strong electronic coupling between the π-conjugated molecules in the gel fibre network, thereby inducing transformations in the physico-optical properties of the discrete solvated chromophores.

In an extension of this work, diverse supramolecular gel structures combining donor and acceptor OPV derivatives were prepared.[70] In this case, a tuneable fluorescence emission was demonstrated by tailoring the electron acceptor groups present in the π-conjugated OPV scaffold (Figure 7.11a). Surprisingly, different HOMO–LUMO gaps could be described in these

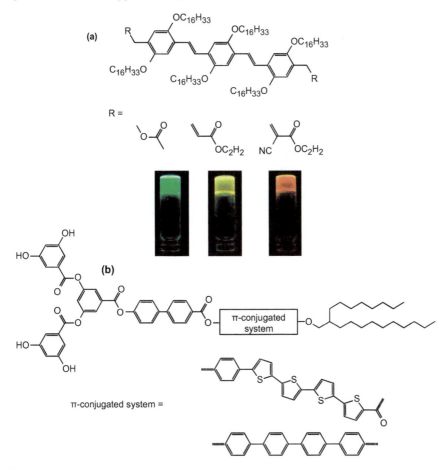

Figure 7.11 In (a) molecular structures of the OPV derivatives containing an insulated ester group, a conjugated ester, and a cyanoester moiety together with photographs of the respective gels in *n*-hexane under illumination at 365 nm. (b) Molecular structure of the phenyl-tetra(thiophene), and tetra(phenylene) derivatives.

systems. For example, the blue emission of the OPV derivative containing an insulated ester group located at 466 nm in dichloromethane is red-shifted to green in the gel state ($\lambda_{max} = 530$ nm), whereas the green and orange emissions of the OPV derivatives containing a conjugated ester and a cyanoester moiety, turned to yellow and red, respectively, in *n*-hexane gels (Figure 7.11a).

Gelation in nonpolar solvents with oligo(thiophene) or oligo(phenylene) derivatives attached covalently to dendritic segments has also been shown.[71] The so-called dendron rod–coil molecules containing phenyl-tetra(thiophene) or tetra(phenylene) segments also showed a fluorescence emission red-shift upon gellification in a toluene:tetrahydrofuran solution (30:1) (Figure 7.11b). Indeed, enhanced electronic properties were asserted by these organogelators as a result of improved π-orbital overlap in the gel state.

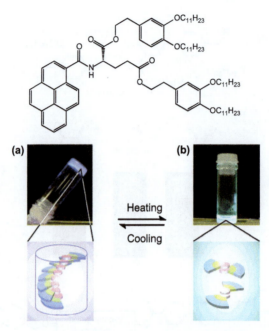

Figure 7.12 Molecular structure of the pyrene-based gelator and photographs acquired under UV light of: (a) the gel state, and (b) the sol phase.[72] Below the micrographs, schematic illustrations showing the self-assembly of the pyrene-based gelator in the two states are shown.

Thereafter, Kamikawa and Kato designed pyrene-based gelators containing dendritic oligopeptides that were capable of efficient gelation of a wide variety of organic solvents[72] (Figure 7.12). The hydrogen bonds of peptide residues together with the π–π interactions between the pyrene moieties facilitated the self-assembly, and prompted the gelation process. Self-assembled structures consisting of helical columnar assemblies were characterised by XRD, UV–visible absorption and circular dichroism (CD) spectroscopies. Surprisingly, the pyrene-based gelator shown in Figure 7.12 exhibited a drastic emission colour change between the gel and sol phases under UV irradiation. The gel state displayed a blue monomer emission, with bands located at 396 and 412 nm, whereas the sol phase displayed a bright-green emission located around 480 nm, a band characteristic of pyrene excimers (Figures 7.12a and b). The remarkable luminescent properties of pyrene derivatives incorporating alkynyl units when they form gels in which gelled DMF, toluene, or cyclohexane has also been shown.[73] The use of laser scanning confocal microscopy to characterise the photoluminescence showed that it was independent of superstructure size and that fibre entanglement prevents the efficient formation of excimers. Apart from their optical properties, it was shown that the xerogels could behave as active layers in organic field-effect transistors.

Apart from pyrene, OPVs, and other fluorescent π-conjugated organogelators, interesting energy-transfer processes have been described in organogels of acenes derivatives.[74,75] For example, white-light emission in organogels and dry films formed by blending three acene derivatives with distinct fluorescence emissions has been demonstrated.[76] The gel of the anthracene derivative shown in Figure 7.13 exhibits a blue fluorescent emission in DMSO with a band located at 412 nm, (Figure 7.13a), while the two other tetracene derivatives display green (502 nm) and orange-red (555 nm) fluorescence emissions in tetrahydrofuran (Figures 7.13b and c). However, when a DMSO solution containing a mixture of the three compounds was prepared at 120 °C and subsequently cooled to room temperature, an organogel that was translucent to visible light was synthesised (Figure 7.13d). Surprisingly, tunable emissive properties were achieved in this mixed-gel phase by controlling both; the UV-light excitation, and the concentration of the tetracene derivatives. In principle,

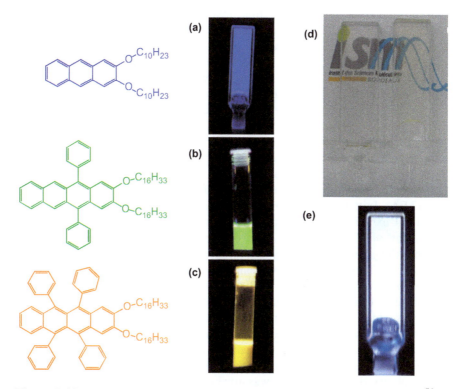

Figure 7.13 Molecular structures of the anthracene and tetracene derivatives.[76] (a) Photograph of the anthracene derivative gel in DMSO, and (b) (c) photographs of the tetracene derivatives in tetrahydrofuran. (d) Photograph showing two quartz cuvettes containing an anthracene derivative gel (left) and the mixed gel (right) under daylight. (e) Picture of the mixed gel under UV light. All UV light photographs are acquired under an excitation wavelength of 365 nm.

the excitation and energy transfer from the anthracene derivative to the tetracene derivatives was modulated by varying the proportions of the three components. By employing 0.012 equivalents of both tetracene derivatives in the final mixed gel; a white-light-emitting organogel was achieved (Figure 7.13e). The white light emission is therefore the result of the partial excitation energy transfer from the anthracene derivative, which acts as a light-harvesting matrix, to the tetracene derivatives, which behave as energy acceptors. Importantly, this kind of energy transfer cascade has also been observed in other organogelators, *e.g.* cholesteryl-based perylene bisimide gelators,[77] or donor–acceptor OPV-based organogelators,[78] which will be specifically featured in the next section, amongst others. Apart from this example, borondipyrromethene units incorporated into a branched aromatic moieties with amide groups and six alkyl chains gives a molecule that forms highly luminescent gels in nonane.[79] The amide groups were responsible for the formation of this superstructure.

A ground-breaking study in this area was the incorporation of photochemically active units in the molecular structure of the gelator molecules. In this case, gelator molecules can respond and be sensitive to one or more external stimuli, thus yielding new smart materials. Even though this is not the central theme of the present chapter, we will present a few examples of stimuli-sensitive supramolecular gels with effects on the photoactivity of the constituent chromophore moieties. For a comprehensive account of stimuli-sensitive organogelators, the readers are referred to Chapter 3 of this book.

In one of the earliest examples, a 1,10-phenanthroline derivative with two cholesteryl side groups was described as a proton-sensitive gel[80] (Figure 7.14). The 1,10-phenanthroline-based gel formed in 1-propanol showed interesting fluorescence behaviour upon protonation of the basic sites with trifluoroacetic acid (TFA). Initially and under UV-light irradiation, the 1,10-phenanthroline-based gel exhibited a bright purple colour, whereas upon addition of two equivalents of TFA a greenish-yellow colour was observed (Figure 7.14a–c). On heating, the greenish-yellow gel, a light blue sol was observed under UV-light irradiation (Figure 7.14d). Fluorescence spectroscopy studies indicated that the fluorescence maximum located at 396 nm for the neutral gel phase quenched completely when small amounts of TFA were added. Surprisingly, when the TFA was added a broad longer-wavelength band located at 530 nm appeared, giving rise the characteristic greenish-yellow colour of the protonated gel form under UV light. The colour change between phases was explained by the energy transfer from the neutral gelator to the protonated form. The greenish-yellow colour observed in the gel state was attributed to the exceptional organisation of the neutral gelator in the gel fibre network, favouring an effective energy-transfer process. Whereas, in the sol phase, the energy-transfer process was not efficient, leading to a blue emission under UV-light irradiation.

Another route that facilitates optical changes in the gel state is the incorporation of photochromic moieties in the organogelator structure. In particular, photochromic bisthienylethene derivatives have been shown to be particularly appealing molecules for the bottom-up design of multicolour

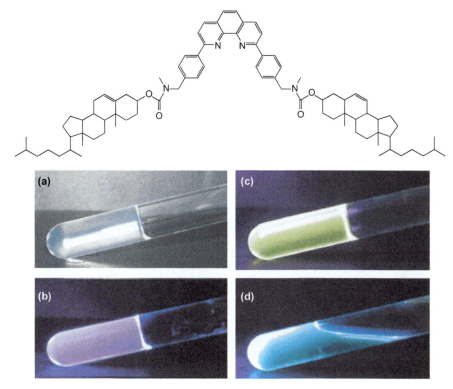

Figure 7.14 Molecular structure of the 1,10-phenanthroline-based gelator.[80] (a) Micrograph of the gel under visible light, and (b), (c), (d) under UV-light irradiation. (c) Shows the fluorescence of the gel after addition of TFA, and (d) after a heating process was applied to the gel shown in (c).

photochromic materials and multiswitchable organogelators.[81] For example, Tian and coworkers have designed a fluorescent photochromic gelator based on a bisthienylethene core bridging two naphthalimide-cholesteryl boundary groups[82] (Figure 7.15). Excellent reversible photochromic properties were characterised for this low molecular weight gelator in its gel and liquid phases, thus facilitating the description of a multistimuli-responsive system. Surprisingly, when a yellow gel of the bisthienylethene derivative was irradiated with UV light at 365 nm, a photoinduced ring closing was prompted, and a new absorption band at round 537 nm appeared, turning the colour of the gel from yellow to red (Figures 7.15a and b). This change in the absorption spectra was attributed to the extended π-electron delocalisation favoured in its cyclic form, and it was switchable by subsequent visible-light irradiation. Importantly, a difference in the luminescence spectra between the two gel states was established, while a strong fluorescence emission was observed in the cyclic form (460 nm), a weak fluorescence emission was characterised for the acyclic phase.

The use of metal ions in molecular gelators can give rise to materials with very interesting optical properties, a subject that has received attention in

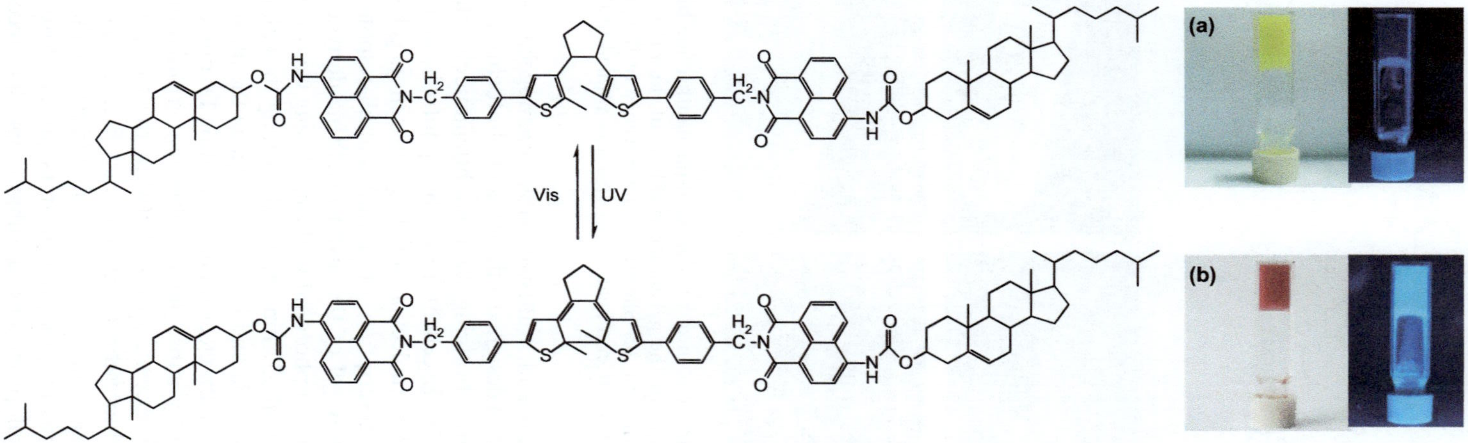

Figure 7.15 Structure of the bisthienylethene derivative and its photochromic behavior, together with images showing: (a) the open gel form under visible light (left) and under UV light (right), and (b) the corresponding pictures for the closed gel phase.

excellent reviews elsewhere.[83–85] A particularly interesting family of gels involve complexes of platinum(II), which show phosphorescence that can be favoured by the presence of short metal to metal contacts in the colloids. The use of hydrogen bonds for the aggregation of platinum(II) complexes led to the formation of long fibres in dodecane, where the interaction between amides also supported the close contact between metal atoms and therefore enhancement of the optical properties.[86] A conceptually similar but simpler complex – bearing a planar tridentate ligand and a monodentate one on the remaining coordination site – was also shown to have interesting luminescent behaviour.[87] A gel of the compound was made by diffusing hexane into a chloroform solution, yielding a material that was highly luminescent when irradiated with UV light, and was shown to have a photoluminescence quantum yield of up to 90%. Films of the material also revealed electroluminescence, making them interesting candidates for use in organic light-emitting diodes. The generality of this approach to phosphorescent materials has recently been shown by the preparation of a family of derivatives.[88]

7.3.3 Photoinduced Energy Transfer

In plants, solar energy in the form of light can be collected and efficiently harvested thanks to the precisely organised assemblies of chlorophyll moieties attached to a peptide scaffold. All the moieties that participate in the charge-separation process in the photosynthetic reaction centre are arranged through noncovalent interactions that lead to large particular arrangements that facilitate a directional and an effective energy transfer required for the photosynthesis process.[89–92] Noteworthy are the peptides that play a crucial role in assisting the packing, separation and the orientation of the chlorophyll moieties in the arrangement. Inspired by nature, scientists have tried to mimic nature's processes with synthetic chromophore-based molecules. Effective excitation energy-transfer processes have been described by noncovalently bound supramolecular architectures formed in solution from π-conjugated molecules.[93–96] However, a major hurdle for future applications of robust synthetic systems is to design molecules with self-capacity to organise under controlled molecular packaging and with precise orientation in both fluid and solid phases. The idea of employing chromophore-based gelators has gained considerable attention, not only for facilitating an anisotropic organisation of chromophore moieties, but also because the gel state is a promising route towards a suitable close-packed multimolecule-layered structure formation. In addition, these hierarchically self-assembled scaffolds can be employed to entrap different donors or acceptors, thereby providing new strategies to design light-harvesting systems. The challenge is to identify suitable moieties that enable a good spectral overlap between the emission of the donor chromophore and the absorption of the acceptor, thereby favouring electron transfer.

One of the earlier examples of energy transfer in gel-based donor scaffolds is that of L-glutamate derivatives containing porphyrin and pyrene groups[97] (Figure 7.16a). It was found that the pyrene-containing molecule could gel

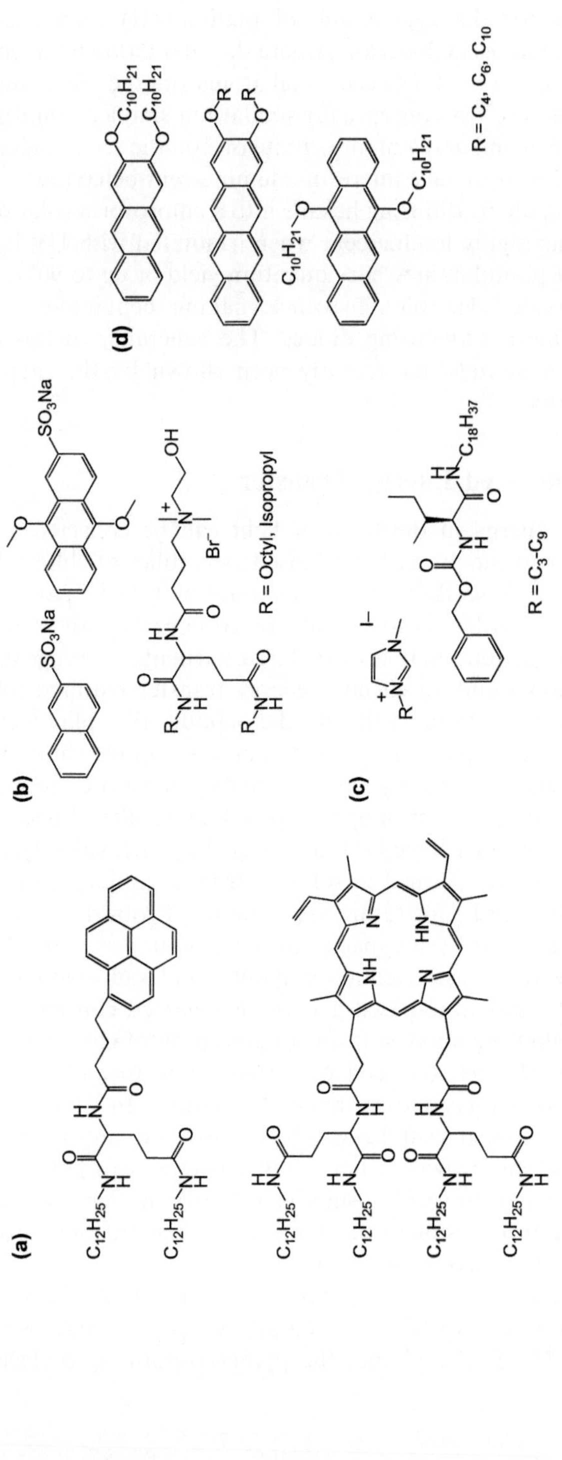

Figure 7.16 From (a) to (d) chemical structures of chromophores and organogelators.

benzene and cyclohexane, while the porphyrin-containing compound failed to form physical gels in either solvent. Spectroscopic studies conducted in mixtures of both chromophore derivatives indicated the formation of highly oriented aggregates in benzene. Cofacial and chiral chromophore aggregates were proposed from these studies. Indeed, a thorough fluorescent spectroscopic analysis indicated that the excimer emission of the pyrene-based gelator overlapped well with the absorption of porphyrin-containing compounds. Therefore, energy transfer studies in mixed gel phases corroborated an efficient energy-transfer process from the pyrene units to the free-base porphyrin moieties. An efficient energy transfer was also reported between anionic naphthalene and anthracene derivatives electrostatically entrapped in cationic glutamate-based hydrogels[98] (Figure 7.16b). It was demonstrated that the addition of an equimolar amount of anionic naphthalene derivative to an aqueous dispersion of glutamate derivative enabled the formation of a two-component hydrogel phase. Upon doping with an anionic anthracene derivative, quenching of the naphthalene fluorescence was demonstrated and a concomitant rise of the anthracene emission was observed. The energy-transfer process was described as the result of the favourable excitation energy migration among the well-organised naphthalene derivatives to the anthracene acceptors. More relevantly, a pronounced quenching was only described in the gel state, in solution only a weak fluorescence emission was detected from the acceptor chromophore, demonstrating the relevance of the gel network in the organisation of the chromophores, and especially in the excitation energy-transfer process.

The gel electrolyte approach was also employed to synthesise a dye-sensitised solar cell based on imidazolium-derived molten salts and iodine[99] (Figure 7.16c). In this case, the gelled molten salt was prepared with a benzyl acetate-based gelator. The authors proved that the gel state fixed a three-dimensional network where efficient charge-transport processes were possible, even upon thermal annealing.

The versatility of an anthracene-based organogel as a medium towards the formation of efficient energy-transfer systems has been shown. for 2,3-*n*-di-decyloxyanthracene as the excitation energy donor and 5,12 or 2,3-*n*-dialkoxytetracene derivatives as the acceptors (Figure 7.16d).[74] The doping of the energy donor scaffold with 2,3-*n*-alkyltetracene derivatives facilitated the occurrence of efficient energy-transfer processes. Importantly, this work showed that the acceptors required an appropriate molecular structure in order to be incorporated and associated to the donor scaffold. Hence, the tetracene derivatives with longer alkyl chains and with a 2,3-substitution turned out to be more effective for the energy-transfer process. Remarkably, it was shown that 0.05% doping of the gelator with 2,3-di-*n*-decyloxytetracene acceptor was sufficient to accomplish 35% of the maximum energy transfer, and with only 1% doping the complete quenching was achieved. In sharp contrast, the 5,12-substituted and the butyl tetracene derivative proved to be less efficient.

The efficient exciton migration observed in fluorescent OPV-based gelators (see Section 7.3.2) has contributed greatly to the idea of constructing OPV-derivatised gel networks that act as donor scaffolds to suitable acceptor moieties. For example, a good fluorescence resonance energy transfer (FRET) from an OPV-based organogelator network to an entrapped acceptor dye such as Rhodamine B has been demonstrated[46,49] (Figure 7.17). Upon excitation at 380 nm, a precise energy transfer from the OPVs conforming the solid gel matrix to the entrapped dye was demonstrated by the quenching of the donor emission and the concomitant increase of the acceptor emission at 625 nm (Figure 7.17c). In addition, the energy-transfer process could be thermally controlled by the reversible sol–gel transition of the OPV organogelators. When the gels were heated above the gel melting temperature, a complete quench in emission was described (Figure 7.18). From the distinct OPV derived organogelators used, the OPV with the carboxylic acid groups proved to be more efficient in the energy-transfer process, probably due to the favoured localisation and the proximity of the acceptor dye to the carboxylated donor gel network (Figure 7.18). Significantly, the FRET process solely occurred from the self-assembled donor network provided by the gel and not directly from the individual donor molecules in solution (Figure 7.18).

More efficient light-harvesting gels were described by the same group upon combining donor and acceptor OPV-based organogelators[70] (Figure 7.19). The tight analogy between the donor and acceptor moieties proves to be crucial for an efficient energy-transfer process. As discussed in the previous section, tuneable red-green-blue (RGB) emissions were demonstrated by tailoring the electron-acceptor groups present in the π-conjugated OPV scaffold (Figure 7.19). We have seen that these OPV derivatives show red-shift in their emission in the gel state. However, apart from these observations, an efficient energy transfer was reported between the gelling donor derivative containing an insulated ester group and the nongelling acceptor derivative with dicyano groups. It was found that the addition of only 2.62 mol% of the dyciano OPV derivative to a gel of the OPV with the insulated ester group in *n*-decane induced a 90% quenching of donor emission and a concomitant acceptor emission (Figure 7.19a). The inset in Figure 7.19a compares the emission of the dicyano derivative on direct excitation and under indirect excitation, *i.e.* upon the energy-transfer process. A threefold increase in the emission intensity was described due to the efficient energy-transfer process between the self-assembled donor gel network and the entrapped dicyano acceptor. Importantly, after conducting the same doping experiments in dichloromethane, where no gel formation occurred, insignificant emissions were observed, hence proving once more the crucial role of the gel state in the energy-transfer process (Figure 7.19b).

Encapsulation of a self-assembled semiconductor molecular wire in a gel-forming OPV matrix has also been investigated.[100] In this approach, the hydroxyl-containing OPV organogelator shown in Figure 7.17 was co-assembled with an acceptor phenylenevinylene and pyrolylenevinylene based

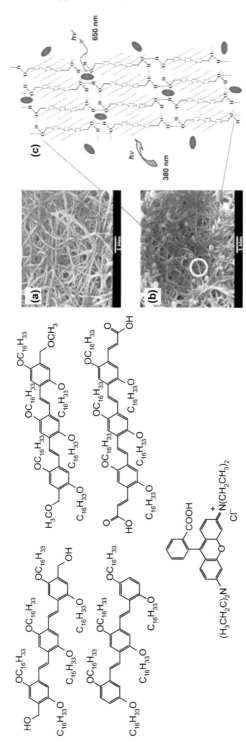

Figure 7.17   Molecular structures of the OPV-based gelators and of Rhodamine B dye.[46] (a) and (b) SEM images of the gel formed by the hydroxyl-containing OPV-derivative in the absence of Rhodamine B, and with the presence of Rhodamine B, respectively. (c) Suggested schematic view of the self-assembly of Rhodamine B dispersed in the OPV gel, the dark ellipsoid forms represent the trapped Rhodamine B molecules. (Reproduced with permission from Wiley-VCH.)

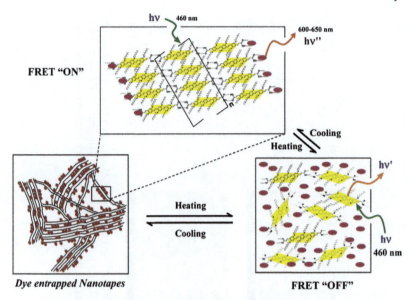

Figure 7.18   Schematic representation of the FRET process in self-assembled donor
networks formed from the carboxylic-containing OPV derivative. The
red ellipsoid forms represent the trapped Rhodamine B molecules.
(Reproduced with permission from American Chemical Society).[49]

copolymer (Figure 7.20). The idea was to use the acceptor copolymer wire as an
efficient excitation energy trap that could work at very low concentrations.
Indeed, fast exciton funnelling and efficient energy transfer were reported for
this system, even with less than 2 mol% of the copolymer added to the donor
gel-forming matrix. The reason for that is the optimal spectral overlap between
the emission of the self-assembled donor and the absorption band of the molecular acceptor wire. It was demonstrated that the possibility of a direct excitation of the acceptor was nearly absent at the excitation wavelength of the
donor. For example, upon illumination with UV light, the bright yellow
emission of the OPV derivative with the hydroxyl groups turned to red with the
encapsulation of 1.53 mol% of the copolymer (Figure 7.20a–c). A direct excitation of the acceptor without the donor scaffold network resulted in a negligible fluorescent emission. Surprisingly, temperature-dependence studies of
the coassembled gel containing 1.53 mol% of copolymer indicated a reversible
and variable temperature-controlled emission, which was associated to the
different aggregates generated during the process (Figure 7.20d). The red
emission colour was attributed to the energy-transfer process from the donor
scaffold to the acceptor, and the blue emission to the single molecules in solution. Most intriguingly, the residual green emission from self-assembled aggregates together with the blue emission of the single molecules lead to a white
emission that was found at the halfway point between the gel and the solution
phases.

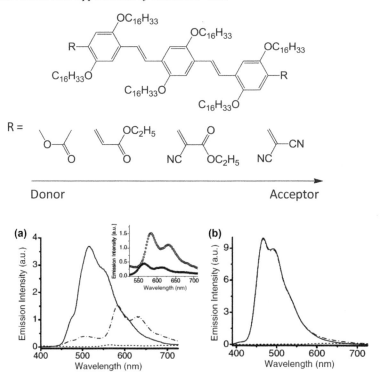

Donor   Acceptor

Figure 7.19   Molecular structures of OPV-based organogelators that combine donor
and acceptor groups. (a) Emission spectra of the gel formed from the
OPV with the insulated ester group alone (solid line), upon doping with
2.62 mol% of the dicyano derivative (dot-dashed line), and the emission
of the dicyano derivative alone (dashed line). In (b) spectra of the same
samples but acquired from solution (in dichloromethane). The inset in (a)
shows the emission of the dicyano moieties in the doped gel on direct
excitation (filled circles) and upon indirect excitation (empty circles).
(Reproduced with permission from Wiley-VCH.)[70]

This innovative organogel-polymer approach has been further exploited for
the generation of novel donor–acceptor systems suitable for charge transport
and charge generation. For example, Thelakkat and coworkers have used a
perylene bisimide gel network as an acceptor conductive system self-assembled
in an amorphous p-donor conductive polymer matrix based on a tetraphenyl
benzidine derivative[101] (Figure 7.21). By using this gel embedded polymer
matrix approach an interpenetrating network structure with multiple electron
donor–electron acceptor connections was successfully accomplished, with the
scope of defining good charge percolation pathways at the nanoscale. In addition, the authors optimised the formation of smooth films of high optical
quality, for the preparation and characterisation of prototype solar cells. Two
different approaches were studied, a two-step process including two subsequent
depositions, or a single-step, where a blend of both compounds was directly
dried on a surface. The results indicated that the solar cells prepared with the

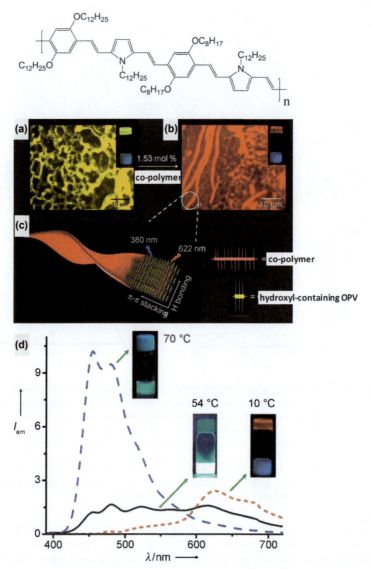

Figure 7.20 Molecular structure of the acceptor phenylenevinylene and pyrolylene-vinylene based copolymer.[100] (a) and (b) fluorescence microscopy images of the gel formed with the hydroxyl OPV derivative in absence of the copolymer and with copolymer, respectively. The insets show photographs of the respective gels. (c) Schematic representation of the encapsulation of the copolymer in a self-assembled donor structure. (d) Temperature dependence of the energy transfer between the donor OPV gel scaffold and the copolymer in a cyclohexane gel. The illumination in all images was at 365 nm.
(Reproduced with permission from Wiley-VCH.)

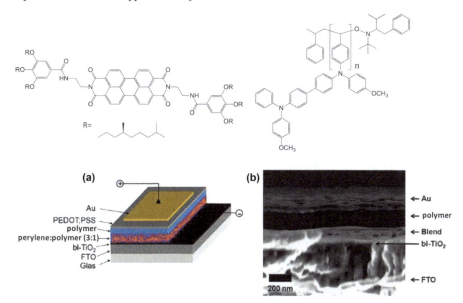

Figure 7.21 Molecular structures of the perylene bisamide organogelator and of the tetraphenyl benzidine-based polymer.[101] (a) Schematic side view of the solar cell configuration where FTO is fluorine doped tin oxide, PEDOT is poly(3,4-ethylenedioxythiophene), and PSS is poly(styrene sulfonate). (b) A cross-sectional SEM image of a real solar cell device. (Reproduced with permission from American Chemical Society.)

single-step approach showed better performances than those prepared with the two-step approach, probably because the former enabled a better interface active area between the donor and acceptor matrix. In addition, during solar cell construction, it was found that the roughness of the perylene–polymer blend was too irregular to evaporate a metal contact on it, so a hole-transporting overlayer made only of the tetraphenyl benzidine-based polymer was required in the solar cell (around 150 nm in thickness) (Figure 7.21a). Furthermore, different thicknesses and compositions were studied, where best performances were described for blends comprising a perylene-polymer ratio 3 : 1 in weight percentage. It is also worthy of note that control experiments conducted with a nongelator but electronically active perylene derivative exhibited very low efficiency values, hence demonstrating the significance of the self-assembled gel network in the final device performance.

The idea of synthesising hybrid gel systems (which will be discussed more fully below) attempting the incorporation of other good acceptor moieties to π-extended organic-based donor gel networks has recently been launched. For example, Zhu and coworkers have recently described a blend where a new extended TTF-based organogelator incorporating L-glutamide-derived lipid units was combined with fullerene (C_{60}).[102] Preliminary results indicated that stable and rapid photocurrents were generated with these porous gel network blends after their exposure to light, and that high energy-conversion efficiencies were accomplished. This work highlights the interesting effects that can be observed by the incorporation of novel functional components into fibrous gel matrices, which, in fact, could be extended to a variety of gel-based electronic materials.[103,104]

7.4 Gel-Derived Materials as Electrical Conductors

Suitable electron–phonon coupling interactions are achieved in fibre–gel networks between noncovalently stacked chromophores or other π-conjugated moieties, as shown in previous sections. Therefore, the design of electrically active π-electron moieties capable of self-assembling into gel-based fibre networks can certainly lead to new functional materials with very appealing features for organic electronics and device fabrication. While this approach has gained a great deal of interest among scientist, insights into the strengths and weaknesses have been evaluated by designing archetypal devices such as solar cells (see preceding sections) or organogelator-based field-effect transistors (OFETs). In this section, we shall demonstrate how the gel-chemistry approach can favour the construction of one-dimensional assemblies of electrically active π-electron molecules through noncovalent interactions, and how this methodology can be employed to generate new conducting molecular assemblies from the bottom-up. There are a number of organogelators with electronic conductivity and charge-carrier mobility, however, a perspective view of exemplary gel-based electronic materials will be given, with special attention to monocomponent assemblies in this section (multiple component assemblies are discussed subsequently).

In one of the earlier examples Bechgaard's and Schaumburg's groups reported a gel-forming compound incorporating a tetrathiafulvalene (TTF) core with two branched polyalcohol segments, which formed gels in mixtures of ethanol-water or dimethylformamide-water[105] (Figure 7.22). AFM and TEM characterisation of xerogel samples revealed a nanowire texture, which upon doping with iodine prompted the formation of TTF cation radicals, as demonstrated by UV-visible absorption spectroscopy. The gelation of organic solvents by bis-thiophene derivatives that self-assemble thanks to urea groups gives materials that exhibit charge mobility.[106] The technique pulse-radiolysis time-resolved microwave conductivity was used on solid samples of the materials because it avoids problems arising from domain boundaries and impurities. The compound with two thiophene units showed much higher mobility than the one with a single heterocycle, because of greater overlap and delocalisation. In fact, the mobility was claimed to be close to that of much larger oligothiophenes, an effect attributed to the excellent overlap between neighbouring molecules in the supramolecular fibres. Shinkai and coworkers also described a TTF-based gelator incorporating a tridodecyl oxy-benzamide group as a gel-forming segment that could effectively gellify various hydrocarbon solvents[107] (Figure 7.22). Surprisingly, the authors proved by TEM analysis that aniso-tropically arranged nanowire domains could be achieved with gels containing a higher concentration of the TTF-based organogelator (Figure 7.22a). To gain

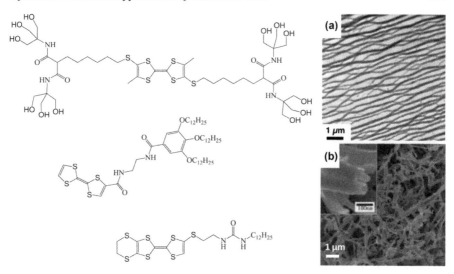

Figure 7.22   Molecular structures of TTF-based organogelators. (a) TEM image of anisotropically aligned fibres formed from physical gels of the benzamide derivatised TTF-based gelator. (b) SEM images of the TTF-TCNQ fibrous-gel network. The inset is a high-magnification SEM image of nanoscaled tube-like structures.

insight into the molecular packing of the TTF units in the self-assembled nanowires, XRD analyses, theoretical calculations, and cyclic voltammetry studies were performed. The cyclic voltammograms obtained indicated higher oxidation states of the fibrous gel network when compared with nonaggregated solutions of TTF-based organolgelator, thereby providing good evidence for a strong π–π interaction between the TTF units in the fibrous gel network. Furthermore, upon iodine doping a characteristic absorption band at around 1750 nm appeared in the NIR-UV-visible absorption spectra, clearly indicating the formation of a mixed-valence state. To achieve a mixed-valence state the formation of conductive pathways from TTF-based donor systems is essential; however, the mixed-valence state is only favoured with a good close-packed molecular arrangement of the π functional moieties. A promising TTF packing in gel networks was also achieved by Zhu and coworkers with the incorporation of one single urea group into the molecular structure of a TTF-based orga-nogelator[108] (Figure 7.22). In this case, the authors demonstrated the formation of a charge transfer (CT) complex in the gel state by the incorporation of an electron-acceptor molecule such as 7,7,8,8-tetracyanoquinodimethane (TCNQ). Surprisingly, and for the first time, entangled nanoscaled tube-like structures were generated from the TTF-TCNQ fibrous gel network (Figure 7.22b).

Conductivity values from physical measurements performed on TTF-based organogelators were first evaluated by Becher and coworkers on spin-coated films of the TTF derivative incorporating branched polyalcohol groups[109] (Figure 7.22). The authors demonstrated the increase in the conductivity

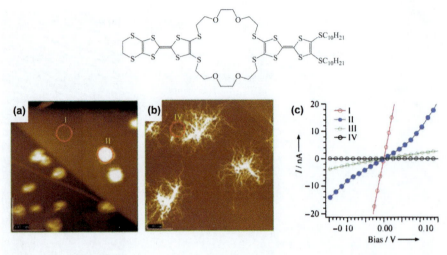

Figure 7.23   Chemical structure of a bis-TTF derivative incorporating a cage-annulated crown ether macrocycle.[111] (a) and (b) AFM images of nanodots and bundles of nanowires of the bis-TTF derivative on graphite, respectively. (c) graphs showing *I–V* sweeps of: (I) graphite, (II) single nanodot in air, (III) single nanodot in vacuum, and (IV) a bundle of nanowires.
(Reproduced with permission from Wiley-VCH.)

value upon iodine doping of the xerogel. The initial conductivity value changed from 10^{-6} S cm^{-1} to 10^{-4} S cm^{-1} after iodine doping. In a subsequent study, the same group carried out conducting AFM (C-AFM) measurements[110] on doped xerogel films of a bis-TTF derivative incorporating a cage-annulated crown ether macrocycle[111] (Figure 7.23). This study showed the formation of electrically active nanostructures including size-controllable nanodots and nanowires. By precisely locating the conducting AFM tip on top of oxidised nanodots or bundles of nanowires, conductance values could be determined. While the conductance of an oxidised nanodot increased from 20 nS to 86 nS under ambient conditions, because of the effects of oxygen and/or water, the conductance of an oxidised bundle of nanowires was 0.03 nS, four orders of magnitude smaller. It was assumed that the high conductance of single doped nanodots reflected the high concentration of the conduction carriers in these assemblies.

In contrast to other TTF-based organogelators, we demonstrated that interesting electronic properties can be displayed by doped xerogels of an octadecyl-amide functionalised TTF derivative upon annealing[112] (Figure 7.24). The xerogel, when doped to form the mixed-valence state, exhibited a bulk conductivity of $3–5 \times 10^{-3}$ Ω^{-1} cm^{-1} at room temperature. Upon heating, a drastic drop in the resistance value was observed at 77 °C, attributed to an irreversible phase transition from one initial phase (referred to as the α-phase) of the doped xerogel to a thermally converted one (the β-phase).

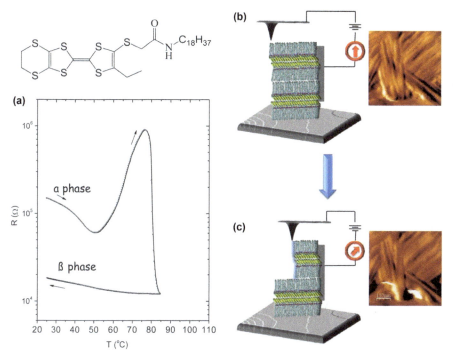

Figure 7.24 Molecular structure of the octadecyl-amide functionalised TTF derivative.[112] (a) Graph showing the resistance–temperature dependence and the conversion from α-phase to β-phase. (b) and (c) schematic illustrations of the proposed peeling process for a doped xerogel film. The insets are AFM images of the control removal of material in a doped xerogel sample.[116]
(Reproduced with permission from Wiley-VCH.)

The room-temperature resistivity of the β-phase was less than that of the α-phase by a factor of about 10 (Figure 7.24a). The phase transition from α-phase to β-phase was proved by four-probe measurements and EPR characterisation. Furthermore, current-sensing AFM studies of the doped β-phase indicated an apparently Ohmic behaviour of the nanowires, and the EPR linewidth indicated a metallic state in the material. We also described four different conducting phases in total from the same doped xerogel sample after different heat and/or doping treatments,[113] and demonstrated that the solvent used to form these physical gels can play an important role in the conductivity of related doped materials.[114,115] In a subsequent study, an unprecedented controlled peeling of doped xerogel films of the octadecyl-amide TTF-based gelator was also accomplished with the conducting probe of an AFM[116] (Figures 7.22b and c). Remarkably, this study showed the importance of the electric field in the stability of organic supramolecular conductive systems at the nanoscale, an aspect that should be further considered in order to achieve optimal organic-based functional devices from organogelators.

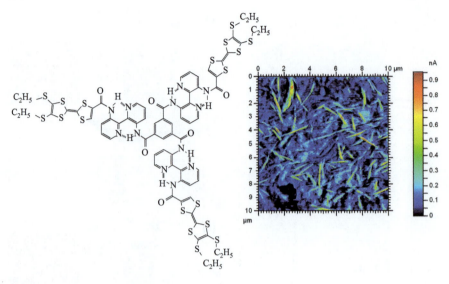

Figure 7.25   Molecular structure of the TTF-based organogelators with C_3 symmetry, and C-AFM image of a doped xerogel film.[117] Note the higher conductivity values attributed to the thicker nanofibre assemblies.

Recently, more complex TTF-based organogelators with C_3 symmetry have been described[117] (Figure 7.25). In this case, the gelator properties in chlorinated solvents were reported by an unprecedented redox-active C_3 symmetric tris-TTF organogelator incorporating a central 1,3,5-benzenetricarbonyl unit. Upon iodine doping of a xerogel sample, the formation of a mixed-valence state was demonstrated by IR-NIR spectroscopic measurements. Current-sensing AFM measurements indicated the formation of a complex material containing thick and thin conductive nanofibre structures, the former being more conductive than the thinner ones, probably as a consequence of the better ordering and/or more effective interfibre contacts. Importantly, this study introduced the formation three collinear conductive pathways into a single organogelator structure. Further, the incorporation of short ethyl chains appended to the TTF cores opened up the possibility of electronic contact between the nanoscaled fibres in the network.

A mixed-valence perchlorate (ClO_4^-)-doped TTF organogelator has been prepared from an octadecylthio TTF-diamide derivative[118] (Figure 7.26). Upon doping with perchlorate, the rod-like appearance of the neutral xerogel changed to a multiply coiled structure with a spiral configuration, *i.e.* double and triple helices were reported by the incorporation of ClO_4^- ions into the assembly. Furthermore, the conductivity properties as a function of temperature were measured for the ClO_4^--doped TTF organogelator in micrometer-gap electrodes. Nonlinear *I–V* characteristics were determined over the temperature range measured (70–300 K), and upon cooling a decrease in the conductance was verified. In this case, a room-temperature resistance of 4.59×10^{-5} S cm^{-1}

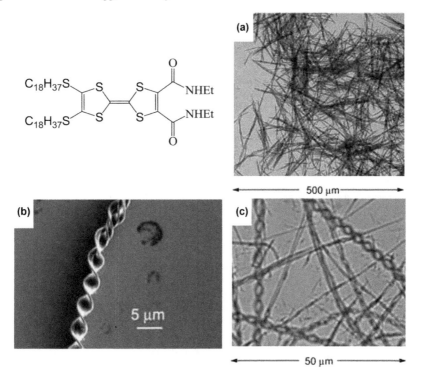

Figure 7.26 Molecular structure of the octadecylthio TTF-diamide organogelator.[118]
(a) Optical microscope image of the rod-like appearance of the neutral xerogel. (b) and (c) SEM and optical images of the doubly coiled structure of the ClO_4^--doped xerogel, respectively.
(Reproduced with permission from the RSC.)

was confirmed for a ClO_4^--doped nanofibre, with a thermal activation energy value of 0.108 eV.

Recently, the gelation ability and the modulation of the conductivity values of four different TTF-based xerogels was reported[119] (Figure 7.27). The authors designed organogelators with two different gel-forming segments, one containing a tridodecyl oxy-phenyl substituent or a cholesteryl group. Moreover, one or two TTF moieties were incorporated in the molecular structure of the gelator (Figure 7.27). From the four distinct TTF-based organogelators, higher conductivity values were determined for the two TTF-derivatives incorporating two TTF moieties. In these samples, the conductivity values were one order of magnitude higher than that of the doped xerogels incorporating only one TTF moiety (from $2-4 \times 10^{-5}$ S cm^{-1} to $2-4 \times 10^{-4}$ S cm^{-1}). Furthermore, the better gelation performance was described by the TTF organogelators with a cholesteryl side group, independently of the number of TTF moieties, presumably on account of the higher rigidity of this group as compared to the tridodecyl oxy-phenyl substituent.

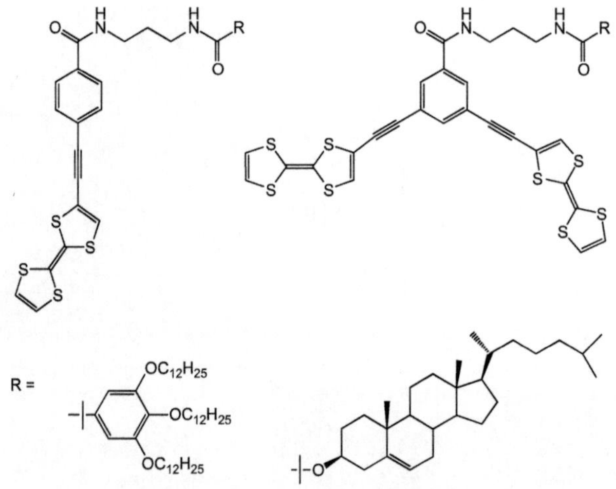

Figure 7.27   Chemical structures of some TTF-based gelators.[119]

Low molecular weight gelators designed with extended functional π-conjugated oligomers have also attracted a great deal of attention, for example, organogelators incorporating π-conjugated oligothiophenes moieties. Oligothiophenes are attractive semiconductor moieties because of their excellent ability to form conductive π-stacks systems and their relatively simple chemical modification.[120–123] Recently, the synthesis of a quaterthiophene gelator functionalised with a barbituric acid segment and a tridodecyl oxy-phenyl end group was reported[124] (Figure 7.28). The formation of conductive self-assembled nanorods and nanotape-like aggregates was demonstrated through complementary hydrogen bonds with a flexible bis(melamine) receptor (Figure 7.28). By controlling the temperature of the coaggregated solution, a selective self-assembly towards nanorod or nanotape supramolecular objects was reported. Flash-photolysis time-resolved microwave conductivity (FP-TRMC) measurements performed with a laser pulse $\lambda = 355$ nm, showed the formation of long-lived charge carriers with maximum transient conductivities of 1.0×10^{-4} cm^2 V^{-1} s^{-1} for the quaterthiophene assembly and of 0.67×10^{-4} cm^2 V^{-1} s^{-1} for the coassembled structure. The hole mobility of the coassembled nanotape-like structure was calculated to be 0.57 cm^2 V^{-1} s^{-1}, whereas a value of 1.3 cm^2 V^{-1} s^{-1} was estimated for the nanorods. The hole mobility of the nanotape-like structures is one order of magnitude higher than the mobility observed for lamellarly organised poly(hexylthiophene)s. This high mobility value was attributed to the interchain transportation of charge carriers within a lamella. Similar effects have also been seen in a quaterthiophene gelator.[125]

A hairpin-shaped molecule containing two sexithiophene moieties employed as charge-conducting segments has been prepared and studied[126] (Figure 7.29). To determine the charge mobility of this thiophene-based organogelator, OFETs from thin films were fabricated using an assembling solvent (toluene) or nonassembling solvents (chlorobenzene or o-dichlorobenzene). All OFET

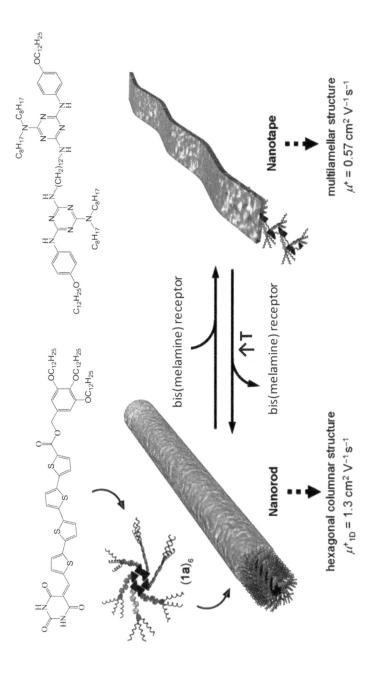

Figure 7.28 Molecular structures of the quaterthiophene gelator and the bis(melamine) receptor, together with a schematic representation of the interconversion between nanorod and nanotape-like aggregates. (Reproduced with permission from Wiley-VCH.)[124]

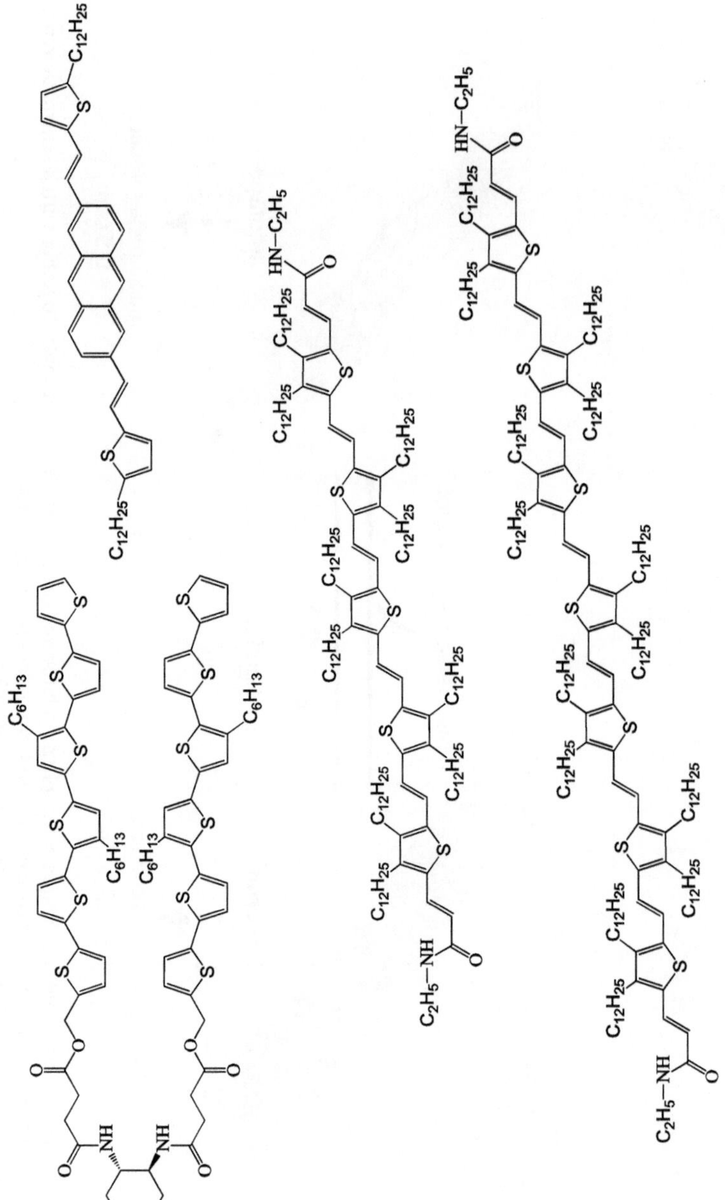

Figure 7.29 Chemical structures of different organogelators.

devices measured indicated typical p-type semiconductor behaviour. In addition, self-assembled fibres deposited from a toluene solution exhibit a hole mobility of 3.46×10^{-6} cm^2 V^{-1} s^{-1}, while films obtained from nonassembling solvents had values one order of magnitude lower. These results are very interesting because they prove the strong influence of the gel chemistry through the extended fibre length with improved π-orbital overlap continuity to the charge-carrier mobility.

Recent studies have also highlighted the importance of other organic linear π-systems towards the creation of conductive fibrous gel networks.[127] For example, two oligo(thienylenevinylene) (OTV) based gelators with five and seven thienylenevinylene moieties have been reported recently[128] (Figure 7.29). Both compounds form gels in a variety of nonpolar solvents, nonetheless the oligomer with increased conjugation proved to be a supergelator, *i.e.* only tiny amounts were needed to guarantee physical gel formation. In addition, current-sensing AFM analysis of the respective xerogels demonstrated the formation of conductive films, which upon iodine doping drastically increased their conductivity values. The extended conjugation length together with the strong gelation ability of the extended OTV derivative resulted in an efficient charge carrier generation upon doping, thereby leading to the highest electrical conductivity reported for an organogelator with an electrical conductivity value of 4.8 S cm^{-1}.

On the other hand, a thienylvinylene derivative incorporating an anthracene moiety was also described,[129] (Figure 7.29) who reported a simple method to prepare nano- or microfibers of the anthracene-based molecule through a gelation process. Furthermore, and for the first time, they fabricated free-standing fibre FETs or film-based FET devices with the anthracene-based gelator. When film-type fibrous gel networks were measured, an estimated hole mobility in the range of 0.02–0.05 cm^2 V^{-1} s^{-1} was determined. In contrast, single nanofibres showed higher hole mobilities (0.48 cm^2 V^{-1} s^{-1}) with a current on-off ratio of 10^5. This higher mobility was attributed to the highly ordered structure of the single nanofibre structure, as showed from XRD analysis. Surprisingly, an extremely high mobility (8.7 cm^2 V^{-1} s^{-1}) was determined from a single nanofibre transistor with the width as narrow as 70 nm.

Graphite-based gelators have also been described from polyaromatic hydrocarbons. For example, gels and nanotubular structures have been described from amphiphilic hexabenzocoronene (HBC) derivatives.[130] In an early study, Aida and coworkers described an amphiphilic HBC derivative incorporating two dodecyl chains on one side and two triethylene glycol segments on the other of the aromatic system, which formed physical gels in tetrahydrofuran (THF)[131] (Figure 7.30). Filtration of gel-diluted samples enabled the characterisation of the redox-active tubular fibre structures by spectroscopic techniques, TEM, SEM, and AFM measurements. Moreover, upon oxidation with nitrosonium tetrafluoroborate (NOBF$_4$), an electrical resistivity of 2.5 MΩ was found for this amphiphilic methyl-derivatised HBC-based gelator (Me-HBC) material. In a subsequent study, the incorporation of trinitrofluorenone (TNF), an electron-accepting group, into the donor HBC

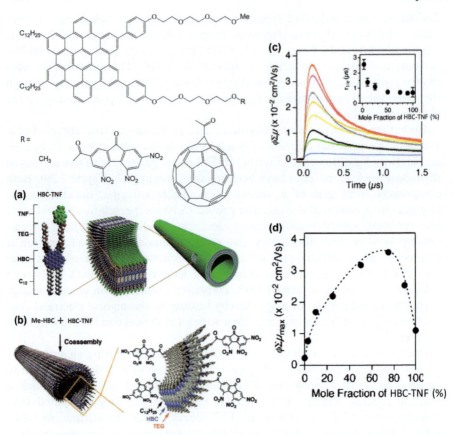

Figure 7.30 Molecular structures of HBC-based gelators.[131–134] (a) Schematic illustration of the self-assembly of HBC-TNF towards the formation of a nanotube structure with a coaxial donor–acceptor configuration. (b) Schematic illustration of the Me-HBC and HBC-TNF coassembled structure. (c) FP-TRMC profiles of cast films of the nanotubes with varying mole fractions of HBC-TNF: 0 (blue), 3 (green), 10 (yellow), 25 (orange), 50 (purple), 75 (red), 90 (gray), and 100% (black), upon photoirradiation at 355 nm. Inset shows a plot of $\tau_{1/e}$ values *versus* mole fractions of HBC-TNF. (d) Plot of $\phi\sum\mu_{max}$ values *versus* mole fractions of HBC-TNF.
(Reproduced with permission from the American Chemical Society, ref. 133)

amphiphile structure enabled the formation of a coaxial donor–acceptor nanotubular structure (HBC-TNF)[132] (Figure 7.30a). HBC-TNF nanotubes showed a quick photoconductive response as a result of the spatial separation of charge carriers along the structure. A photocurrent and dark current of 4.2 nA and 0.07 pA were determined for HBC-TNF nanotubes, respectively. A significant enhancement in photoconductivity was achieved by coassembling the Me-HBC derivative into a HBC-TNF nanotube[133] (Figure 7.30b). FP-TRMC studies indicated that the incorporation of an appropriate amount of

Me-HBC derivative into HBC-TNF nanotubes greatly enhanced the photo-conductivity properties (Figures 7.30c and d). Recently, an ambipolar charge carrier mobility was described with a HBC derivative incorporating a fullerene electron-acceptor group[134] (Figure 7.30). By means of a FET device setup, the authors determined an electron and hole mobility values of 1.1×10^{-5} and 9.7×10^{-7} cm^2 V^{-1} s^{-1}, respectively. In addition, tailoring of the p-n heterojunction was possible by coassembling Me-HBC with the fullerene HBC derivative.

The use of liquid crystals (LCs) to align conducting fibres is an interesting approach for the preparation of materials with anisotropic charge transport. In this regard, the influence of the LC component towards the alignment of an electroactive TTF-based organogelator has been studied.[135] While a pentyl-cyanobiphenyl LC favoured the formation of a randomly dispersed fibrous gel network ($T_{iso-lc} < T_{sol-gel}$), a phenyl benzoate LC derivative enabled the formation of aligned fibrous aggregates of the TTF-based organogelator ($T_{iso-lc} > T_{sol-gel}$). Moreover, molecular conductive wires could be generated from these LC physical gels after iodine doping of the xerogels. Therefore, the incorporation of electroactive moieties in LC physical gels may be a promising approach towards new advances in molecular electronic circuitry.

7.5 Gels as Organising Material for Nanocomposites and Nanohybrids with Optical, Electrical and Magnetic Activity

In many ways, the gelation of solvents with functional molecules is an ideal way to prepare nanocomposites. Compared with other processing methods, the formation of a gel gives a very high probability of the formation of fibres whose widths lie in the nanometre-size range. The dispersion of any other colloid in a solvent during the formation of the supramolecular fibres will often lead to composite materials that are not accessible through other routes because of the phase separation of the components. This statement holds true for all types of material, the gel itself, the aerogel or the xerogels. Provided that there is a strong interaction between the suspended colloid and the molecule forming the gel the two will be bound with the smaller component distributed along the length of the other, as shown schematically on the left-hand side of Figure 7.31 for the case of nanoparticles on supramolecular gel fibres. Conversely, if for the same case no strong interaction is present then the fibres will form all the same and the nanoparticulate colloid will adhere in the spaces.

Composite materials comprising inorganic nanoparticles and π-conjugated molecules have their optoelectronic properties varied greatly by the supramolecular organisation of the components, which influences the way devices based on these materials behave.[136] Gels are highly appropriate for applications involving transportation of charge because domains with high aspect ratios, including fibres and tapes, favour the movement and separation of charges.[106] Although Chapter 8 of this book deals with the templating effects that gels can

Nanoparticle adherence to fibre –
strong fibre-particle interaction

Nanoparticle fibre segregation –
weak fibre-particle interaction

 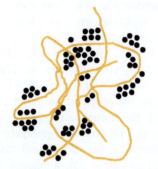

Figure 7.31   Representation of contrasting outcomes of molecule-driven gelation in the presence of nanoparticles. On the left, the particulate material (represented as a black dot) interacts strongly with the supramolecular fibres (represented as an orange line) and on the right the two components interact weakly so that the nanoparticles collect in the voids in a situation reminiscent of nanometre-scale phase segregation.

have in general, here we will focus on the changes in properties when composites are formed.

The preparation of a composite of gelator and gold nanoparticles can be achieved by a chemical exchange reaction of thiols on the surface of the metal with thiols present in the self-assembling molecule.[137] A derivative of *trans*-1,2-diamidocyclohexane bearing alkyl chains terminated with thiol groups acts as a gelator, and when heated with octylthiol passivated gold nanoparticles undergoes an exchange reaction, to give, ultimately, a composite material in which the metal colloid forms lines with a morphology similar to the pure gelator. The role of the thiol groups was confirmed by performing parallel experiments in which the gelator did not possess these moieties, leading to a material with nanophase separation, and therefore poor particle–nanofibre interaction. It was foreseen that this kind of network would open new possibilities for electronic and optical materials. Apart from these properties, the viscoelastic behaviour of gels is affected significantly by nanoparticles as a result of changes in fibre morphology through interaction of the coated particles with the supramolecular fibres.[138,139] Indeed, suitable complementary hydrogen-bonding groups can lead to very specific attachment of nanoparticles to gel fibres.[140]

The combination of an OPV (oligo(p-phenylenevinylene)) gelator and one weight per cent of gold nanoparticles coated with a derivative of the same OPV bearing a disulfide moiety leads to a material that displays very different optical properties to the pure OPV.[141] Proportions of 20:1 were possible, but the critical gel concentration (CGC) for the 100:1 and the 750:1 mixtures was lower than that of the pure OPV, indicating a synergic effect of the particles on the network of fibres. The presence of an OPV unit in the capping group of the nanoparticles is essential, since simple aliphatically capped particles do not show the effect. TEM images of the composite showed the nanoparticles running along the sides of tapes formed by the organogelators at a distance consistent with the length of the capping unit being intercalated in the organogels part. The fluorescence of the OPV gelator is quenched significantly by the presence of even 1% of the nanoparticles. The decay of luminescence is also faster in the composite, presumably because the diffusion of the electronic excitations through the fibres to the particles (that occurs on the nanosecond timescale) is very efficient. The nature of the particles is essential to providing this efficient pathway: The noble-metal colloids coated with aliphatic groups show weak quenching of the OPV luminescence. The self-assembly of the composite was facilitated by the functionalisation of the nanoparticles with ligands capable of interaction with the gelator through noncovalent interactions similar to those that hold the fibres of the main component together (aromatic stacking and hydrogen bonds between the hydroxyl groups). The metal particles are very close to the stacked chromophores, which makes electronic communication efficient.

Appropriately substituted nanoparticles can help structure gels and increase electrical conductivity.[142] The combination of a gelator with a tetrathiafulvalene (TTF) unit that allows electrical conductivity (when doped) and gold nanoparticles passivated with tetrathiafulvalene units bearing amide groups compatible with the same kind of unit in the gelator leads to transparent gels upon cooling hot solutions in apolar organic solvents when the metal colloid is present in 1% weight. Even at this low proportion the morphology of the gel fibres is affected dramatically: The presence of the amide-coated nanoparticles gives rise to long well-defined fibres, as demonstrated by atomic force microscope imaging (Figure 7.32). On the contrary, when the nanoparticles were coated with a TTF unit bearing only alkyl chains the corresponding nanocomposites comprise very short fibres and areas where nanoparticles agglomerate. Importantly, in the transmission electron microscope images of the composites the gold colloid is only observed attached to the gel fibres in the case of the amide, while with alkyl chains a clear phase separation takes place. After the materials were partially oxidised using iodine vapours to give the mixed-valence material (with neutral and cation radical TTFs on average) it became clear that the nanoparticles are able to induce the more conducting phase of fibres of the TTF derivative, as proven by electron paramagnetic resonance spectroscopy. Furthermore, current-sensing atomic force microscopy showed that the material with amide-containing nanoparticles gave an apparently much more conducting material, probably because of the increased interfibre connections that must be promoted by the additive. The nanocomposites containing nanoparticles with no hydrogen bonding units were very inhomogeneous materials with areas of high conductivity, possibly caused by agglomerated nanoparticles. The bulk conductivity in these materials is modified as a result of the changes in fibre structure by the nanoparticles, and the examples show how a small amount of additive can help structure the gel and give rise to optimised electrical properties.

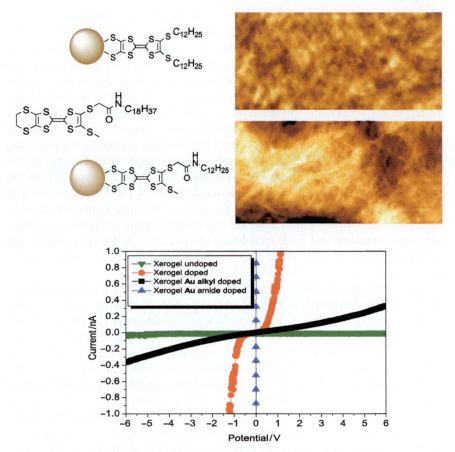

Figure 7.32 The tetrathiafulvalene-based gelator that forms nanocomposites with gold nanoparticles coated with similar structures containing only alkyl chains or an amide function.[142] The AFM images shown correspond to the composite's xerogels containing 1 weight per cent of the nanoparticles. Below, the current–voltage sweeps from conducting AFM measurements on the undoped xerogel, the iodine-doped xerogel, and the same for the nanocomposite materials.

The use of precursors to electrically conducting material and magnetic nanoparticles is an easy and potentially interesting way to generate nanocomposites comprised of components with these often synergic properties. The preparation of this kind of material without recourse to the gel route is quite challenging, particularly for the formation of crystals. The combination of the same gelator used in the previous example of nanocomposite in hexane with an oleate-coated iron oxide nanoparticle provided transparent gels (Figure 7.33) with up to 50% by weight of the inorganic colloid.[143] The high degree of loading was possible because of the great solubility of the inorganic colloid in hydrocarbon solvents.

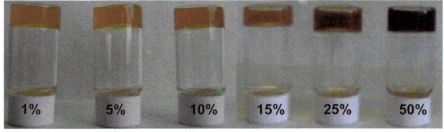

Figure 7.33 The tetrathiafulvalene-based gelator and representation of the oleate-coated iron oxide nanoparticles used for the formation of nanocomposites and a photograph of the gels formed in hexane with different weight per cents of the inorganic colloid.[143]

The transmission electron microscopy (TEM) images of the xerogels of these samples on a holey carbon grid (Figure 7.34) revealed some fascinating effects. At low concentrations of the inorganic colloid, TEM images showed that the nanoparticles (which appear as dark dots) were mainly located at the edges of the nanofibres of the gelator, although a few nanoparticles can be spotted that are apparently not in contact with the purely organic material. The structure of the nanocomposites was clearly different when the mixture contains more than 10% of the inorganic colloid. Two important observations are worthy of note for the mixtures with high proportions of iron oxide: First, because of the high content of absorbing inorganic colloid in these samples the fibres cannot be imaged, and the presence of clear tracks in the images infers the presence of the organic fibres. Secondly, in the clear tracks no nanoparticles were seen, suggesting that the inorganic and organic components phase separate.

Closer inspection of the TEM images of the xerogels containing 15 and 50% by weight of nanoparticles showed the apparent noncovalent "misunderstanding" between the two components. When the xerogel contained 15% of particles obvious separation of the two components was seen. At even higher concentrations the inorganic spheres started to organise and self-assemble, suggesting a more long-range organisation in between the organic fibres of the final hybrid gel. The iron oxide nanoparticles self-assembled in a very different way to a solution of pure colloid drop cast over a TEM holey carbon grid. The hybrid xerogel gave rise small and large domains where mono- and bilayers of nanoparticles were seen, as a result of the separation and confinement conditions in the material. The situation observed in this two-component composite contrasted with the cases (like the previous one presented here) where

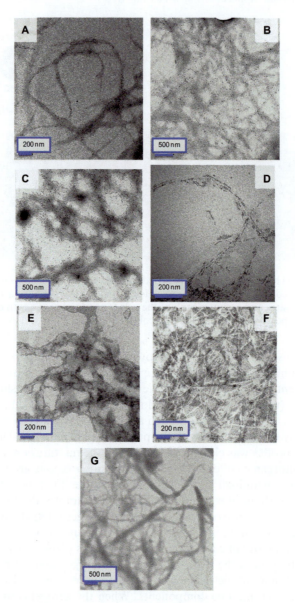

Figure 7.34   TEM images of the nanocomposite xerogels formed from a TTF-
                  containing organogelator and different proportions of oleate-coated
                  iron oxide nanoparticles. (A) Xerogel containing 1% in weight of
                  nanoparticles, (B) 5%, (C) 10%, (D) 15%, (E) 25% (F) 50%, and (G)
                  the xerogel of the TTF organogelator from hexane as reference. All the
                  TEM images were acquired by casting hot solutions of the gel-forming
                  solution onto holey carbon grids.
                  (Reproduced with permission of the RSC.)[143]

clear noncovalent interactions take place between the components. For this reason, it is revealing to see the effects that this kind of phase separation had on the conducting properties of the doped materials.

The doped gels containing 1, 5, 15 and 25% by weight of iron oxide nano-particles were subject to current sensing (CS) atomic force microscopy (AFM) on highly oriented pyrolytic graphite (HOPG). Relatively uniform areas with clear fibre-like morphology are imaged in the current maps from these AFM experiments across the sample when the percentage of inorganic colloid is under 10% by weight, where poor phase segregation exists (Figure 7.35). When the percentage of inorganic component was 10% or more, dark regions became apparent in the CS-AFM image, corresponding to poorly conducting areas from regions containing the iron oxide. All the current images in Figure 7.35 were measured using the same bias voltage (1 V), and yet bright areas of highly conducting material are observed even in the latter composites, where the size of the domains was limited. Compared with either the pure organogelator or

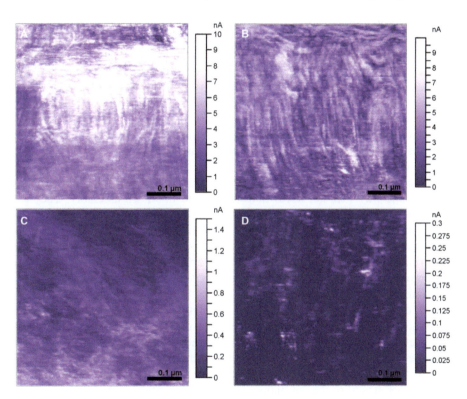

Figure 7.35 Current-sensing AFM image of a doped xerogel of a TTF-containing organogelator and different proportions of oleate-coated iron oxide nanoparticles; 1% (A), 5% (B), 10% (C) and 15% (D) by weight of nanoparticles. The images - recorded on HOPG when applying 1 V bias voltage – show how the conductivity is reduced upon adding nano-particles, presumably because interfibre connections are limited. (Reproduced with permission of the RSC.)[143]

other xerogel samples of this type these nanocomposites containing 1 or 5 weight per cent of nanoparticles show relatively large areas of parallel fibres. Higher concentrations of the nanoparticles do affect the alignment of the fibres of the organogelator. It is an interesting empirical observation that improved alignment of the fibres in the xerogel samples are achieved with small amounts of nanoparticles, although the reason for this alignment is unclear at present. The current *versus* potential curves measured during the CS-AFM indicated that increasing amount of nanoparticles results in a decrease of the bulk conductivity (even at 1% loading) presumably because interfibre connections are disturbed by the presence of the inorganic colloid. Nonetheless, these composites show a unique way of combining magnetic and conducting components in a single material.

The incorporation of fluorescent quantum dots into organogels can also be readily achieved.[144] Toluene-soluble core–shell CdSe/ZnS quantum dots surface capped with hexadecylamine were combined with a pseudopeptidic macrocycle gelator in that solvent giving transparent gels with the same thermal stability and optical transparency as the parent organogel. The material is fluorescent thanks to the embedded semiconductor nanocrystals, and this fluorescence is virtually identical to that observed in solution, no matter how concentrated the gelator. TEM measurements indicated that the quantum dots reside at the edge of the gel fibres. The sensing capacity of the gels for nitric oxide was probed by monitoring the fluorescence change after injection of the molecule, giving a sensitivity from 0.05 to 0.5 (vol%) although the response time was rather slow compared with the solution. However, the composite is an interesting material for this purpose, because the gel is not affected upon complexation of the sensed material.

Apart from nanoparticles, nanotubes can also be used to interact in a synergistic way with gel fibres that are formed upon self-assembly. They can even help reinforce gels, aiding the formation of gel networks, as is the case of carbon nanotubes with OPV derivatives.[145] The formation of hybrids or composites can also affect electrical properties. A TTF derivative (Figure 7.36) bearing solubilising groups on one "side" and on the other amide groups to promote aggregation to give a gel and pyrene groups to interact with carbon nanoforms by π–π interactions leads to the formation of nanocomposite gels with carbon nanotubes.[146] When the gelator was mixed with increasing amounts of single-walled carbon nanotubes (from 0.01% to 1% by weight with respect to the gelator) in 1,2-dichlorobenzene the gels were reinforced (even after a few days no solvent was released), although heterogeneous gels appeared when the nanotube ratio was over 0.1% by weight. It was shown that the nanotubes structure the gel thanks to the incorporation of pyrene units in the organogelator, as shown by TEM images of the materials. The pure organic material formed a complex network of fibres, while the composite comprised much tighter bundles of fibres. This change also affects the conducting character of the doped material; the nanotubes increase slightly the conductivity of the material compared with the organic system. This effect was seen in the CS-AFM measurements of the sample, which showed the quite uniform

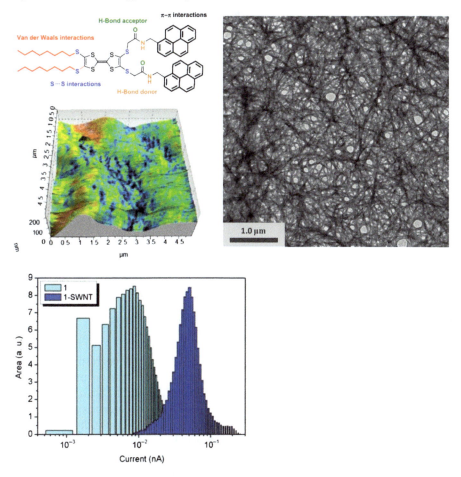

Figure 7.36 The chemical structure of a gelator that forms a nanocomposite gel with carbon nanotubes, as shown in the TEM image on the right.[146] The contour plot shows the current-sensing atomic force microscope image overlaid on the topography of a doped sample where the highly conducting regions are shown to appear all over the surface, indicating high interfibre connectivity. The graph shows a comparison of the conductivity of the doped xerogel with and without carbon nanotubes.

nature of the composite. The measurements allowed a comparison of the current flow through the materials, and it can be seen in the graph in Figure 7.36 that the conductivity is clearly improved in the carbon nanotube-containing material. This enhancement was caused by an extremely small quantity (0.1% w/w) of the nanotubes.

While most examples of composites are demonstrated for organogels, great opportunity lies in the formation of molecular hydrogels for applications in optics and sensing. The combination of an amine-functionalised clay and carboxylate dye molecules in water leads to hydrogels with potential for light harvesting.[147] The amine groups appended to the sheets of the magnesium

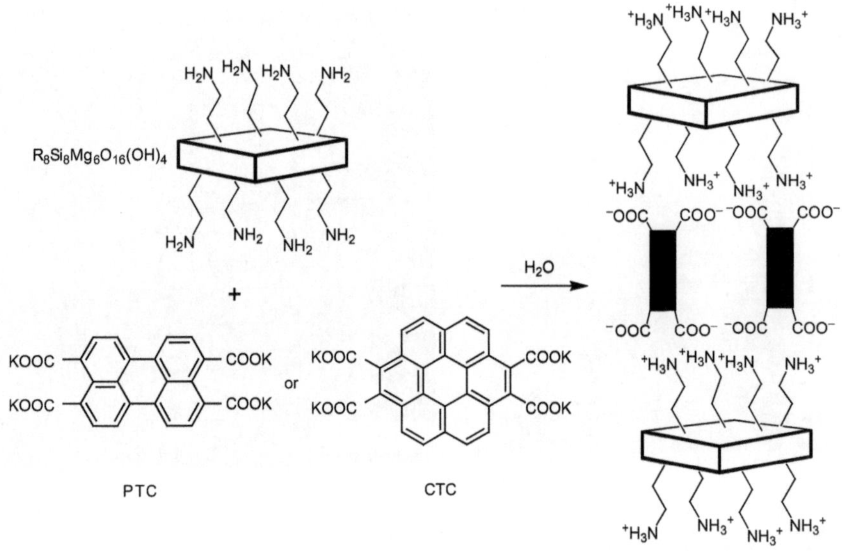

Figure 7.37   Structures of perylene and coronene derivatives that form hydrogels when combined with a lamellar amino-derived clay.[147]

silicate of general formula $R_8Si_8Mg_6O_{16}(OH)_4$ (where R corresponds to the alkylamine groups) are protonated, leading to their exfoliation in water, and allowing the formation of complexes (Figure 7.37) with both a perylene tetracarboxylate derivative (PTC) and a coronene tetracarboxylate (CTC) derivative through electrostatic interactions primarily. The individual complexes form hydrogels at approximately ten weight per cent upon sonication of the components in water. The fluorescence of the molecules is quenched under these conditions because of stacking in the complexes. However, when an appropriate combination of PTC and CTC is used for the formation of the gels Förster-type resonance energy transfer takes place from the coronene (donor) to the perylene (acceptor) derivative, as witnessed by a green fluorescence. The dynamic characteristic of these noncovalent hybrid systems is very promising for the preparation of responsive materials whose state can be probed through their luminescent properties.

7.6   Orthogonal Assembly for Functional Materials in Gels

The term "orthogonal assembly" is meant to infer the simultaneous formation of two or more networks of molecules that assemble selectively with their own kind,[148,149] perhaps the most obvious and relevant example being the preparation of donor and acceptor stacks.[150] It is widely recognised that nanophase separation of donors and acceptors in photovoltaic materials, but also related photoconducting materials, can be optimised through appropriate control of

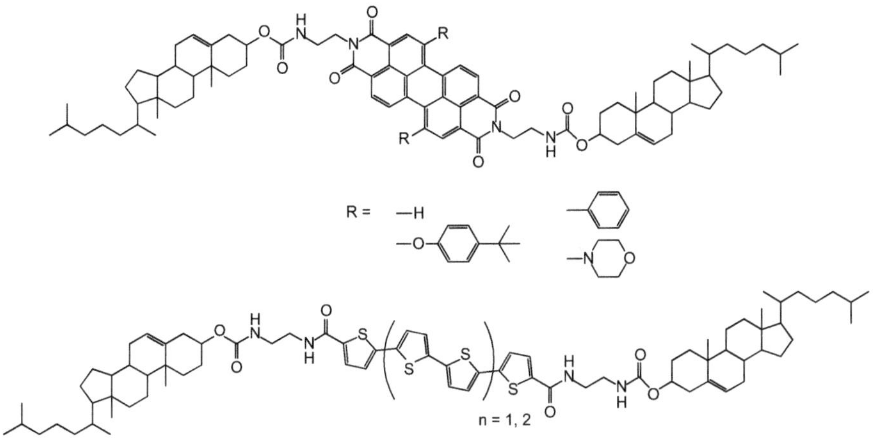

Figure 7.38 Molecular structures of cholesteryl-containing perylene and thiophene derivatives.

dimensions.[151,152] However, apart from the scientific interest and challenge in designing and making this kind of system, one could envisage several other applications of this kind of system in electronics, optics and sensing. Some nice examples of the approach have been proven in molecular-gel systems, although outside the remit of the coverage in this chapter, and we refer the interested reader to those for inspiration.[153–155]

In terms of electro-optically functional materials, perylene-stacked assemblies formed in the gel phase have proved to be particularly appealing visible-light-harvesting system. For example, Shinkai and coworkers described energy-transfer cascade processes in mixtures of perylene bisimide organogelators containing cholesteryl end groups[77] (Figure 7.38). Binary, ternary and quaternary perylene-based gels were synthesised in a mixed solvent of p-xylene and 1-propanol (3:1). A thorough characterisation of different mixed gels by fluorescence spectroscopy permitted the determination of a stepwise efficient energy-transfer process. The energy-transfer process was most efficient for binary gels containing the hydrogen-based and the *tert*-butyl benzene perylene derivative. Notably, no energy transfer was reported in the solution phase, therefore reinforcing the importance of chromophore gel networks for the excited energy-transfer process. In a subsequent study from the same group, p-type charge carrier organogelators based on cholesteryl-containing thiophene derivatives were mixed with the hydrogen-based perylene bisimide organogelator to form binary gels in chlorobenzene[156] (Figure 7.38). The results indicated that the binary gel formed with the cholesteryl-tetra(thiophene) derivative self-organised with the two constituents self-assembled independently, *i.e.* the two aggregated structures did not interfere with each other and truly orthogonal self-assembly had taken place. However, better binary self-sorted organogel systems were described with the cholesteryl-hexa(thiophene) derivative. In this case, a donor–acceptor coupling interaction was demonstrated in the entangled fibrous gel network, enabling photoelectrical conversion upon visible-light irradiation.

Hydrogen bonding can play a very important role in the development of segregated stacks of naphthalene diimide- and dioxynaphthalene-derived gelators.[157] The two aromatic units interact with each other through a charge-transfer type π–π interaction, and kinetically this is the force that is apparent in methylcyclohexane. However, within a few hours the system converts into the assembly where each unit is combined with itself. Hydrogen bonding was seen as the driving force for this process, as the units each had two amide groups, but spaced at different distances in each compound. The solvent is critical in the process, because in tetrachloroethylene the charge-transfer interaction dominates. The work shows quite beautifully how subtle changes in structure and solvent can bring about a desirable segregated state or a less useful coassembled structure, and it provides useful information for those wishing to use organogels as a way to process molecules for applications where parallel networks are required.

7.7   Conclusions and Outlook

The assembly of functional units using the gel as a vehicle to obtain nanoscale fibres has been proven beyond reasonable doubt. Hydrogen bonds are a reliable source of noncovalent cement for the aggregation of the molecules to give long fibres that can conduct energy and charge. The method overcomes many of the shortcomings of using crystalline materials, and the dogma that disorder may not allow the formation of efficient conductors has been proven wrong. The possibility of using these materials as conducting wires and sensors is a very real possibility that should be pursued with the most promising candidates. Certainly one of the major reasons evoking a rapid development in the research area described in this chapter is the simplicity of the gel approach towards the bottom-up fabrication of functional supramolecular architectures. A significant amount of knowledge is now available in order to control energy-transfer processes and efficient charge-carrier mobilities in fibrous gel networks. However, a precise control of the molecular order in the gel state is still required, not just to achieve enhanced properties, but to underpin future emerging technologies. The liquid-crystal approach provides convenient template networks to assemble organogelators. Nonetheless, to control the molecular order and properties of liquid crystals is not perfectly achieved. Frequently, additional gadgets are required that give electric fields with large voltages. Notwithstanding this, an important asset in gel chemistry is the fact that discrete chemical modifications in an organogelator's structure can easily lead to tunable functional properties. This is an extremely appealing feature because a large number of different applications can yet appear. An ultimate goal would be to tailor make a simple synthetic molecule that upon defined conditions can potentially lead to one or another efficient property. Under this perspective, the gel approach represents the best-case scenario for this challenge to become true, even among other nondynamic approaches such as controlled crystallisations and so forth.

On the other hand, long-term stability and electronic robustness of gels and xerogel films is an issue that needs further investigation. We have seen that a controlled peeling process can be conducted on doped xerogel samples with a conductive AFM configuration.[116] Furthermore, a clear understanding of the gelator assembly at the nanoscale level is still required, which undoubtedly will lead to new advanced functions. For example, it is expected that in organic solar-cell technologies the gel approach would lead to novel and effective molecular bulk heterojunctions with optimal charge separation, where donor–acceptor interfaces would be considerably extended. Low-cost and flexible bulk heterojunctions photovoltaic devices with high performance are anticipated. Relevantly, very few examples exist in the literature where gel-based organic solar cells are considered and relatively discrete device performances have been achieved so far. The postassembly covalent linking of gel materials, which may also lead to stable aerogels from molecular components, is an interesting possibility in order to improve the applicability of gel systems. It has already been shown that gels formed by porphyrin derivatives incorporating diacetylene units in the side chains next to the hydrogen-bonding group can be used to trap the organisation by polymerisation.[158] Disulfide formation[159] or more traditional sol–gel chemistry can also be used.[160] The proof that this paradigm could be used to generate electronically useful materials for information processing or sensing functions is an attractive prospect. The role of molecular metallogels is likely to be increasingly exploited in this sense.[83–85]

Clearly, many challenges remain in this nascent field of research. Therefore, and even though great advance has been made, new breakthrough research and further knowledge are required to properly design future emerging technologies with gel-based materials. Nonetheless, the promise offered by the gel approach in the coming years is immense, a myriad of applications extending beyond the ones shown here are likely.

Acknowledgements

We are indebted to the institutions that fund our research in this area under projects RYC-2011-08071, 2009 SGR 158 and CTQ2010-16339.

References

1. J. W. Steed, P. A. Gale (ed.), *Supramolecular Chemistry: From Molecules to Nanomaterials*, Wiley-VCH, Weinheim, 2012, ISBN: 978-0-470-74640-0.
2. A. Dawn, T. Shiraki, S. Haraguchi, S. Tamaru and S. Shinkai, *Chem. Asian J.*, 2011, **6**, 266.
3. T. J. Barton, L. M. Bull, W. G. Klemperer, D. A. Loy, B. McEnaney, M. Misono, P. A. Monson, G. Pez, G. W. Scherer, J. C. Vartuli and O. M. Yaghi, *Chem. Mater.*, 1999, **11**, 2633.

4. E. Gomar-Nadal, J. Puigmartí-Luis and D. B. Amabilino, *Chem. Soc. Rev.*, 2008, **37**, 490.
5. D. B. Amabilino, S. De Feyter, R. Lazzaroni, E. Gomar-Nadal, J. Veciana, C. Rovira, M. M. Abdel-Mottaleb, W. Mamdouh, P. Iavicoli, K. Psychogyiopoulou, M. Linares, A. Minoia, H. Xu and J. Puigmartí-Luis, *J. Phys. Condens. Matter*, 2008, **20**, 184003.
6. J. Puigmartí-Luis, A. Minoia, A. Pérez del Pino, G. Ujaque, C. Rovira, A. Lledós, R. Lazzaroni and D. B. Amabilino, *Chem. Eur. J*, 2006, **12**, 9161.
7. W. Wang and L. Chi, *Acc. Chem. Res.*, 2012, **45**, 1646.
8. J. R. Heath, *Annu. Rev. Mater. Res.*, 2009, **39**, 1.
9. Ph. Leclère, M. Surin, P. Jonkheijm, O. Henze, A. P. H. J. Schenning, F. Biscarini, A. C. Grimsdale, W. J. Feast, E. W. Meijer, K. Müllen, J. L. Brédas and R. Lazzaroni, *Eur. Polym. J.*, 2004, **40**, 885.
10. D. B. Amabilino and J. Puigmartí-Luis, *Soft Matter*, 2010, **6**, 1605–1612.
11. D. Gonzalez-Rodriguez and A. P. H. J. Schenning, *Chem. Mater.*, 2011, **23**, 310.
12. A. R. Hirst, B. Escuder, J. F. Miravet and D. K. Smith, *Angew. Chem. Int. Ed.*, 2008, **47**, 8002.
13. D. Das, T. Kar and P. K. Das, *Soft Matter*, 2012, **8**, 2348.
14. A. C. Pierre and G. M. Pajonk, *Chem. Rev.*, 2002, **102**, 4243.
15. R. W. Pekala, *J. Mater. Sci.*, 1989, **24**, 3221.
16. P. J. M. Carrott, L. M. Marques, M. M. Carrott and L. Ribeiro, *Micropor. Mesopor. Mater*, 2012, **158**, 170.
17. A. P. Katsoulidis, J. He and M. G. Kanatzidis, *Chem. Mater.*, 2012, **24**, 1937.
18. S. A. Al-Muhtaseb and J. A. Ritter, *Adv. Mater.*, 2003, **15**, 101.
19. S. T. Mayer, R. W. Pekala and J. L. Kaschmitter, *J. Electrochem. Soc.*, 1993, **140**, 446.
20. P. Iavicoli, H. Xu, L. N. Feldborg, M. Linares, M. Paradinas, S. Stafström, C. Ocal, B. Nieto-Ortega, J. Casado, J. T. López Navarrete, R. Lazzaroni, S. De Feyter and D. B. Amabilino, *J. Am. Chem. Soc.*, 2010, **132**, 9350.
21. D. Demus, J. W. Goodby, G. W. Gray, H.-W. Spiess and V. Vill, (eds), *Handbook of Liquid Crystals*, Wiley-VCH, Weinheim, 1998.
22. T. Kato, (ed.), *Liquid Crystalline Functional Assemblies and Their Supramolecular Structures, Structure and Bonding*, Springer, Berlin, 2008.
23. T. Kato, *Science*, 2002, **295**, 2414.
24. T. Kato, N. Mizoshita and K. Kishimoto, *Angew. Chem. Int. Ed.*, 2006, **45**, 38.
25. T. Kato, T. Yasuda, Y. Kamikawa and M. Yoshio, *Chem. Commun.*, 2009, 729.
26. M. Funahashi, H. Shimura, M. Yoshio and T. Kato, *Struct. Bond.*, 2008, **128**, 151.
27. Y. Suzuki, N. Mizoshita, K. Hanabusa and T. Kato, *J. Mater. Chem.*, 2003, **13**, 2870.

28. P. Cirkel, T. Kato, N. Mizoshita, H. Jagt and K. Hanabusa, *Liq. Cryst.*, 2004, **31**, 1649.
29. T. Kato, N. Mizoshita and K. Kishimoto, *Angew. Chem. Int. Ed.*, 2006, **45**, 38.
30. K. Yabuuchi, Y. Tochigi, N. Mizoshita, K. Hanabusa and T. Kato, *Tetrahedron*, 2007, **63**, 7358.
31. N. Mizoshita, T. Kutsuna, K. Hanabusa and T. Kato, *Chem. Commun.*, 1999, 781.
32. T. Kato, Y. Hirai, S. Nakaso and M. Moriyama, *Chem. Soc. Rev.*, 2007, **36**, 1857.
33. T. Kato, T. Kutsuna, K. Yabuuchi and N. Mizoshita, *Langmuir*, 2002, **18**, 7086.
34. T. Kato, T. Kutsuna, K. Hanabusa and M. Ukon, *Adv. Mater.*, 1998, **10**, 606.
35. N. Mizoshita, H. Monobe, M. Inoue, M. Ukon, T. Watanabe, Y. Shimizu, K. Hanabusa and T. Kato, *Chem. Commun.*, 2002, 428.
36. R. H. C. Janssen, V. Stumpflen, D. J. Broer, C. W. M. Bastiaansen, T. A. Tervoort and P. Smith, *J. Appl. Phys.*, 2000, **88**, 161.
37. N. Mizoshita, K. Hanabusa and T. Kato, *Adv. Funct. Mater.*, 2003, **13**, 313.
38. N. Mizoshita, K. Hanabusa and T. Kato, *Adv. Mater.*, 1999, **11**, 392.
39. N. Mizoshita, K. Hanabusa and T. Kato, *Displays*, 2001, **22**, 33.
40. N. Mizoshita, K. Hanabusa and T. Kato, *Adv. Mater.*, 2005, **17**, 692.
41. N. Mizoshita and T. Kato, *Adv. Funct. Mater.*, 2006, **16**, 2218.
42. M. Moriyama, N. Mizoshita, T. Yokota, K. Kishimoto and T. Kato, *Adv. Mater.*, 2003, **15**, 1335.
43. J. F. Hulvat, M. Sofos, K. Tajima and S. I. Stupp, *J. Am. Chem. Soc.*, 2005, **127**, 366.
44. S. Yagai, Y. Nakano, S. Seki, A. Asano, T. Okubo, T. Isoshima, T. Karatsu, A. Kitamura and Y. Kikkawa, *Angew. Chem. Int. Ed.*, 2010, **49**, 9990.
45. A. Ajayaghosh and S. J. George, *J. Am. Chem. Soc.*, 2001, **123**, 5148.
46. A. Ajayaghosh, S. J. George and V. K. Praveen, *Angew. Chem. Int. Ed.*, 2003, **42**, 332.
47. S. J. George, A. Ajayaghosh, P. Jonkheijm, A. P. H. J. Schenning and E. W. Meijer, *Angew. Chem. Int. Ed.*, 2004, **43**, 3421.
48. S. J. George and A. Ajayaghosh, *Chem. Eur. J*, 2005, **11**, 3217.
49. V. K. Praveen, S. J. George, R. Varghese, C. Vijayakumar and A. Ajayaghosh, *J. Am. Chem. Soc.*, 2006, **128**, 7542.
50. A. Ajayaghosh, C. Vijayakumar, R. Varghese and S. J. George, *Angew. Chem. Int. Ed.*, 2006, **45**, 456.
51. A. Ajayaghosh, R. Varghese, S. J. George and C. Vijayakumar, *Angew. Chem. Int. Ed.*, 2006, **45**, 1141.
52. H. Yu, H. Kawanishi and H. Koshima, *J. Photochem. Photobiol., A*, 2006, **178**, 62.
53. K. Murata, M. Aoki, T. Suzuki, T. Harada, H. Kawabata, T. Komori, F. Ohseto, K. Ueda and S. Shinkai, *J. Am. Chem. Soc.*, 1994, **116**, 6664.

54. S. van der Laan, B. L. Feringa, R. M. Kellogg and J. van Esch, *Langmuir*, 2002, **18**, 7136.
55. S. A. Ahmed, X. Sallenave, F. Fages, G. Mieden-Gundert, W. M. Muller, U. Muller, F. Vogtle and J.-L. Pozzo, *Langmuir*, 2002, **18**, 7096.
56. N. Koumura, M. Kudo and N. Tamaoki, *Langmuir*, 2004, **20**, 9897.
57. S. Yagai, T. Nakajima, K. Kishikawa, S. Kohmoto, T. Karatsu and A. Kitamura, *J. Am. Chem. Soc.*, 2005, **127**, 11134.
58. Y. Ji, G. C. Kuang, X. R. Jia, E. Q. Chen, B. B. Wang, W. S. Li, Y. Wei and J. Lei, *Chem. Commun.*, 2007, 4233.
59. J. Eastoe, M. Sánchez-Domínguez, P. Wyatt and R. K. Heenan, *Chem. Commun.*, 2004, 2608.
60. T. Suzuki, S. Shinkai and K. Sada, *Adv. Mater.*, 2006, **18**, 1043.
61. J. H. Kim, M. Seo, Y. J. Kim and S. Y. Kim, *Langmuir*, 2009, **25**, 1761.
62. M. De Loos, J. van Esch, R. M. Kellogg and B. L. Feringa, *Angew. Chem. Int. Ed.*, 2001, **40**, 613.
63. J. J. D. de Jong, P. Ralph Hania, A. Pugžlys, L. N. Lucas, Maaike de Loos, M. Kellogg Richard, B. L. Feringa, K. Duppen and J. H. van Esch, *Angew. Chem. Int. Ed.*, 2005, **44**, 2373.
64. C. Geiger, M. Stanescu, L. H. Chen and D. G. Whitten, *Langmuir*, 1999, **15**, 2241.
65. A. Ajayaghosh, S. J. George and A. P. H. J. Schenning, *Top. Curr. Chem.*, 2005, **258**, 83.
66. F. Würthner, C. Bauer, V. Stepanenko and S. Yagai, *Adv. Mater.*, 2008, **20**, 1695.
67. T. Ishi-I and S. Shinkai, *Top. Curr. Chem.*, 2005, **258**, 119.
68. C.-C. Lu and S.-K. Su, *J. Chin. Chem. Soc.*, 2009, **56**, 115.
69. S. J. George and A. Ajayaghosh, *Chem. Eur. J*, 2005, **11**, 3217.
70. A. Ajayaghosh, V. K. Praveen, S. Srinivasan and R. Varghese, *Adv. Mater.*, 2007, **19**, 411.
71. B. W. Messmore, J. F. Hulvat, E. D. Sone and S. I. Stupp, *J. Am. Chem. Soc.*, 2004, **126**, 14452.
72. Y. Kamikawa and T. Kato, *Langmuir*, 2007, **23**, 274.
73. S. Diring, F. Camerel, B. Donnio, T. Dintzer, S. Toffanin, R. Capelli, M. Muccini and R. Ziessel, *J. Am. Chem. Soc.*, 2009, **131**, 18177.
74. A. D. Guerzo, A. G. L. Olive, J. Reichwagen, H. Hopf and J.-P. Desvergne, *J. Am. Chem. Soc.*, 2005, **127**, 17984.
75. J.-P. Desvergne, A. G. L. Olive, N. M. Sangeetha, J. Reichwagen, H. Hopf and A. Del Guerzo, *Pure Appl. Chem.*, 2006, **78**, 2333.
76. C. Giansante, G. Raffy, C. Schäfer, H. Rahma, M.-T. Kao, A. G. L. Olive and A. Del Guerzo, *J. Am. Chem. Soc.*, 2011, **133**, 316.
77. K. Sugiyasu, N. Fujita and S. Shinkai, *Angew. Chem. Int. Ed.*, 2004, **43**, 1229.
78. A. Ajayaghosh, C. Vijayakumar, V. K. Praveen, S. S. Babu and R. Varghese, *J. Am. Chem. Soc.*, 2006, **128**, 7174.

79. F. Camerel, L. Bonardi, G. Ulrich, L. Charbonniere, B. Donnio, C. Bourgogne, D. Guillon, P. Retailleau and R. Ziessel, *Chem. Mater.*, 2006, **18**, 5009.
80. K. Sugiyasu, N. Fujita, M. Takeuchi, S. Yamada and S. Shinkai, *Org. Biomol. Chem.*, 2003, **1**, 895.
81. H. Tian and S. Wang, *Chem. Commun.*, 2007, 781.
82. S. Wang, W. Shen, Y. Feng and H. Tian, *Chem. Commun.*, 2006, 1497.
83. F. Fages, *Angew. Chem. Int. Ed.*, 2006, **45**, 1680.
84. M.-O. M. Piepenbrock, G. O. Lloyd, N. Clarke and J. W. Steed, *Chem. Rev.*, 2010, **110**, 1960.
85. A. Y.-Y. Tam and V. W.-W. Yam, *Chem. Soc. Rev.*, 2013, **42**, 1540.
86. F. Camerel, R. Ziessel, B. Donnio, C. Bourgogne, D. Guillon, M. Schmutz, C. Iacovita and J.-P. Bucher, *Angew. Chem. Int. Ed.*, 2007, **46**, 2659.
87. C. A. Strassert, C.-H. Chien, M. D. Galvez Lopez, D. Kourkoulos, D. Hertel, K. Meerholz and L. De Cola, *Angew. Chem. Int. Ed.*, 2011, **50**, 946.
88. N. K. Allampally, C. A. Strassert and L. De Cola, *Dalton Trans.*, 2012, **41**, 13132.
89. W. Kühlbrandt and D. N. Wang, *Nature*, 1991, **350**, 130.
90. W. Kühlbrandt, D. N. Wang and Y. Fujiyoshi, *Nature*, 1994, **367**, 614.
91. G. McDermott, S. M. Prince, A. A. Freer, A. M. Hawthornthwaite-Lawless, M. Z. Papiz, R. J. Cogdell and N. W. Isaacs, *Nature*, 1995, **374**, 517.
92. X. Hu, A. Damjanovic, T. Ritz and K. Schulten, *Proc. Natl. Acad. Sci. USA*, 1998, **95**, 5935.
93. X. Li, L. E. Sinks, B. Rybtchinski and M. R. Wasielewski, *J. Am. Chem. Soc.*, 2004, **126**, 10810.
94. F. J. M. Hoeben, I. O. Shklyarevskiy, M. J. Pouderoijen, H. Engelkamp, A. P. H. J. Schenning, P. C. M. Christianen, J. C. Maan and E. W. Meijer, *Angew. Chem., Int. Ed.*, 2006, **45**, 1232.
95. J. Zhang, F. J. M. Hoeben, M. J. Pouderoijen, A. P. H. J. Schenning, E. W. Meijer, F. C. De Schryver and S. De Feyter, *Chem. Eur. J*, 2006, **12**, 9046.
96. F. J. M. Hoeben, M. Wolffs, J. Zhang, S. De Feyter, P. Leclère, A. P. H. J. Schenning and E. W. Meijer, *J. Am. Chem. Soc.*, 2007, **129**, 9819.
97. T. Sagawa, S. Fukugawa, T. Yamada and H. Ihara, *Langmuir*, 2002, **18**, 7223.
98. T. Nakashima and N. Kimizuka, *Adv. Mater.*, 2002, **14**, 1113.
99. W. Kubo, T. Kitamura, K. Hanabusa, Y. Wada and S. Yanagida, *Chem. Commun.*, 2002, 374.
100. A. Ajayaghosh, V. K. Praveen, C. Vijayakumar and S. J. George, *Angew. Chem. Int. Ed.*, 2007, **46**, 6260.
101. A. Wicklein, S. Ghosh, M. Sommer, F. Würthner and M. Thelakkat, *ACS Nano*, 2009, **3**, 1107.
102. X. Yang, G. Zhang, D. Zhang and D. Zhu, *Langmuir*, 2010, **26**, 11720.

103. P. Xue, R. Lu, L. Zhao, D. Xu, X. Zhang, K. Li, Z. Song, X. Yang, M. Takafuji and H. Ihara, *Langmuir*, 2010, **26**, 6669.

104. R. J. Kumar, J. M. MacDonald, T. B. Singh, L. J. Waddington and A. B. Holmes, *J. Am. Chem. Soc.*, 2011, **133**, 8564.

105. M. Jørgensen, K. Bechgaard, T. Bjørnholm, P. Sommer-Larsen, L. G. Hansen and K. Schaumburg, *J. Org. Chem.*, 1994, **59**, 5877.

106. F. S. Schoonbeek, J. H. van Esch, B. Wegewijs, D. B. A. Rep, M. P. de Haas, T. M. Klapwijk, R. M. Kellogg and B. L. Feringa, *Angew. Chem. Int. Ed.*, 1999, **38**, 1393.

107. T. Kitahara, M. Shirakawa, S.-i. Kawano, U. Beginn, N. Fujita and S. Shinkai, *J. Am. Chem. Soc.*, 2005, **127**, 14980.

108. C. Wang, D. Zhang and D. Zhu, *J. Am. Chem. Soc.*, 2005, **127**, 16372.

109. T. L. Gall, C. Pearson, M. R. Bryce, M. C. Petty, H. Dahlgaard and J. Becher, *Eur. J. Org. Chem.*, 2003, 3562.

110. T. W. Kelley, E. L. Granstrom and C. D. Frisbie, *Adv. Mater.*, 1999, **11**, 261.

111. T. Akutagawa, K. Kakiuchi, T. Hasegawa, S.-i. Noro, T. Nakamura, H. Hasegawa, S. Mashiko and J. Becher, *Angew. Chem. Int. Ed.*, 2005, **44**, 7283.

112. J. Puigmartí-Luis, V. Laukhin, A. P. del Pino, J. Vidal-Gancedo, C. Rovira and D. B. Amabilino, *Angew. Chem. Int. Ed.*, 2007, **46**, 238.

113. J. Puigmartí-Luis, E. E. Laukhina, V. N. Laukhin, Á. Pérez del Pino, N. Mestres, J. Vidal-Gancedo, C. Rovira and D. B. Amabilino, *Adv. Func. Mater*, 2009, **19**, 934.

114. J. Puigmartí-Luis, Á. Pérez del Pino, V. Laukhin, L. N. Feldborg, C. Rovira, E. Laukhina and D. B. Amabilino, *J. Mater. Chem.*, 2010, **20**, 466.

115. D. Canevet, Á. Pérez del Pino, D. B. Amabilino and M. Sallé, *J. Mater. Chem.*, 2011, **21**, 1428.

116. C. Munuera, J. Puigmartí-Luis, M. Paradinas, L. Garzón, D. B. Amabilino and C. Ocal, *Small*, 2009, **5**, 214.

117. I. Danila, F. Riobé, J. Puigmartí-Luis, A. Pérez del Pino, J. D. Wallis, D. B. Amabilino and N. Avarvari, *J. Mater. Chem.*, 2009, **19**, 4495.

118. S. Ahn, Y. Kim, S. Beak, S. Ishimoto, H. Enozawa, E. Isomura, M. Hasegawa, M. Iyoda and Y. Park, *J. Mater. Chem.*, 2010, **20**, 10817.

119. X.-J. Wang, L.-B. Xing, W.-N. Cao, X.-B. Li, B. Chen, C.-H. Tung and L.-Z. Wu, *Langmuir*, 2011, **27**, 774.

120. F. Garnier, A. Yassar, R. Hajlaoui, G. Horowitz, F. Deloffre, B. Servet, S. Ries and P. Alnot, *J. Am. Chem. Soc.*, 1993, **115**, 8716.

121. F. Garnier, R. Hajlaoui, A. Yassar and P. Srivastava, *Science*, 1994, **265**, 1684.

122. A. Dodabalapur, L. Torsi and H. E. Katz, *Science*, 1995, **268**, 270.

123. R. E. Martin and F. Diederich, *Angew. Chem. Int. Ed.*, 1999, **38**, 1350.

124. S. Yagai, T. Kinoshita, Y. Kikkawa, T. Karatsu, A. Kitamura, Y. Honsho and S. Seki, *Chem. Eur. J*, 2009, **15**, 9320.

125. P. Pratihar, S. Ghosh, V. Stepanenko, S. Patwardhan, F. C. Grozema, L. D. A. Siebbeles and F. Wurthner, *Beilstein J. Org. Chem.*, 2010, **6**, 1070.

126. W.-W. Tsai, I. D. Tevis, A. S. Tayi, H. Cui and S. I. Stupp, *J. Phys. Chem. B*, 2010, **114**, 14778.

127. S. Prasanthkumąr, A. Saeki, S. Seki and A. Ajayaghosh, *J. Am. Chem. Soc.*, 2010, **132**, 8866.

128. S. Prasanthkumar, A. Gopal and A. Ajayaghosh, *J. Am. Chem. Soc.*, 2010, **132**, 13206.

129. J.-P. Hong, M.-C. Um, S.-R. Nam, J.-I. Hong and S. Lee, *Chem. Commun.*, 2009, 310.

130. Y. Yamamoto, G. Zhang, W. Jin, T. Fukushima, N. Ishii, A. Saeki, S. Seki, S. Tagawa, T. Minari, K. Tsukaghosi and T. Aida, *Proc. Natl. Acad. Sci. USA*, 2005, **102**, 10801.

131. J. P. Hill, W. Jin, A. Kosaka, T. Fukushima, H. Ichihara, T. Shimomura, K. Ito, T. Hashizume, N. Ishii and T. Aida, *Science*, 2004, **304**, 1481.

132. Y. Yamamoto, T. Fukushima, Y. Suna, N. Ishii, A. Saeki, S. Seki, S. Tagawa, M. Taniguchi, T. Kawai and T. Aida, *Science*, 2006, **314**, 1761.

133. Y. Yamamoto, T. Fukushima, A. Saeki, S. Seki, S. Tagawa, N. Ishii and T. Aida, *J. Am. Chem. Soc.*, 2007, **129**, 9276.

134. Y. Yamamotoa, G. Zhanga, W. Jina, T. Fukushima, N. Ishii, A. Saeki, S. Seki, S. Tagawa, T. Minari, K. Tsukagoshi and T. Aida, *Proc. Natl. Acad. Sci. USA*, 2009, **106**, 21051.

135. T. Kitamura, S. Nakaso, N. Mizoshita, Y. Tochigi, T. Shimomura, M. Moriyama, K. Ito and T. Kato, *J. Am. Chem. Soc.*, 2005, **127**, 14769.

136. B. C. Sih and M. O. Wolf, *Chem. Commun.*, 2005, 3375.

137. M. Kimura, S. Kobayashi, T. Kuroda, K. Hanabusa and H. Shirai, *Adv. Mater.*, 2004, **16**, 335.

138. S. Bhattacharya, A. Srivastava and A. Pal, *Angew. Chem. Int. Ed.*, 2006, **45**, 2934.

139. A. Pal, A. Srivastava and S. Bhattacharya, *Chem. Eur. J*, 2009, **15**, 9169.

140. L.-S. Li and S. I. Stupp, *Angew. Chem. Int. Ed.*, 2005, **44**, 1833.

141. J. van Herrikhuyzen, S. J. George, M. R. J. Vos, N. A. J. M. Sommerdijk, A. Ajayaghosh, S. C. J. Meskers and A. P. H. J. Schenning, *Angew. Chem. Int. Ed.*, 2007, **46**, 1825.

142. J. Puigmartí-Luis, A. Pérez del Pino, E. Laukhina, J. Esquena, V. Laukhin, C. Rovira, J. Vidal-Gancedo, A. G. Kanaras, R. J. Nichols, M. Brust and D. B. Amabilino, *Angew. Chem. Int. Ed.*, 2008, **47**, 1861.

143. E. Taboada, L. N. Feldborg, A. Pérez del Pino, A. Roig, D. B. Amabilino and J. Puigmartí-Luis, *Soft Matter*, 2011, **7**, 2755.

144. P. D. Wadhavane, M. A. Izquierdo, F. Galindo, M. I. Burguete and S. V. Luis, *Soft Matter*, 2012, **8**, 4373.

145. S. Srinivasan, S. S. Babu, V. K. Praveen and A. Ajayaghosh, *Angew. Chem. Int. Ed.*, 2008, **47**, 5746.

146. D. Canevet, A. Pérez del Pino, D. B. Amabilino and M. Sallé, *Nanoscale*, 2011, **3**, 2898.
147. K. V. Rao, K. K. R. Datta, M. Eswaramoorthy and S. J. George, *Angew. Chem. Int. Ed.*, 2011, **50**, 1179.
148. M. M. Safont-Sempere, G. Fernandez and F. Wuerthner, *Chem. Rev.*, 2011, **111**, 5784.
149. S. L. Li, T. X. Xiao, C. Lin and L. Y. Wang, *Chem. Soc. Rev.*, 2012, **41**, 5950.
150. J. van Herrikhuyzen, A. Syamakumari, A. P. H. J. Schenning and E. W. Meijer, *J. Am. Chem. Soc.*, 2004, **126**, 10021.
151. A. Pivrikas, H. Neugebauer and N. S. Sariciftci, *Solar Energy*, 2011, **85**, 1226.
152. Z. X. Wang, F. J. Zhang, J. Wang, X. W. Xu, J. Wang, Y. Liu and Z. Xu, *Chinese Bull. Sci.*, 2012, **57**, 4143.
153. A. Brizard, M. Stuart, K. van Bommel, A. Friggeri, M. de Jong and J. van Esch, *Angew. Chem. Int. Ed.*, 2008, **47**, 2063.
154. D. García-Velázquez and R. Luque, *Chem. Eur. J*, 2011, **17**, 3847.
155. X. Z. Yan, D. H. Xu, X. D. Chi, J. Z. Chen, S. Y. Dong, X. Ding, Y. H. Yu and F. H. Huang, *Adv. Mater.*, 2012, **24**, 362.
156. K. Sugiyasu, S.-i. Kawano, N. Fujita and S. Shinkai, *Chem. Mater.*, 2008, **20**, 2863.
157. A. Das, M. R. Molla, B. Maity, D. Koley and S. Ghosh, *Chem. Eur. J.*, 2012, **18**, 9849.
158. M. Shirakawa, N. Fujita and S. Shinkai, *J. Am. Chem. Soc.*, 2005, **127**, 4164.
159. C. S. Love, V. Chechik, D. K. Smith, I. Ashworth and C. Brennan, *Chem. Commun.*, 2005, 5647.
160. T. Kishida, N. Fujita, K. Sada and S. Shinkai, *Langmuir*, 2005, **21**, 9432.

CHAPTER 8

Molecular Gels as Templates for Nanostructured Materials

TANMOY KAR AND PRASANTA KUMAR DAS*

Department of Biological Chemistry, Indian Association for the Cultivation of Science, Jadavpur, Kolkata – 700 032, India
*Email: bcpkd@iacs.res.in

8.1 Introduction

The self-assembly of organic building blocks can create a wide variety of supramolecular arrangements having diversity in their morphology and function as found in many natural and synthetic systems.[1–5] In these self-assembled structures, the organic molecules are held together by a combination of weak forces, such as hydrogen bonding, π–π stacking and van der Waals interactions, *etc.*[1–18] Over the last decade, researchers from various fields of chemistry have been trying to develop novel methods through which the shape and size of nanostructured materials can be controlled at the micro- or even nanoscopic level. In this regard, the use of self-assembled systems has received immense attention as templates for the generation of anisotropic nanostructures and composite materials.[19–22] The term template can be defined as the pattern or structure that can be used to reproduce another material with a complementary shape/size and architecture. In supramolecular chemistry, templates are generally used as a driving force for bringing together molecular components.[23] Such components afterwards react among themselves and go on to form products in which the template molecule can remain intertwined or from which the template can be easily removed. The use of templates in supramolecular chemistry has been found to be useful for developing new receptor molecules,

RSC Soft Matter No. 1
Functional Molecular Gels
Edited by Beatriu Escuder and Juan F. Miravet
© The Royal Society of Chemistry 2014
Published by the Royal Society of Chemistry, www.rsc.org

and also the construction of complex molecular architectures that would otherwise have been impossible. Previously it was thought that templates were usually single molecules or ionic species, however, recent advancements of supramolecular chemistry have shown that much larger self-assembled structures (like self-assembled fibrillar networks (SAFIN)) of gels from low molecular weight gelators (LMWGs), micelles, vesicles, organic crystals, *etc* can also function in an analogous manner.[24–28] All of these organic superstructures have been employed as templates for the creation of various nanomaterials and processes are known as transcription.[24] With this process the morphology of the organic template can be transferred to an inorganic material that is currently unattainable through any other method. The transcription process consists of several steps (Scheme 8.1).

First, the organic template having self-aggregating properties is brought into contact with an inorganic precursor or small particles of the actual inorganic material that will be finally formed. This step usually takes place in solution in the presence of a catalyst, and then leads to the deposition of the inorganic material on the inner or outer surface of the organic template. This step resulted in the formation of an organic–inorganic hybrid material that may exhibit interesting properties. The organic template material may then be removed to get the pure inorganic material whose morphology is directly related to the organic template. The removal of the organic template can be achieved by heat treatment,[29] microwave irradiation,[30] or washing with organic solvents.[31] The procedure of template removal by heat treatment is known as calcination and it

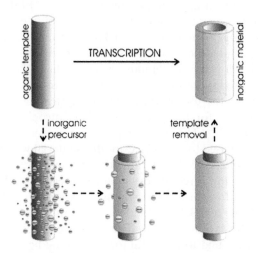

Scheme 8.1 Schematic representation of the transcription process: the organic template (top left) is brought into contact with the inorganic precursor that leads to the deposition of inorganic material on the surface of the template; the latter can subsequently be removed to yield the transcribed inorganic material (top right).
Reproduced with permission from ref. 24, Wiley-VCH.

is the most widely used method.[29] Removing the organic template by dissolving it with an appropriate solvent would seem to be the simplest and quickest method for obtaining the transcribed, inorganic material. The ease with which the original template structure can be removed is one of the advantages of using organic instead of inorganic templates.

On the other hand, the SAFINs of LMW gels have also been used for the stabilisation and organisation of many nanoparticles in two- and three-dimensional (2D/3D) ordered assemblies as these composite materials would find potential applications in developing electronic and optical devices (see Chapter 7).[32] The gel fibres are much larger than the nanomaterials and they could act as a host on which the nanoparticles are embedded during fibre formation. Due to these properties, low molecular weight organo-gelators (LMOGs) can be successfully applied toward templating desired nanoparticle composites. Additionally, gels in the swollen stage provide a large free space between the 3D crosslinked networks that acts as a nanoreactor for the nucleation and growth of nanoparticles. Intriguingly, gelators comprising of appropriate functional moiety capable of reducing metal ions can also be used for the *in situ* synthesis and stabilisation of nanoparticles even at room temperature. Interestingly, this fact simplifies the preparation of soft nanohybrids and rather eliminates the use of hazardous external reducing agents and extreme conditions.

In this chapter we will discuss i) the chemical reactions involved in the sol–gel chemistry necessary for the transcription process, ii) the imperative choice of the gelator molecule for successful transcription of the template structure, iii) the importance of positively charged moieties and hydrogen-bonding groups, either covalently or noncovalently attached in successful transcription, iv) organisation of nanomaterials into 2D/3D architectures using SAFINs of LMW gels and v) development of novel soft nanocomposites by hybridising LMW gels with metal nanoparticles, carbon nanotube (CNT) and graphene-based nanomaterials with examples from the recent literature.

8.2 Sol–Gel Chemistry

In order to understand the involvement of supramolecular gel templates in transcription for preparing nanomaterials, at first it is essential to know the chemistry involved in sol–gel processes.[33] Sol–gel chemistry is primarily based on two reactions; one is hydrolysis and the other is condensation of precursors. Most works in the sol–gel field have been performed by the use of alkoxides as precursors. Alkoxides provide a convenient source for "inorganic" monomers that in most cases are soluble in a variety of solvents, especially alcohols. Alcohols enable a convenient addition of water to start the reaction. Another advantage of the alkoxide route is the possibility to regulate rates by controlling hydrolysis and condensation by chemical means. With an alkoxide as a precursor, sol–gel chemistry can be simplified in terms of the following reaction equations.

8.2.1 Hydrolysis of Metal Alkoxides

Hydrolysis of alkoxides is greatly influenced by the addition of a catalyst (acid or base). Acetic acid and benzylamine are the two most commonly used catalysts for transcription of LMOG-based template structures as both of them are readily soluble in a variety of organic solvents as well as water. Under acidic conditions, the mechanism involves a rapid protonation of an alkoxide group followed by the nucleophilic attack of a water molecule onto electrophilic metal centre. The reaction may proceed either by S_N^1 or S_N^2 as shown in Scheme 8.2. Subsequently, transfer of a proton from the water to a negatively charged alkoxy group of the metal and release of the resulting ROH molecule.

Under basic conditions, hydroxyl ions formed by the dissociation of water molecule, attack the silicon atom. Hydrolysis then occurs by the displacement of the alkoxide anion, a process that may be aided by hydrogen bonding of the alkoxide anion with the solvent (Scheme 8.3). It is important to realise that concurrent with the hydrolysis steps, trans-esterification and re-esterification reactions will also take place.

8.2.2 Condensation

As mentioned, polymerisation to form siloxane bonds occurs by either an alcohol-producing or a water-producing condensation reaction. However, the mechanisms of both reactions are analogous. The first step of acid-catalysed condensation reaction involves protonation of the silanol species (Scheme 8.4). As a result, the silicon centre becomes more susceptible to nucleophilic attack. The silanols existing as monomers or weakly branched oligomers are the most basic silanol species, and are most likely to be protonated. These results in

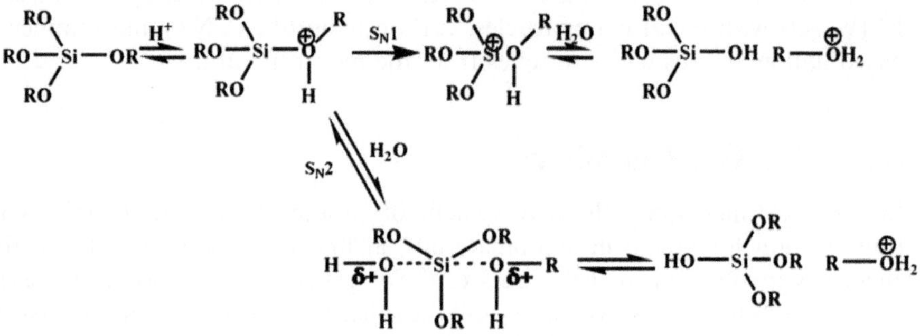

Scheme 8.2 Two possible mechanisms for the acid-catalysed hydrolysis step.

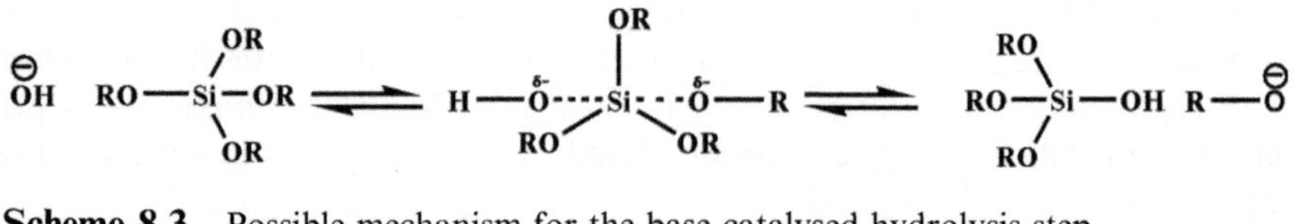

Scheme 8.3 Possible mechanism for the base-catalysed hydrolysis step.

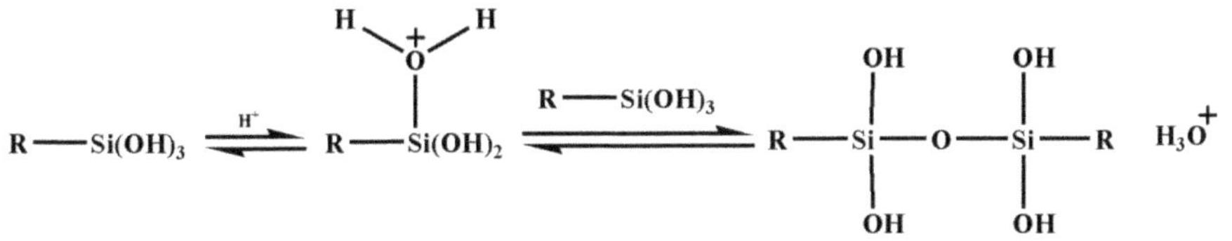

Scheme 8.4 Possible mechanism for the acid-catalysed condensation step.

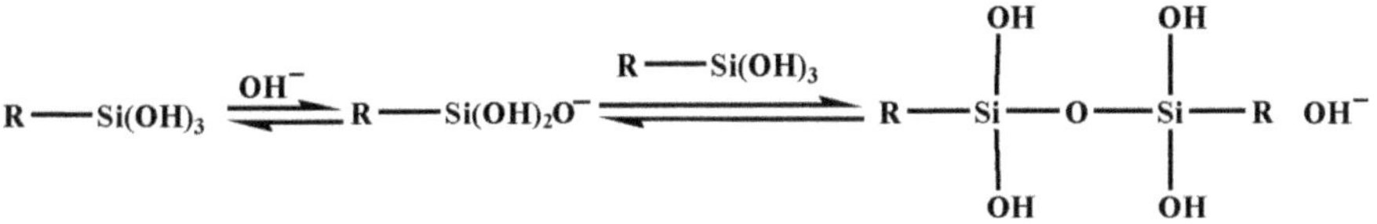

Scheme 8.5 Possible mechanism for the base-catalysed condensation step.

preferential reactions involving condensation between neutral species and protonated silanols situated on monomers, end groups of chains, *etc.*

For base-catalysed condensation, nucleophilic, deprotonated silanol reacts with neutral silicate species (Scheme 8.5). This reaction takes place above the isoelectric point of silica, where surface silanols are deprotonated. As the acidity of silanols depends on the other substituents on the silicon atom, this isoelectric point may vary from pH 1.5–4.5 depending on the extent of condensation of the silicate species. Replacement of basic OR and OH moieties with -OSi moieties reduces the electron density on the Si, thus increasing the acidity of the protons of the remaining silanols. Hence, reactions between larger, more highly condensed species, containing acidic silanols, and smaller, less weakly branched species are favoured. The condensation rate is highest near neutral pH, where both species are present in significant concentrations. The presence of these charged species in both the acid- and base-catalysed condensation steps is essential for successful transcription.

8.3 Organic Templates for Creation of Inorganic Nanomaterials

Synthesising inorganic materials of different morphologies including fibres, rods, ribbons, helices and tubes employing the SAFINs of LMW gels as template is the most widely known method of developing gel-nanocomposites.[19–22] Most of the early examples of templated formation of inorganic materials, demonstrated the synthesis of SiO_2, TiO_2, or other common oxides, however, there is a rapidly increasing number of materials (for example, $ZnSO_4$, CdS, *etc.*) that can nowadays be shaped into fascinating structures with the help of supramolecular gels.

8.3.1 Silica Nanomaterials

Successful exploitation of a supramolecular organogel as a template for the preparation of inorganic nanostructures was reported for the first time by

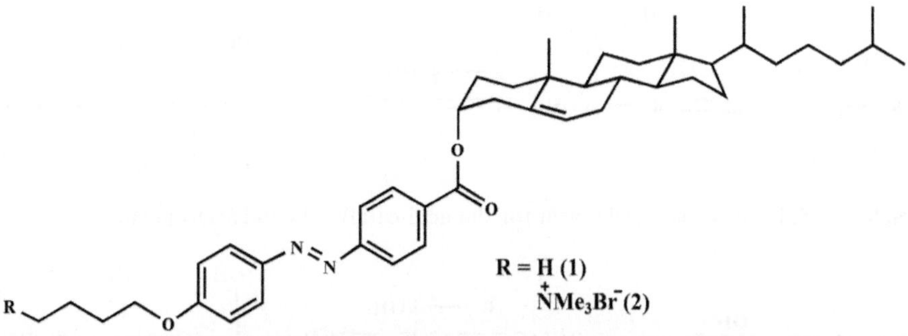

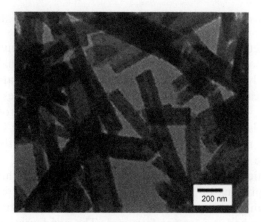

R = H (1)
$\overset{+}{N}Me_3Br^-$ (2)

Chart 8.1

Figure 8.1 TEM image of silica with tubular structure (after calcination) obtained by transcription of an acetic acid gel of **2**.
Reproduced with permission from ref. 35, The Royal Society of Chemistry.

Shinkai and coworkers. The chlolesterol-based gelator **1** (Chart 8.1) was found to gelatinise a number of organic solvents including tetraethoxysilane (TEOS) and this gel was polymerised under acidic or basic conditions.[34] However, the obtained silica from the polymerisation of TEOS under acidic or basic conditions did not exhibit any sign of being templated by the organogel fibres of **1**.[35]

Consequently, gelator **2** (Chart 8.1) comprising a quaternary ammonium group like in a conventional cationic surfactant that are generally used in sol–gel polymerisation systems was used as a template. Interestingly, polymerisation using TEOS under acidic condition resulted in the formation of tubular silica with inner diameters of 10–200 nm as observed in the transmission electron microscopic (TEM) image (Figure 8.1).[35] These results indicate that the fibrils of organogelator **2** carrying a positive charge acted as a template in the sol–gel process to create the tubular structure. This sol–gel polymerisation was carried out in acetic acid, which is also an excellent catalyst for the TEOS polycondensation reaction. Under such acidic conditions (pH 2–7) anionic silica species ($\equiv$Si–O–) are present in solution. Consequently, the oligomeric silica species preferentially get adsorbed onto the positively charged gelator fibrils. Once this deposition started, the polymerisation further proceeded along the surface of the template. Removal of the gelator molecules through calcination resulted in the isolation of the hollow silica fibres. Subsequently, transcription caused by electronic interactions between positively charged gelator fibres and anionic silica species can also be modulated by the presence of other cationic charges (*e.g.*, Me_4NCl).[36]

The influence of cationic charge at the fibre surface on the formation of the tubular silica was further established by carrying out the polymerisation reaction in organogels prepared by mixing **1** and **2** at different molar ratios that also resulted in the formation of tubular silica.[36] But surprisingly, the tubular silica obtained at a lower molar composition fraction was found to be also helical in nature. The "dilution" of the cationic charges present in the gel fibres clearly played a decisive role in whether or not the chirality present in the template is transcribed into the silica material. At high cationic charge concentration, rapid adsorption of anionic silica precursors took place at the fibre surface that led to the formation of tubular silica and did not display any of the chirality present in the gel fibres. Conversely, a very low concentration of positive charges did not result in rapid transcription as the charge density on the fibre surface is insufficient for preferential deposition of silica particles. However, a moderate concentration of cationic charges resulted in excellent transcription (helical silica) due to more controlled adsorption of negatively charged silica precursors at the fibre surface. Thus, it can be said that control over the deposition speed is the key to the accuracy of the transcription process.

Similarly, Jung *et al.* employed an organogel system formulated by mixing a positively charged and a neutral gelator based on chiral diaminocyclohexane for efficient sol–gel transcriptions (**3–6**) (Chart 8.2).[37] A combination of neutral and cationic gelators (**3** with **4** and **5** with **6**) was used for the transcription of helicity. When neutral **3** was mixed with cationic **4**, organogel resulted in the formation of left-handed helical silica, while right- handed helical silica was produced by the mixture of neutral **5** and cationic **6** (Figure 8.2). Here also at lower molar composition fraction irregular granular silica was obtained, illustrating the importance of cationic charge density at the fibre surface for successful transcription.

In a separate work, Shinkai and coworkers employed cationic gemini surfactants (**7 and 8**, Chart 8.3) with C-16 tail and chiral D-/L-tartarate as counterions to create double-helical silica fibres with a tunable pitch by sol–gel polycondensation of TEOS in 1:1 pyridine–water in the presence of benzyl

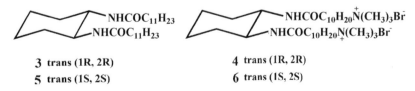

$NHCOC_{11}H_{23}$
$NHCOC_{11}H_{23}$

3 trans (1R, 2R)
5 trans (1S, 2S)

$NHCOC_{10}H_{20}\overset{+}{N}(CH_3)_3Br^-$
$NHCOC_{10}H_{20}\overset{+}{N}(CH_3)_3Br^-$

4 trans (1R, 2R)
6 trans (1S, 2S)

Chart 8.2

Figure 8.2   TEM images of the silica obtained by sol–gel transcription in (A) left-handed **3 + 4** (1 : 1 wt%), and (B) right-handed **5 + 6** (1 : 1 wt%) organogels. Reprinted with permission from ref. 37. Copyright 2000 American Chemical Society.

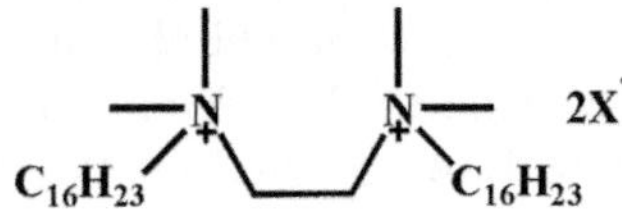

$$\text{C}_{16}\text{H}_{23}\overset{|}{\underset{+}{\text{N}}}\diagdown\diagup\overset{|}{\underset{+}{\text{N}}}\text{C}_{16}\text{H}_{23}\quad 2\text{X}^-\qquad \text{X}^- = \text{L-tartarate (7)}$$

D-tartarate (**8**)

Chart 8.3

amine.[38] Additionally, it was shown that the twist pitch (T) and the width (W) of these ribbons can be regulated by varying the enantiomeric excess (*ee*) in water–pyridine (Figure 8.3). For example, with a gel of pure **7** (100% ee), right-handed helical fibrils were obtained whereas with lower *ee* (50% or below), the formation of double-stranded helical structures was observed.

In a separate example, Ono *et al.* used a cholesterol-based gelator containing benzo-18-crown-6 moiety (**9**, Chart 8.4) capable of binding potassium cation to transcribe the helical organogel fibre structure into the inorganic silica.[39] The benzylamine-catalysed polycondensation of TEOS in the butanol gel of **9** in the absence of K^+ produced granular silica. However, in the presence of a considerable amount of K^+ ion tubular silica was obtained. However, in the presence of other cations like Li^+, Na^+, Rb^+, Cs^+, *etc.* only granular silica was formed due to their low affinity for the crownether in gelator **9**. Thus, the positive charge introduced in a noncovalent manner can aid transcription in a manner analogous to covalently attached positive charges.

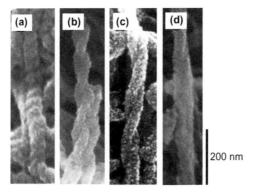

Figure 8.3 Influence of the *ee* (L-**7** in excess) on the helical pitch of double-stranded silica: (a) 100% *ee*; (b) 50% *ee*; (c) 33% *ee*; (d) 20% *ee*.
Reproduced with permission from ref. 38, The Royal Society of Chemistry.

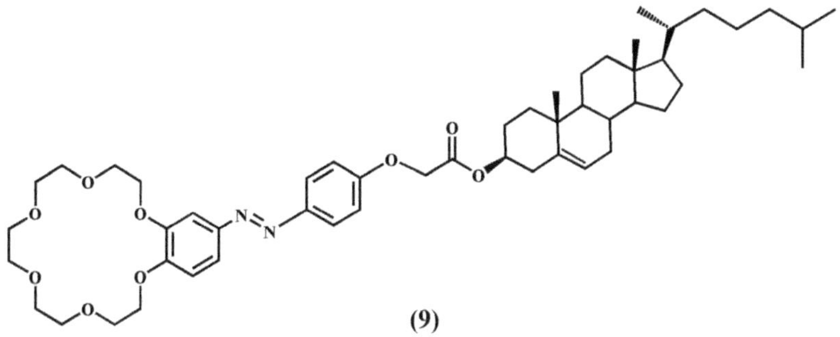

Chart 8.4

Subsequently, a number of azacrown-containing organogelators were reported to obtain transcripted inorganic materials by binding group-I metal ions. The 1-butanol gels of organogelators **10**, **11** and **12**, as well as the aniline gel of **13**, were successfully transcribed into spiral silica in the presence of various metal salts like $AgNO_3$ (**10** and **12**), $CsClO_4$ (**11** and **12**), $Pd(NO_3)_2$ (**13**), and $KClO_4$ (**10** and **11**), respectively (Chart 8.5).[40–43] Complexation of the metal cations by the (aza)crownethers introduced positive charges into the LMOG-based templates, which facilitated the transcription process. Intriguingly, 1-butanol gels of **10** and **11** and acetic acid gels of **12** and **13** could be successfully transcribed into silica even in the absence of metal salts due to the protonation of the nitrogen atom in the azacrownether resulting in a positive charge on the gel fibres. Similarly, phenanthroline containing LMW organogelator (**14**) functioned as transcription directors in acetic acid medium and resulted in the formation of hollow tubular silica (Chart 8.6).[44]

In a search for an organogel system that would act as a neutral template, Jung and Shinkai synthesised compounds 15_n to produce mesoporous silica (Chart 8.7).[45] Here, the transcription occurred due to hydrogen bonding, rather

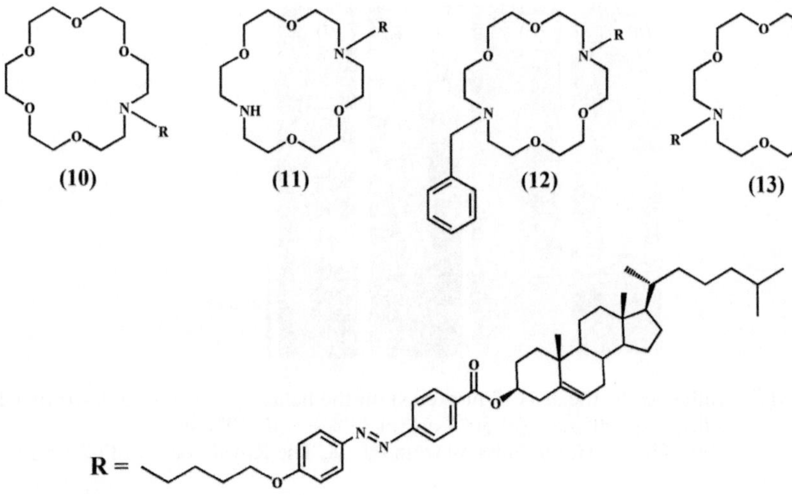

(10)   (11)   (12)   (13)

R =

Chart 8.5

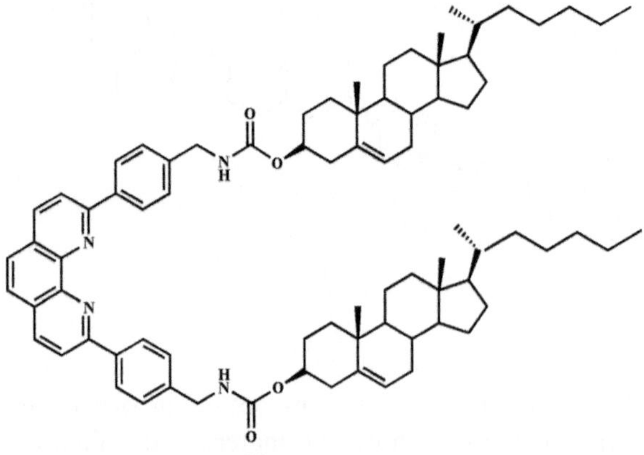

(14)

Chart 8.6

than ionic interactions, between template and silica precursor. The attractive forces between the negatively charged silica precursor species and the amide moieties of the gelator are large enough to result in a successful transcription of the template. Interestingly, different morphologies were observed for the organogels and the resultant silica products of gelators 15_n for different values of n. In the cases of $n = 2$ or 4, the organogels possessed a curved lamellar morphology, which resulted in silica with a structure like a paper roll (Figure 8.4) and for odd numbers of n ($n = 3$ or 5), the morphology of the organogel was fibrous in nature and resulted in hollow silica fibres.

In another study, they synthesised sugar-based compounds **16–19** to enable transcription through the action of hydrogen-bond donating moieties (Chart 8.8).[46] The amino group in the gelators (**16, 18** and **19**) was not only

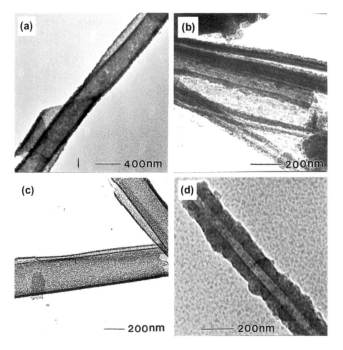

Chart 8.7

Figure 8.4 TEM pictures of the silica obtained from **15** (a) $n=2$, (b) $n=3$, (c) $n=4$ and (d) $n=5$ in butan-1-ol after calcination.
Reproduced with permission from ref. 45, The Royal Society of Chemistry.

expected to stabilise the organogels due to the intensified intergelator hydrogen bonds but also to bind with TEOS through hydrogen bonding. The silica obtained from **16** were tubular in nature with 20–30 nm outer diameter and 350–700 nm in length, whereas the silica obtained from **17** were of conventional granular structure. It is hardly conceivable that the aromatic amino group is protonated in the presence of the benzylamine, used as a catalyst. These results indicate that the tubular structure of the silica was successfully transcribed by the hydrogen-bonding interactions between the amino group of **16** and TEOS (or oligomeric silica species).

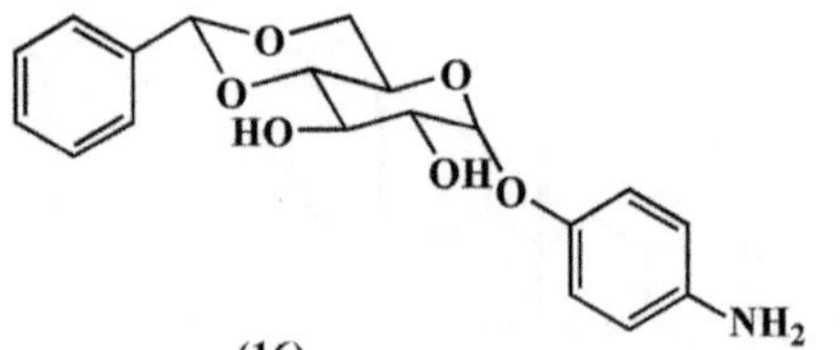

(16)

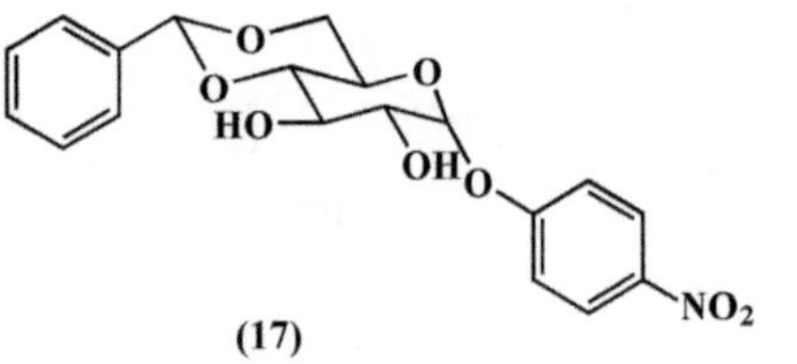

(17)

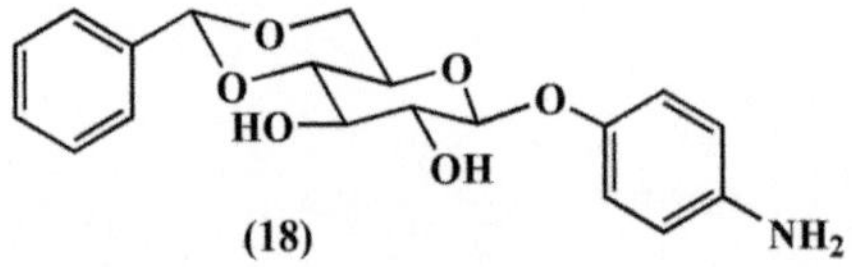

(18)

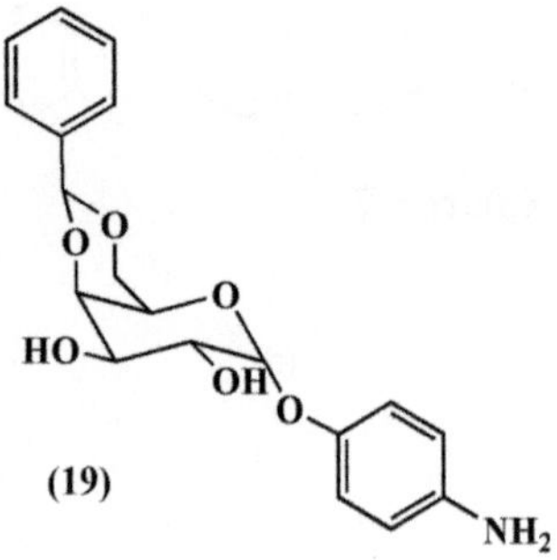

(19)

Chart 8.8

In contrast, β-glucose-type organogel **18** resulted in a tubular silica nano-structure with larger outer diameter of 150–200 nm. Surprisingly, the TEM images of the silica obtained from **18** revealed that the silica consisted of 50–100 nm inner diameters and 150–200 nm outer diameters. Furthermore, the silica in the inner part of the tube was composed of microtubes with diameters of 5–10 nm and eventually led to the formation of a lotus-like structure.

In another example Lu and coworkers used small amphiphilic peptide I_3K (Ac-IIIK-NH$_2$, I = isoleucine; K = lysine) for the generation of silica-based inorganic nanostructures.[47] This peptide I_3K readily self-assembled in water at neutral pH to form nanotubes with diameters of ∼10 nm and lengths of a few micrometres. Consequently, the highly stable I_3K nanotubes were used as templates for silicification from the hydrolysis of organosilicate precursors using TEOS. The lysine groups on the inner and outer nanotube surfaces helped to catalyse the silicification leading to the formation of silica nanotubes. TEM analysis revealed the formation of hollow silica nanotubes templated by the I_3K nanotubes.

In a different study Liu and coworkers reported the formation of uniform helical silica nanotubes (SiNTs) in water using D- or L-glutamic acid based bola-amphiphiles (**20, 21**, Chart 8.9).[48] Most importantly it was found that the as-formed SiNTs have supramolecular chirality in the inner walls. The hydrogelators *N,N*-hexadecanedioyl-di-L-glutamic acid (L-HDGA) and its enantiomer D-HDGA rigidified water forming helical nanotubes.

The handedness of the nanotubes is determined by the chirality of the bola-amphiphile as L-HDGA formed a right-handed and D-HDGA left-handed helical nanotube, respectively. Such chiral hydrogels were used as template to synthesise SiNTs using TEOS and 3-aminopropyltriethoxysilane (APES) in water. The as-prepared silica nanostructures were encapsulated by HDGA

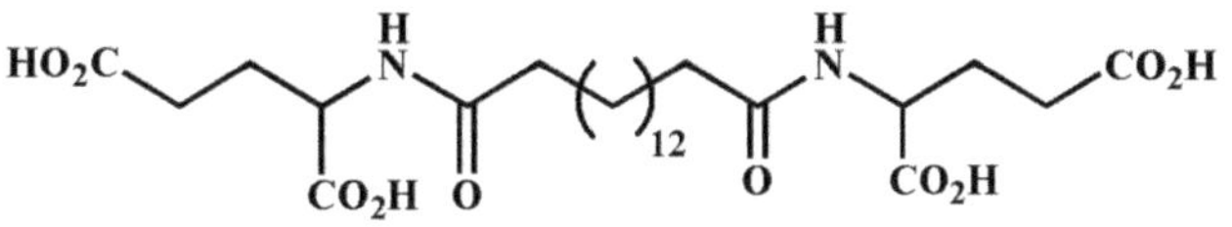

D-HDGA (20)
L-HDGA (21)

Chart 8.9

molecules. Removal of the HDGA from inside resulted in the formation of nanotubes with propylamine groups of APES in them. Interestingly, such nanotubes show chirality (M- and P-chirality) from their inner walls that was established by dispersing SiNTs fabricated from L-HDGA in water containing tetraphenylporphine tetrasulfonic acid (TPPS). TPPS adsorbed within the SiNTs showed both J and H aggregated bands at 490 and 435 nm and a negative cotton effect in a circular dichroism (CD) study suggesting that supramolecular chirality of TPPS was induced (Figure 8.5a). However, HDGA-capped SiNTs upon dispersion into the TPPS solution did not exhibit any kind of aggregation or CD signals. This result indicated that the chirality of the inner wall of the SiNTs was controlled by the chiral gelator molecules. As expected a mirror-imaged CD spectrum was obtained when SiNTs fabricated from D-HDGA hydrogel were used. Similarly, when azobenzene was loaded into the inner walls of the SiNTs, the chirality of the inner walls was transferred to these azobenzene assemblies through π–π stacking, as confirmed from the corresponding CD study (Figure 8.5b).

Nakano and coworkers reported the synthesis of fluorocarbon functionalised silica nanotubes by a perfluoroalkyl chain-containing organogelator (**22**) as a template (Chart 8.10). The sol–gel condensation was carried out using a mixture of TEOS and fluorocarbon functionalised triethoxysilanes (**23a–d**, Chart 8.10) in the presence of organogelator (**22**).[49] The silica obtained after sol–gel polymerisation revealed a homogeneous tubular structure indicating effective transcription of fibrillar structure of **22** to the final silica structure. The presence of functionalised triethoxysilane was found to be crucial for successful transcription as the polycondensation of only TEOS or trialkoxysilanes possessing a long alkyl chain instead of the perfluoroalkyl chain resulted in the formation of granular silica. The fluorocarbon-functionalised trialkoxysilanes were absorbed onto the fibres of gelator **22** through the aggregation of perfluoroalkyl chains between alkoxysilanes and started to condense, resulting in the formation of the fluorocarbon-functionalised silica tube. The chain length of the fluorocarbon functionalised triethoxysilanes also played a role in transcription. Tubular silica structures were obtained with F6-TEOS, F8-TEOS and F10-TEOS, whereas granular silica was obtained with F4-TEOS.

In all the examples discussed above, the structure of the nanomaterials was controlled by the addition of an external organic template that directs the

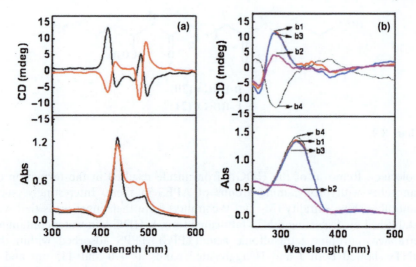

Figure 8.5   (a) CD (top) and UV (bottom) spectra of TPPS adsorbed M-SiNTs (black) and P-SiNTs (red). (b) CD (top) and UV (bottom) spectra of 4-carboxyazobenzene-modified SiNTs: as-prepared Azo-PSiNTs.
Reproduced with permission from ref. 48, The Royal Society of Chemistry.

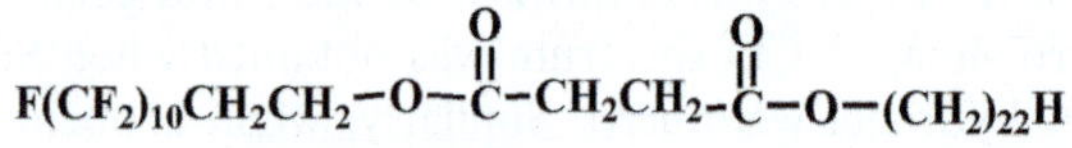

$$F(CF_2)_{10}CH_2CH_2-O-\overset{\overset{\displaystyle O}{\|}}{C}-CH_2CH_2-\overset{\overset{\displaystyle O}{\|}}{C}-O-(CH_2)_{22}H$$

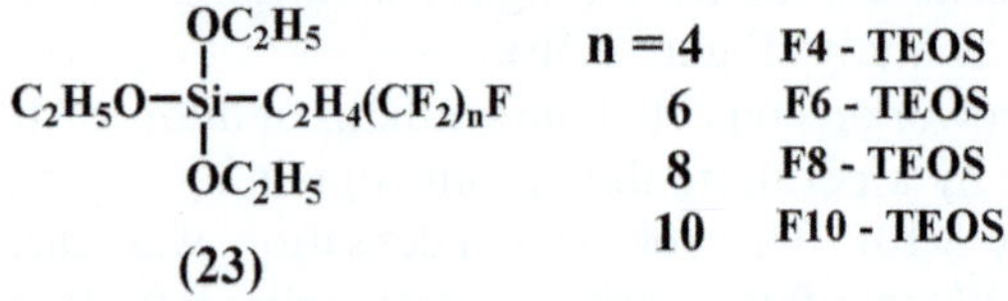

$$(22)$$

$$\begin{array}{l} \quad\quad OC_2H_5 \\ \quad\quad | \\ C_2H_5O-Si-C_2H_4(CF_2)_nF \\ \quad\quad | \\ \quad\quad OC_2H_5 \\ \quad\quad (23) \end{array}$$

n = 4	F4 - TEOS
6	F6 - TEOS
8	F8 - TEOS
10	F10 - TEOS

Chart 8.10

organisation of the inorganic phase. However, in another approach, the precursor of the inorganic was incorporated within the gelator-like structures to obtain templated inorganic nanomaterials. For example, compounds **24** and **25** comprised triethoxysilyl moiety as an integral part of the gelator structure (Chart 8.11).[50] The compound **24** (either as the R,R or S,S enantiomer) gelated a number of organic solvents such as cyclohexane or mesitylene through H-bonds.

However, transcription experiments were carried out under acidic hydrolysis in a purely aqueous medium. In acidic water, the synthesised inorganic/organic hybrid materials displayed opposite macroscopic chirality depending on the enantiomer of **24** employed. The tubular shape is likely to result from a combination of two phenomena: the autoassociation abilities and a self-templating structuration of the hybrid materials by the organic crystalline precursor. In ethanol/water mixtures, tubular silica containing many parallel channels was

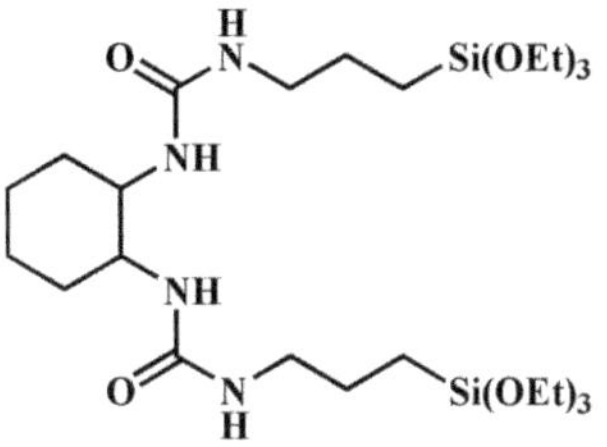

(24)

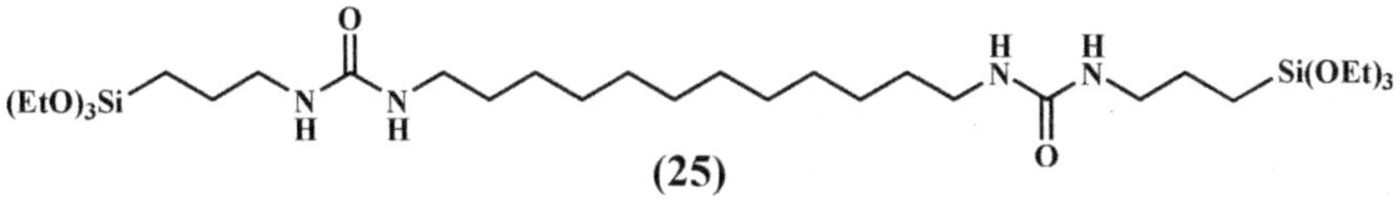

(25)

Chart 8.11

obtained.[51] Although no true gels were used in both cases, the actual template involved stacks of molecules of **24** similar to the stacks present in the gel state. Similarly, compound **25** formed an organogel in toluene. Here also, the polymerisation was carried out in the nongelated solvent, water under acidic condition, and homogeneous lamellar silica was obtained.[52] These examples, although not employing gels, showed that incorporation of the inorganic precursor into a gelator structure is a viable strategy that should be extendable to fibres of true gels.

8.3.2 TiO₂ Nanoparticles

Apart from the preparation of silica nanomaterials, alkoxides of transition metals and other elements (*e.g.*, Ti(O^{i}Pr)$_4$, Ta(OC$_2$H$_5$)$_5$ and O=V[OCH-(CH$_3$)$_2$]$_3$) have been used as the inorganic precursors for the formation of polymeric metal (Ti, Ta, V) oxides or other oxide materials.

In this regard, Hanabusa and coworkers reported the preparation of novel TiO₂ nanomaterials with a hollow-fibre structure like "macaroni" by exploiting the supramolecular fibre network template of organogels of compound (**26**) (Chart 8.12).[53] The sol–gel polymerisation of titanium isopropoxide was carried out in the presence of cationic amphiphilic gelator **26** with the expectation that it would electrostatically interact with titania species in the sol–gel polymerisation process.

Polycondensation of the titanium alkoxide, performed under basic conditions with ammonium hydroxide as catalyst, resulted in the formation of hollow, tubular TiO₂ fibres after calcinations at 450 °C (Figure 8.6). On the contrary, when 2M aqueous hydrochloric acid was used as catalyst, no fibrous structures were observed (Figure 8.6). Under basic condition, the propagating species is considered to be anionic. The electrostatic interaction between oligomeric titania species and cationic gelator was responsible for the formation of TiO₂ hollow fibres. Whereas in strongly acidic medium, cationic titanium

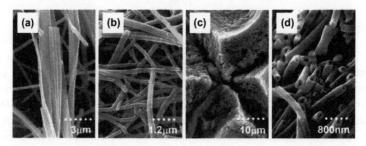

Chart 8.12

$X = PF_6^{\ominus}$ (26) (1R, 1S)

$ClO_4^{\ominus}$ (27) (1R, 1S)

$ClO_4^{\ominus}$ (28) (1R, 1S)

Figure 8.6   SEM images of the dried samples prepared under acidic (a) and basic (b) conditions; SEM images of calcined samples prepared under acidic (c) and basic (d) conditions.
Reprinted with permission from ref. 53. Copyright 2000 American Chemical Society.

species were not adsorbed onto cationic aggregates of **26** and no fibrous structure was observed.

The chiral aggregates of organogelators **27** and **28** (Chart 8.12) containing ClO_4^- as the counteranion were successfully employed as chiral templates in the sol–gel polymerisation.[54] The metal (Ta, V) alkoxides afforded tubular helical Ta_2O_5 and V_2O_5 structures. The helices of metal oxide fibres could be controlled by the chirality of the cationic amphiphile.

In all of the previous examples discussed above, positively charged gelators have been primarily used as template for sol–gel condensation because the corresponding gel fibres will electrostatically attract the negatively charged precursors produced at the initial stage of the polymerisation. In this context, Suzuki and coworkers described the preparation of TiO_2 nanostructures in organogels based on uncharged gelators. The sol–gel polymerisation of $Ti(O^iPr)_4$ in the organogels of **29** (Chart 8.13) containing propylamine as a catalyst resulted in the formation of TiO_2 nanotubes.[55] Propylamines reacted with the -COOH groups in **29**, resulting in the formation of charged nanofibres and sol–gel polymerisation took place on the nanofibres (Figure 8.7). Whereas the negatively charged gelator **30** produced uniform sized TiO_2 nanoparticles as the polymerisation on the negatively charged nanofibres of gelator **30** is inhibited due to the electrostatic repulsion between nanofibres and sol–gel precursors resulting in their inability to act as a template.

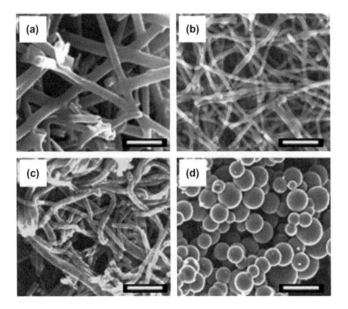

(29)

(30)

Chart 8.13

Figure 8.7 FESEM images of TiO$_2$ prepared in organogels based on **29** in (a) ethanol, (b) 1-butanol, (c) 1,4-dioxane and (d) in ethanol without **29**. Scale bars are 1.5 mm for a and b, 300 nm for c and 6 mm for d.
Reproduced with permission from ref. 55, The Royal Society of Chemistry.

In a separate work Das and coworkers utilised ionogels as a template for the synthesis of TiO$_2$ nanopaticles. Ionogels formed by amino acid based small molecule, **31** (Chart 8.14) was exploited as template to prepare spherical TiO$_2$ nanoparticles.[56] TiO$_2$ nanoparticles were synthesised by adding Ti(O^{i}Pr)$_4$ to the ionogel of **31** in aqueous (10% H$_2$O) BMIMBr and adding an equal amount of water followed by calcination. Formation of uniform-shaped anatase TiO$_2$ nanoparticles of 25–30 nm was confirmed from FESEM and AFM studies. The importance of the supramolecular networks present in the ionogel for the synthesis of TiO$_2$ nanoparticles was highlighted by the fact that aggregated nanoparticle was observed when the same reaction was carried out in the

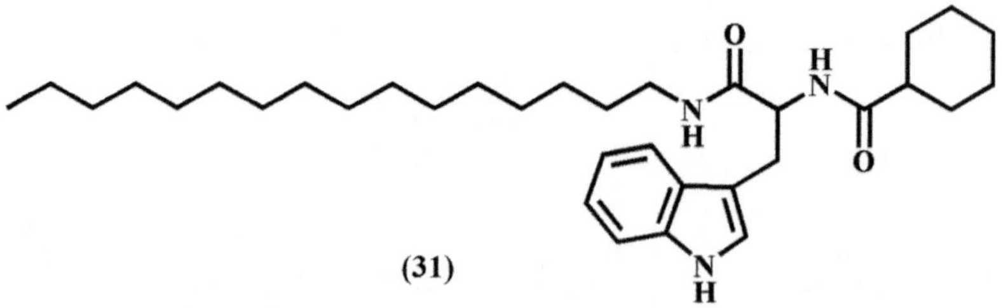

(31)

Chart 8.14

absence of the gelator or below the minimum gelation concentration (MGC) of the ionogelator.

Apart from the synthesis of various metal oxide nanoparticles, LMW gels have also been used for the templated synthesis of a much wider range of inorganic materials. For example, Xue and coworkers introduced a dicholesterol-based organogelator **32** (Chart 8.15) as a template to obtain porous CdS nanofibres having "pearl-necklace" architectures.[57] It was found that Cd^{2+} ions were adsorbed on the organogel fibres by the interaction with ester groups of **32**, which led to the change of the arrangement of the organogelator and also served as nucleation sites for mineralisation. The TEM image revealed the tubular structure of CdS nanofibres with inner diameters of 4–6 nm and 60 nm outer diameters (Figure 8.8).

Similarly, organogelator **33** (Chart 8.15) possessing two binding sites for metals, acted as template for the synthesis of various CuS nanofibres with different helical pitches.[58] The interesting aspect of this templated synthesis was that depending on the sulfur source, CuS nanofibres of different morphologies can be obtained. When H_2S was used as the sulfur source, straight and bent helical CuS nanofibres were formed and only bent CuS nanofibres were produced when thioacetamide was used as the sulfur source.

However, one major limitation of using organogels as template is that the necessary precursors that can be employed are restricted to only metal alk-oxides and this limits the number of metal oxides that can be synthesised by this template method. Other metal salts can be used as precursors if hydrogels are employed as templates. In this regard, a tripodal cholamide-based hydrogel **34** (Chart 8.16) has been employed as a template to synthesise inorganic nanotubes.[59] Various nanotubes of oxides such as SiO_2, TiO_2, ZrO_2, WO_3 and ZnO, as well as nanotubes of sulfates such as the water-soluble $ZnSO_4$ and $BaSO_4$ were prepared using this hydrogel as template. Following similar procedure, nanotubes of CdS, ZnS and CuS from the corresponding metal acetates and Na_2S were prepared using a hydrogel of **34**.

In a separate work, tripeptide-based molecular hydrogels (**35–37**, Chart 8.17) were exploited for immobilising luminescent CdS nanoparticles within the gel phase by incorporation into definite arrays on the gel-nanofibre structures.[60] The size of the CdS nanoparticles remained almost the same before and after deposition on the gel nanofibre. Photoluminescence (PL) measurement of the CdS nanoparticles upon deposition on the gel nanofibres showed a significant

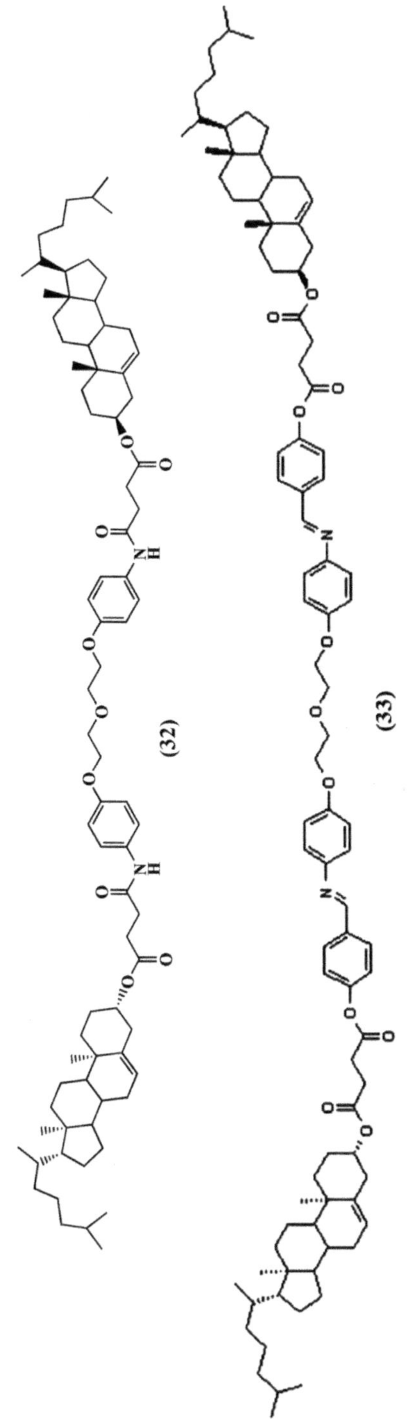

(32)

(33)

Chart 8.15

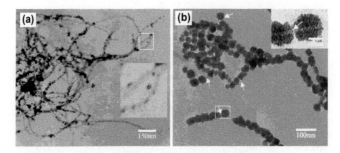

Figure 8.8   TEM micrographs of (a) CdS nanofibres after treatment of composited gel
with H$_2$S: for 2 min. (b) There are many continuous large particles and the
organic ribbon between sections of CdS (indicated by arrow).
Reprinted with permission from ref. 57. Copyright 2004 American
Chemical Society.

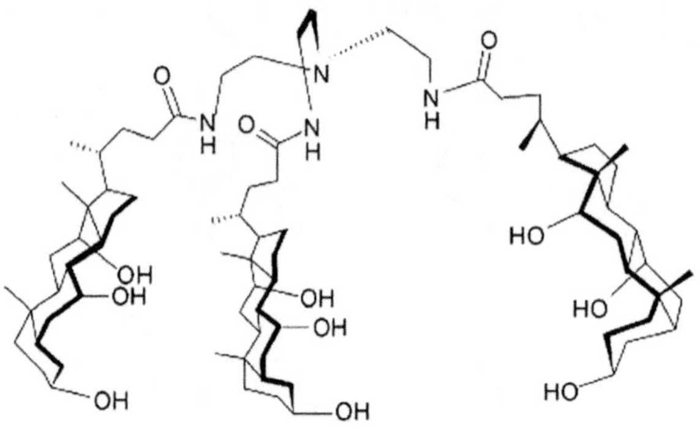

(34)

Chart 8.16

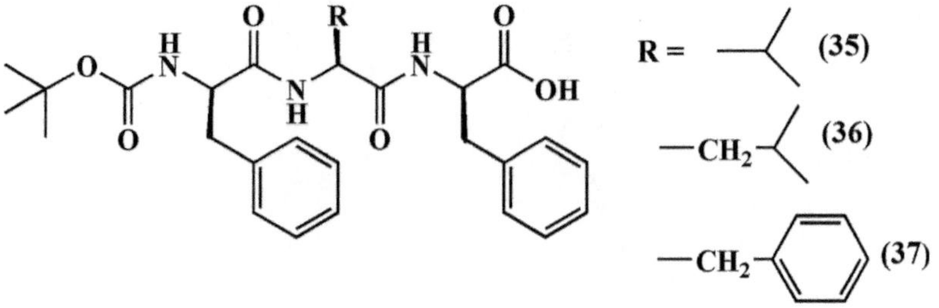

R =   (35)

—CH$_2$ (36)

—CH$_2$ (37)

Chart 8.17

blue shift in the emission spectrum of the nanoparticles. This delineates that the
optoelectronic properties of CdS nanoparticles can be tuned upon deposition
on gel nanofibres.

Stupp and coworkers reported the pH-induced self-assembly and mineralisation of a peptide amphiphile (PA **38**, Chart 8.18) to create a nanostructured

(38)

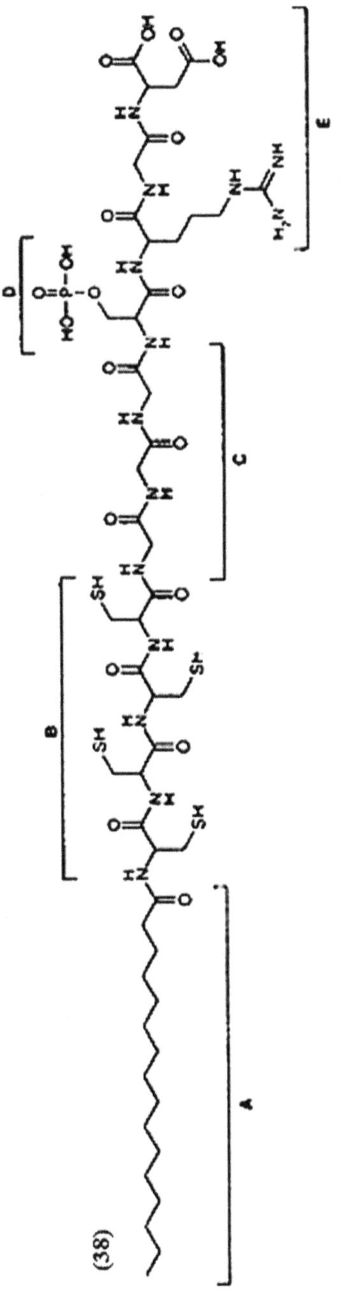

(39)

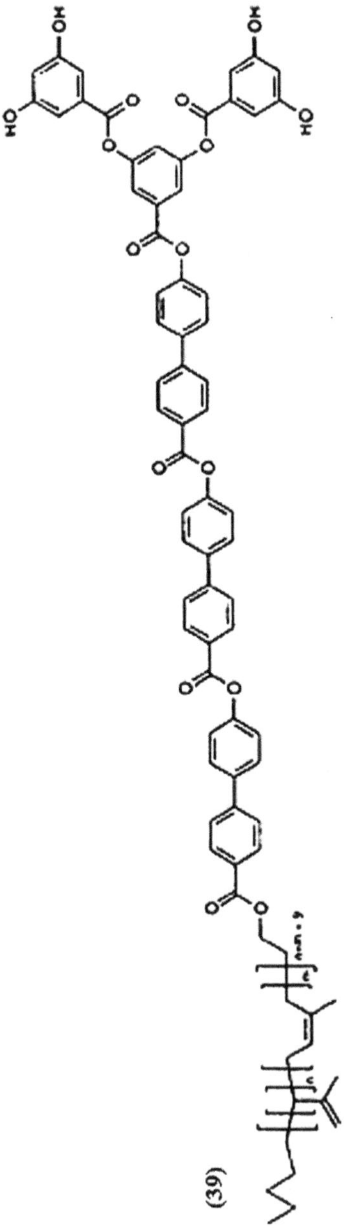

Chart 8.18

composite material that recreates the structural orientation between collagen and hydroxyapatite (HA) observed in bone.[61] The structure of the peptide amphiphile was comprised of five distinct segments such as a hydrophobic alkyl tail with 16 carbon atoms, four consecutive cysteine residues that can form disulfide bonds upon oxidation, three glycine residues as flexible linkers, a phosphorylated serine residue for interacting with calcium ions to direct the mineralisation of hydroxyapatite and the cell-adhesion ligand RGD. The PA formed birefringent gel in water below pH 4 (MGC > 2.5 mg mL^{-1}). The mineralisation property of the PA nanofibres was investigated by mounting a drop of aqueous PA solution on a TEM grid in an atmosphere of HCl vapour, oxidising the cysteines and diffusing $CaCl_2$ and Na_2HPO_4 into the film from opposite sides. In all cases, preferential alignment of the hydroxyapatite crystallographic c-axis with the long axes of the gel fibres was reported. Mineralisation using a nonphosphorylated PA resulted only in amorphous deposits of HA around the gel fibres that indicated the crystallisation-directing effect of the phosphorylated serine group.

Stupp's group later employed a different LMW organogelator (**39**, Chart 8.18) as a template for the mineralisation of polycrystalline CdS.[62] The TEM image of gels in ethyl methacrylate and 2-ethylhexyl methacrylate revealed the formation of helical nanoribbons. The mineralisation was done by exposing a solution of cadmium nitrate and the gelator to hydrogen sulfide gas. The affinity of Cd^{2+} ions for the hydrophilic regions of the ribbons (*i.e.* the hydroxy-containing parts) was the main reason for the observed templating effect. A local supersaturation of Cd^{2+} ions leads to nucleation, growth of the CdS along the fibre surfaces, and, eventually led to the formation of helical structures.

8.4   Gel-Templated Synthesis and Decoration of Metal Nanoparticles

As mentioned in the introduction, not only metal oxide nanostructures, but also metal nanoparticles like Au and Ag have been templated on gel-phase materials. There are two different ways to use LMW gels as template to prepare, stabilise and also to decorate metal nanoparticles (MNPs) into ordered assemblies i) by physically mixing preformed MNPs within the SAFIN of LMWGs and ii) *in situ* synthesis of MNPs in the LMW gels. Importantly, these synthesised gel nanocomposites have diversified applications.

8.4.1   Metal Nanoparticles within SAFIN

Kimura and coworkers first reported the entrapment of AuNPs into a 3D-network structure of an organogelator (**40**) containing two functional groups, a trans-1,2-bis (alkylamide)cyclohexane unit and two thiol groups (Chart 8.19).[63] Octanethiol-stabilised AuNPs (having average diameter 1.7 nm) were prepared separately and then the octanethiol on the surface of AuNPs was replaced by

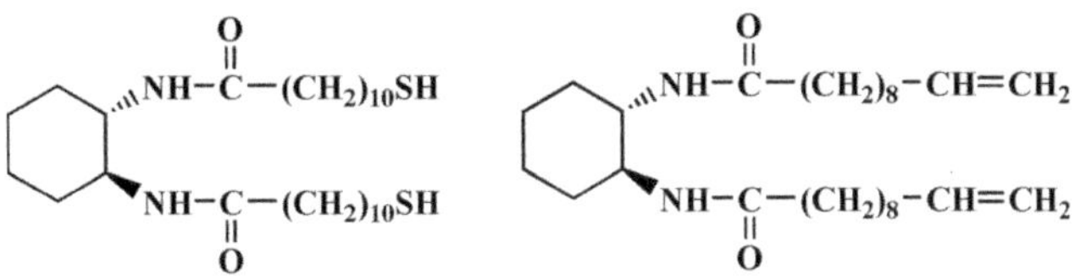

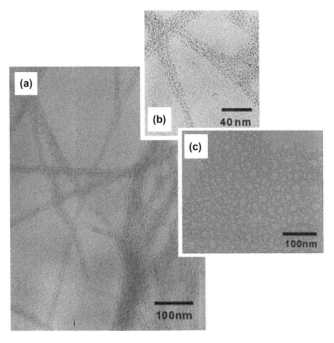

Chart 8.19

Figure 8.9 (a) TEM image and (b) magnified TEM image showing the alignment of AuNPs along the gel fibres, of **40**, (c) TEM image of aggregated AuNPs in presence of **41**.
Reproduced with permission from ref. 63, Wiley-VCH.

the thiol groups of the gelator (**40**) by site-exchange reactions. The AuNP-gel nanocomposites did not show any precipitation even after one month. An interconnected 3D fibrous structure consisting of individual AuNPs with diameters of 10–30 nm was observed from the TEM (Figures 8.9a and b). The importance of the –SH moiety in the stabilisation of AuNPs was confirmed by the absence of 3D fibrillar networks in the organogel formed by structurally analogous molecule, **41** (Chart 8.19) lacking any thiol group (Figure 8.9c).

To the same end, Bhat and Maitra exploited bile-acid-derived hydrogelator **34** (Chart 8.16) to immobilise structurally similar bile-acid-derived thiols capped AuNPs within the hydrogel matrix.[64] The steroidal thiols (**42–44**, Chart 8.20) stabilised AuNPs were synthesised by reducing $HAuCl_4$ by

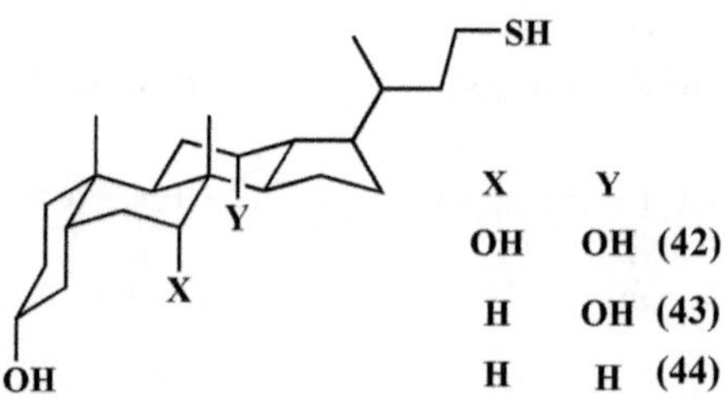

X   Y

OH   OH   (42)

H   OH   (43)

H   H   (44)

Chart 8.20

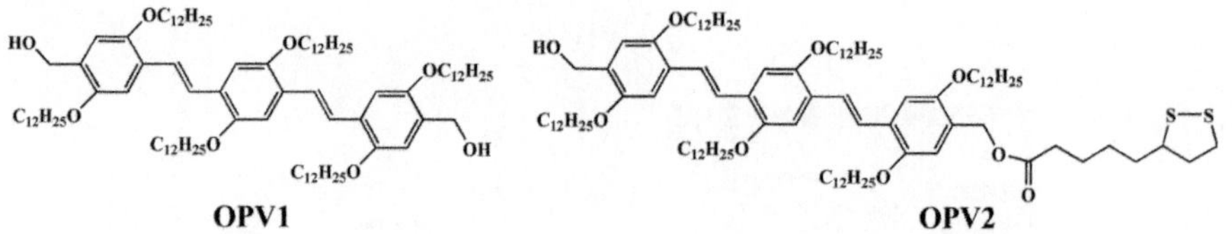

OPV1
(45)

OPV2
(46)

Chart 8.21

NaBH$_4$ in MeOH. This hybrid material was prepared by dissolving AuNPs and the gelator (above MGC) in acetic acid (AcOH) first and then diluting with water. The stability of AuNPs, dispersed within the gel network remarkably improved up to several months in contrast to what was observed in 20% AcOH-water for only 5 h. The TEM images of the AuNP-gel composite showed that steroid-capped nanoparticles were aligned along the gel fibres of **34**. Similarly, Schenning and coworkers used π-conjugated oligo(*p*-phenylenevinylene) (OPV) tapes and AuNPs as a template to develop novel gel-nanocomposites (**45**, **46** Chart 8.21).[65] Hybrid OPV1/OPV2-Au supramolecular tapes were prepared by mixing OPV1 and OPV2–Au in toluene followed by heating above the gelation temperature of OPV1. In the TEM images, pairs of parallel rows of AuNPs were visible, which indicated that the tape-like structures with gold particles on the sides were already formed in solution (Figure 8.10). The proximity of the MNPs to the π-conjugated tapes facilitates electronic communication.

In another report by del Guerzo and coworkers, hybrid gels were developed by simultaneous self-assembly of the 2,3-didecyloxyanthracene (DDOA) gelator (**47**, Chart 8.22) and thiol-capped AuNPs in decanol.[66] This hybrid gel was prepared within minutes and did not show any phase separation. The nanocomposite formation was driven mainly by the supramolecular interaction of the ligands on the nanoparticles and the gelator. The nanoparticles were uniformly distributed over the fibrillar network of the organogelator in *n*-butanol or *n*-decanol. Moreover, the mechanical and thermal stabilities of the DDOA organogels were not significantly altered by the inclusion of AuNPs and they remained as strong, viscoelastic materials.

Bhattacharya and coworkers reported the influence of capping agents on nanoparticle surfaces to tune the morphological features and the viscoelastic

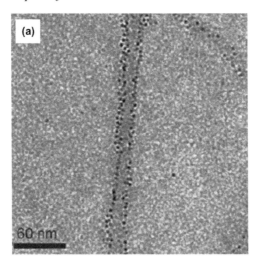

Figure 8.10 TEM images of the OPV1/OPV2–Au (100 : 1) tapes deposited from toluene. Reproduced with permission from ref. 65, Wiley-VCH.

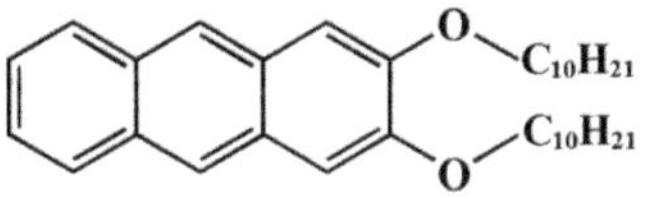

(47)

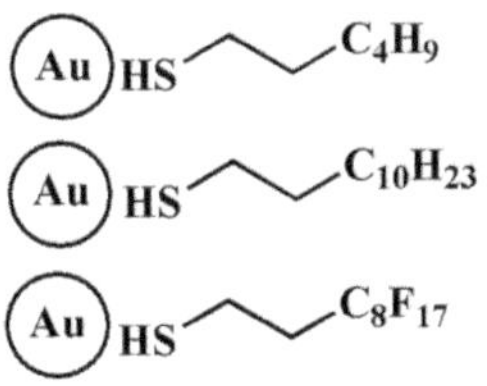

Chart 8.22

properties of gel-AuNP composites.[67] A fatty acid amide of L-alanine (**48**, Chart 8.23) that formed stable gels in various aliphatic and aromatic hydrocarbons was used as the template to develop the hybrid gels.[68] The molecular structure of the capping agents anchored on the nanoparticles' surface was varied such as *n*-alkanethiols (AuC$_{m+2}$, m = 4, 6 and 10), cholesterol-based thiol (AuChol) and *p*-thiocresol (AuPhMe) to probe the gel–nanoparticles interaction. The average diameter of the nanoparticles was 3–6 nm depending on the capping agent used. The incorporation of the AuNPs within the gel had a profound impact on the morphological features of the resulting composite as observed by SEM. The native organogel had a fibrous network, whereas the incorporation of AuC$_{12}$ resulted in the coalescing of the fibres. On the other hand "rolled- tubular" type and platelet aggregates were obtained in the case of AuChol and AuPhMe (Figures 8.11a–d).

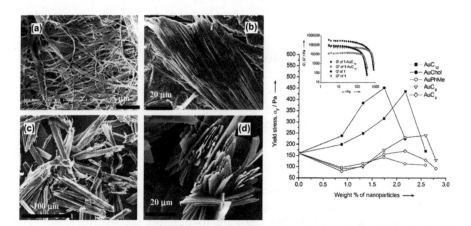

Chart 8.23

Figure 8.11 SEM images of xerogels of (a) **48**, (b) **48**–AuC₁₂ composite, (c) **48**–AuChol composite, and (d) **48**–AuPhMe composite, (e) Plots of yield stress *vs.* AuNP concentration.
Reproduced with permission from ref. 67, Wiley-VCH.

The rheological experiments revealed that incorporation of alkanethiol and cholesterol thiol-capped nanoparticles improved the mechanical strength of the gel due to the increase in the yield stress of the nanocomposites (Figure 8.11e). In the homologous series of *n*-alkanethiol capped nanoparticles, the rigidity of AuC$_{m+2}$-NP composites progressively increased with the chain length of the thiol, suggesting interdigitation of the hydrophobic tail as a probable mechanism of attaining mechanical stability. In contrast, incorporation of thiocresol-based AuNPs made the composite less rigid than the native gel. Therefore, the long *n*-alkyl chain and the cholesteryl group interdigitated better with the gel fibres by superior van der Waals interactions compared to the thiocresol group.

Bhattacharya and coworkers also developed two-component hydrogel systems by mixing stearic acid (SA) or eicosanoic acid with di- or oligomeric amines in specific molar ratios, and the developed hydrogel was used as a medium for the synthesis of silver nanoparticles (AgNPs).[69] The synthesis of gel-AgNP composite was accomplished by reducing $AgNO_3$ using $NaBH_4$ in melted hydrosol of SA-amine, which upon cooling afforded the nanocomposites. These gel-AgNP composites were thermoreversible and the gelation behaviour remained intact even after repeated heating–cooling. The synthesised AgNps were spherical with average diameter 4–7 nm and were preferentially located along the gel fibres in a long assembly. This observation indicated that the gel fibres could act as a host on which the nanoparticles were embedded during fibre formation.

On the other hand, Firestone and coworkers demonstrated that nanostructured, ionic liquid-based gels can be used to template the formation of new nanoparticle morphologies having potential applications in optoelectronics, catalysis, *etc.* They reported the formation of anisotropic gold nanoparticles with varying sizes and morphologies (such as triangular, trigonal prismatic, *etc.*) nanorods by photochemical reduction of tetrachloroaurate ($AuCl_4^-$) ions in the highly constrained aqueous domain of a nanostructured ionogel template (formed *via* self-assembly of the ionic liquid 1-decyl-3-methylimidazolium chloride ($C_{10}mim^+Cl^-$) in water).[70] These materials feature highly constrained water channels indicating their potential utility as templates for the formation of nanosized metal particles. The formation of nonspherical AuNPs was confirmed from the broad, prominent peak at around 590 nm that bears a shoulder at 660–760 nm along with a second band at 796–896 nm in UV/Vis–NIR spectra of the nanocomposite. The SEM images of these nanocomposites revealed the formation of different morphologies of AuNPs such as triangular, hexagonal, trapezoidal, and rectangular shapes, *etc.* (Figure 8.12).

In a separate study, Love and coworkers utilised a dendritic organogelator (**49**, Chart 8.24) comprising a disulfide bridge for the synthesis and stabilisation of AuNPs within a toluene gel.[71] First, a solution of $HAuCl_4/n$-(C_8H_{17})$_4NBr$ in toluene was allowed to stand on a gel for the synthesis of AuNPs. After a few days the gel assumed a uniform yellow colour due to the diffusion of $HAuCl_4$ and the excess of toluene was removed from the top of the gel. When this gel was irradiated with a 100-W mercury lamp, initially the colour of the gel changed from orange-yellow to colourless and then after a few hours from colourless to intense purple. These changes were consistent with the initial reduction of Au(III) to Au(I) and then to Au(0). Uniform spherical-shaped AuNPs having an average diameter of 13 nm were formed with a surface plasmon resonance (SPR) peak at 550 nm. Upon gel melting, it was observed

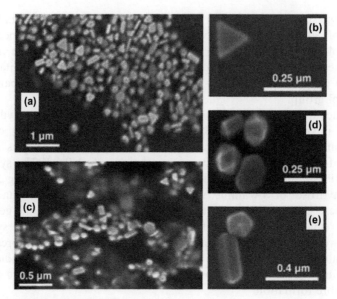

Figure 8.12   SEM images of anisotropic AuNPs (a–e) synthesised in $C_{10}mim^{+}Cl^{-}$
ionogel matrix containing about 16% (w/w) 0.02M aqueous $HAuCl_4$
after UV irradiation for 70 min.
Reproduced with permission from ref. 70, Wiley-VCH.

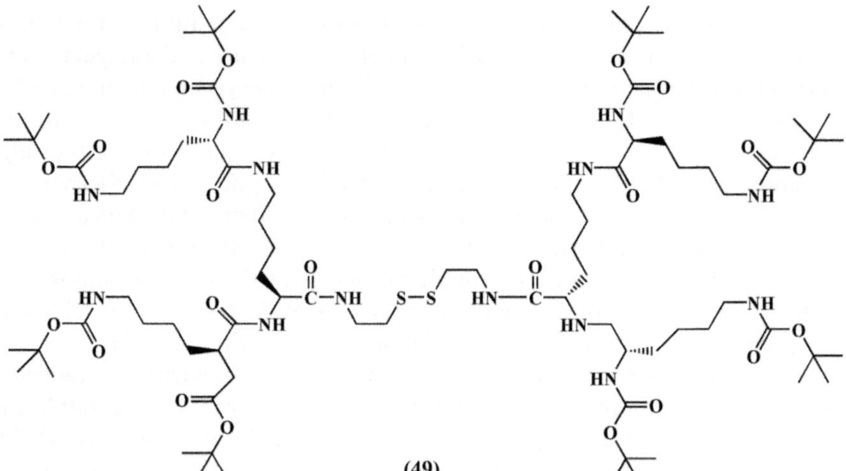

(49)

Chart 8.24

that the purple nanoparticles separated from the medium to form a black
precipitate. This observation once again confirmed the significance of the gel
medium in the synthesis and stabilisation of AuNPs.

Similarly, *N*-terminal protected dipeptide hydrogelator Fmoc-Val-Asp-OH
(**50**, Chart 8.25) was used as a template for the synthesis of fluorescent Ag

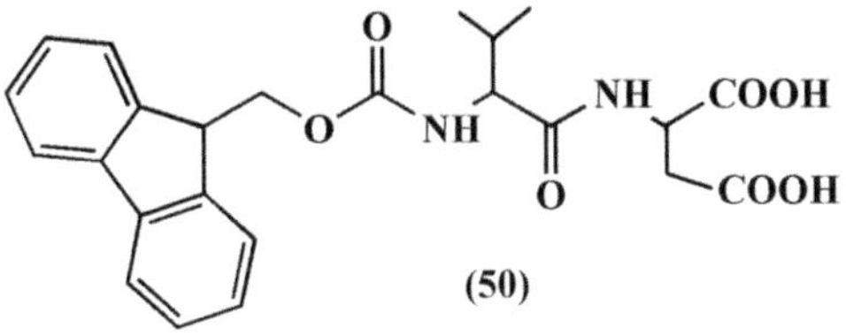

(50)

Chart 8.25

nanoclusters under sunlight by Adhikari and Banerjee.[72] The gelator **50** formed a transparent, stable hydrogel with a MGC of 0.2% w/v. In order to prepare AgNPs within the gel matrix, freshly prepared aqueous AgNO$_3$ was added to a gelator solution (at MGC) in DMSO in a molar ratio of 1:1. An Ag$^+$-encapsulating transparent hydrogel was immediately formed that turned violet upon exposure to bright sunlight. The violet colouration indicates the reduction of silver ions (Ag$^+$) to silver nanoclusters Ag(0) and this colour change was not observed in the absence of sunlight or silver or peptide gelator. Interestingly, when AgNPs were synthesised following the same procedure in an aqueous solution of gelator peptide at a much lower concentration than MGC (not in the gel state) much bigger particles were formed that ultimately precipitated out from solution within 2 h. Thus, it can be envisaged that the 3D-gel network structure provides good stability for Ag nanoclusters. HRTEM images of the nanocomposite revealed the formation of extremely small AgNPs with the dimension of 1–3 nm.

Shen and Xu showed that Ag (I) could serve as a "bridging agent" for triggering sodium carboxymethylcellulose (NaCMC) solution to form a hydrogel.[73] Intriguingly, it was found that the transparent and colourless NaCMC hydrogel gradually turned brown and more stable in the presence of the irradiation of room light (Figure 8.13).

However, no such change in colour was observed when the gel was placed in the dark. FESEM images of the xerogels obtained by freeze-drying the Ag(I) triggered NaCMC hydrogels before and after irradiation of room light showed formation of well-defined 3D nanosized/microsized sheet-like structures in the hydrogel, which were different from that of NaCMC (Figures 8.13a–c). The developed hybrid system also acts as a novel signalling platform for thiol-containing amino acids or small peptides (such as GSH) by affording a visual gel–sol transition in the presence of thiol-containing amino acids or small peptides.

The influence of Ag-salt in supramolecular hydrogelation was also observed in the case of a new set of bis(urea) ligands comprising three pyridyl groups that exhibited strong gelation with silver tetrafluoroborate in THF/water mixtures. Interestingly, the formation of gel by **51** (Chart 8.26) in the presence of Ag indicated the importance of the third central pyridyl moiety in metallogelation.[74] Upon standing for a few weeks AgNPs grow in the gel. A TEM study confirmed that the particles were exclusively formed in the gel fibres. The samples exhibited complex rheological behaviour depending on the initial concentration of AgBF$_4$ and the presence of nanoparticles. After

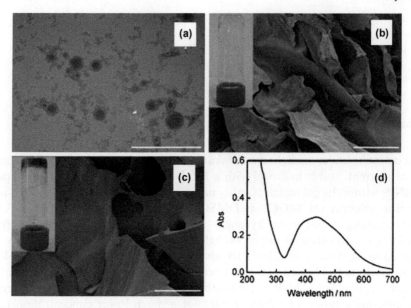

Figure 8.13   FESEM images of NaCMC solution (a), xerogels from the original Ag(I) triggered NaCMC hydrogels before (b) and after (c) then irradiation of room light; absorption spectrum of the irradiated hydrogel (d). Insets are corresponding photographs of the hydrogels. Scale bar for (a) is 50 mm and for (b and c) is 100 mm.
Reproduced with permission from ref. 73, The Royal Society of Chemistry.

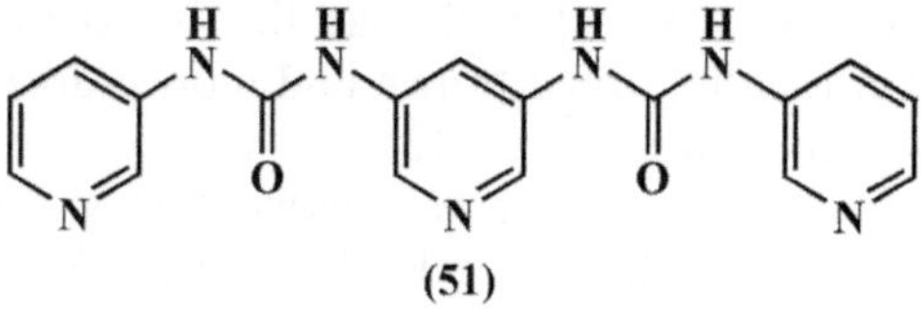

(51)

Chart 8.26

irradiation the gel samples containing 2 and 3 equiv. of $AgBF_4$ showed an increase in $G^{/}$ and $G^{//}$ (storage and loss modulus) after the formation of the nanoparticles.

The increase in $G^{/}$ and $G^{//}$ suggested that the nanoparticles might act as additional nodes to which the gel strands can connect and thereby increased the stability and rigidity in the gel structure as the "hard" nanoparticles fill voids in the gel structure.

8.4.2 *In Situ* Synthesis of Metal Nanoparticles

Until now, LMW gels have been used as a template for the stabilisation and decoration of externally added preformed MNPs and for the synthesis of Au/Ag NPs either by externally added reducing agents or by photochemical means. However, in recent years synthesis of gelator molecules capable of reducing metal ions to corresponding MNPs within the gel matrix in the absence of any reducing agent have attracted immense importance. In this regard, Vemula and John for the first time reported the *in situ* synthesis of AuNPs using LMW hydrogelators comprising of monosubstituted urea-based aryl derivatives with a free terminal amine (**52–54**, Chart 8.27).[75] The free amine group present in the gelator structure was used to reduce HAuCl$_4$ to AuNPs. The amine group also acts as a capping agent to the synthesised AuNPs. The AuNP-embedded gels were pink in colour and showed a characteristic surface plasmon resonance (SPR) peak at 553 nm indicating the formation of gold nanoparticles.

The TEM experiment also revealed the presence of well-dispersed spherical-shaped AuNPs (11–15 nm) on the edges of the gel sheets. Importantly, after reduction of HAuCl$_4$ to AuNP, the gelation properties and also the supramolecular morphologies of the gels remained intact and were stable for several months.

Similarly, Banerjee and coworkers reported the *in situ* synthesis of Au and Ag NPs within the supramolecular gels of small tripeptide-based molecules (**55–57**, Chart 8.28) comprising a redox active tyrosine residue as a template.

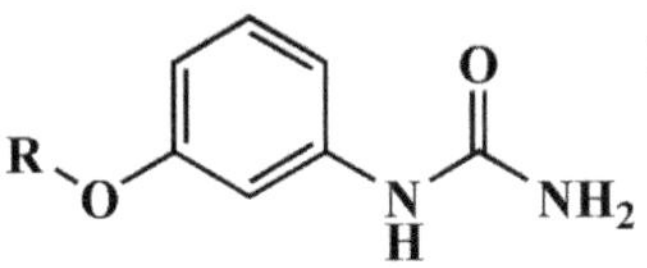

$$R = n\text{-}C_6H_{13} \quad (52)$$
$$n\text{-}C_{10}H_{21} \quad (53)$$
$$n\text{-}C_{14}H_{29} \quad (54)$$

Chart 8.27

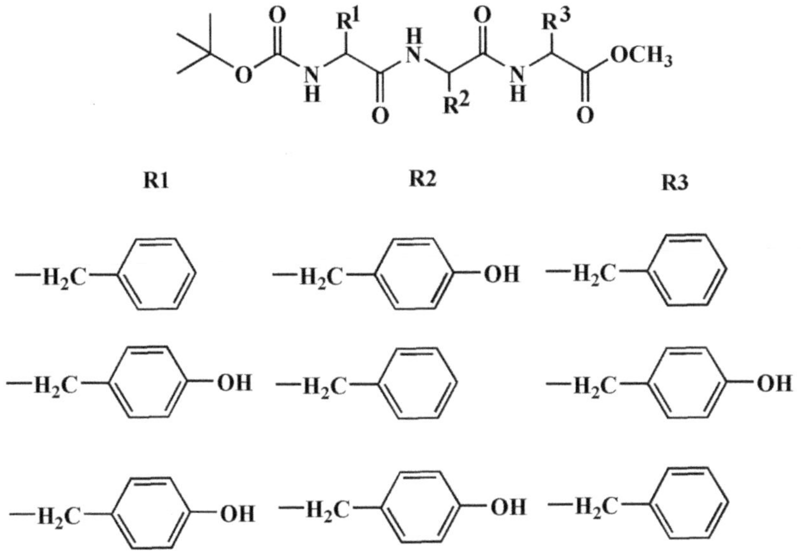

R1	R2	R3	

Chart 8.28

The tripeptide gelators formed 3D supramolecular nanofibrillar networks in the gel state.[76] For the synthesis of AuNPs, gelator peptide was added to chloroaurate solution in toluene and heated above 100 °C to produce a homogeneous solution. However, the yellow colour of the gel did not change even after keeping this for several days. Interestingly, when triethylamine was added into the toluene gel and heated above 100 °C, the yellow colour solution was rapidly changed to a colourless and eventually turned into a violet indicating the formation of AuNPs *via* the oxidation of tyrosine residues. Upon slow cooling, AuNP-embedded violet coloured gel was obtained. Following the same procedure the authors also demonstrated the *in situ* synthesis of AgNPs in tyrosine-containing peptide gels. It was also found that the *in situ* synthesised AuNPs were aligned in a definite array along the gel nanofibre indicating the important role of the nanofibres that enhances the colloidal stability of the AgNPs/AuNPs (Figure 8.14).

In another report, John and coworkers utilised novel ascorbic acid (AscA) based amphiphilic molecules (**58–60**, Chart 8.29) for the *in situ* synthesis of AuNP.[77] AscA is also known for the reduction of metal salts such as chloroauric acid to AuNPs in solution. The synthesised amphiphiles showed excellent gelation abilities in a broad range of solvents at low concentrations with notable thermal stability. The intrinsic metal salt reducing ability of AscA was used to synthesise AuNPs and subsequently to stabilise the nanoparticles within the self-assembled fibrillar network of the gel. TEM studies confirmed the formation of well-dispersed spherical-shaped AuNPs that were distributed throughout the gel matrix.

0.5 µm

Figure 8.14   TEM image indicating alignment of AuNPs along a gel fibre obtained from the peptide **55**–toluene gel.
Reproduced with permission from ref. 76, The Royal Society of Chemistry.

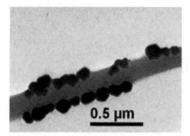

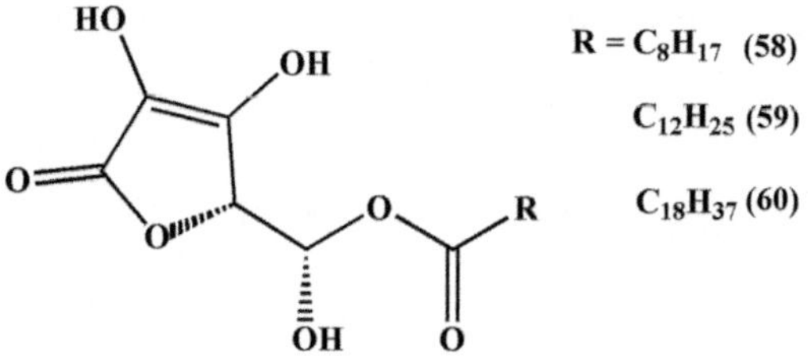

$R = C_8H_{17}$ (**58**)

$C_{12}H_{25}$ (**59**)

$C_{18}H_{37}$ (**60**)

Chart 8.29

Interestingly, similar to the observed chirality induction within silica using a gel template, Zhan and coworkers reported the generation of both right- and left-handed silver nanohelices using the helical ribbons of racemic organogelator 2-acrylamide-dodecane-1-sulfonic acid (ADSA, Figure 8.15A) as a template.[78] Both right- and left-handed silver nanohelices were obtained in chloroform by reducing the adsorbed silver cations (Ag^+) onto the ribbons with sodium borohydride. TEM images of the resultant material showed that opposite-handed helices with widths of 100–200 nm were present (Figure 8.15B).

In a separate study, Li and coworkers reported the preparation of the chiral AgNPs and chiral nanoparticulate films exploiting glutamic acid based organogelators (**61, 62** Chart 8.30).[79] These amphiphilic compounds formed organogels in different organic solvents such as methanol, ethanol or DMSO, *etc*. This organogel matrix was used as a template for *in situ* reduction of silver (I) ions to AgNPs by hydroquinone. The mixture turned yellow after stirring for 3 h and did not change later. The UV-Vis spectrum of the nanoparticles showed a broad absorption band with the maxima at 408 nm, which is a typical SPR of AgNPs. The SEM and TEM images revealed that the nanoparticles were embedded in the fibrous network of the organogel. Interestingly, the CD

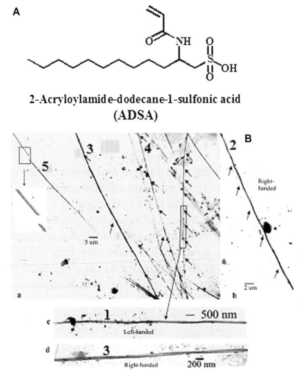

A

2-Acryloylamide-dodecane-1-sulfonic acid
(ADSA)

B

Figure 8.15 (A) Structure of ADSA; (B) TEM images of some typical silver nano-helices (a and b). Enlargements of nanohelix **1** (c) and the upper part of nanohelix **3** (d), respectively.
Reprinted with permission from ref. 78. Copyright 2003 American Chemical Society.

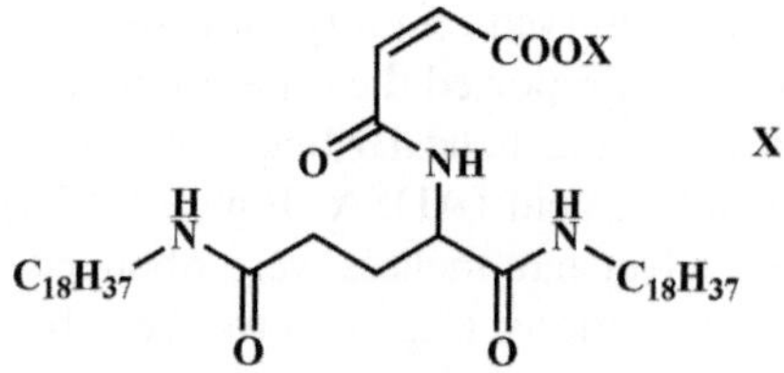

X = H (61)

Ag (62)

Chart 8.30

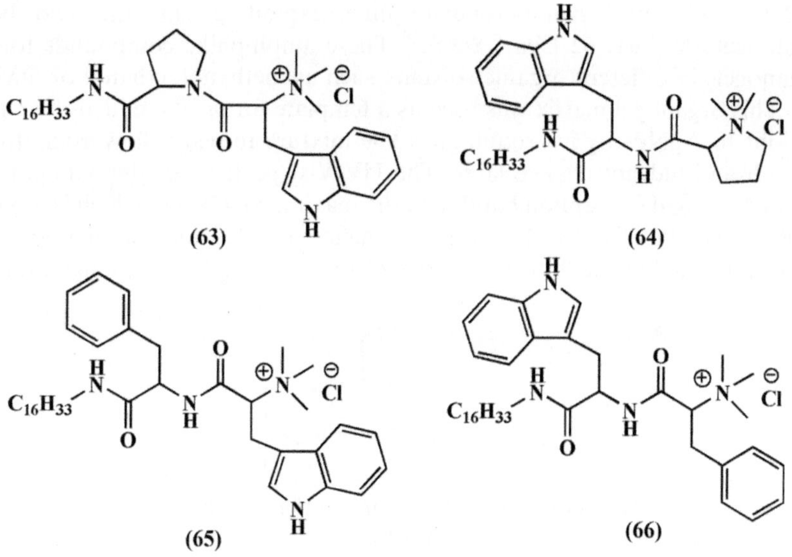

(63)

(64)

(65)

(66)

Chart 8.31

spectrum of the AgNPs exhibited a positive and negative Cotton effect at 442 and 398 nm, respectively with a crossover at 416 nm that indicated that the formed AgNPs were chiral in nature and the chirality of the synthesised AgNPs would probably be related to the chiral nanofibres formed by the gelator.

To this end, LMWGs comprising diversified supramolecular 3D networks, such as fibres, thin sheets, lamellar, *etc.* have also been used as template to regulate the shape/size by of the nanoparticles. For example, Mitra and Das have exploited the different supramolecularmorphologies formed by the tryptophan-based peptide amphiphilic cationic hydrogelators (**64–67**, Chart 8.31) to produce AuNPs having different shape and size such as sheet, wire, octahedral, and decahedral, *etc.* (Figure 8.16).[80] Furthermore, it was found that the gel morphology at MGC is the prime requirement to prepare different nanocrystals of AuNPs.

Similarly, Das and coworkers used the tail-modified amino acid based amphiphiles (**67–69**, Chart 8.32) for the *in situ* synthesis of AuNPs of specific shape without the help of any external reducing agents.[81] The shape and size of the *in situ* synthesised AuNPs was investigated by taking their TEM images that

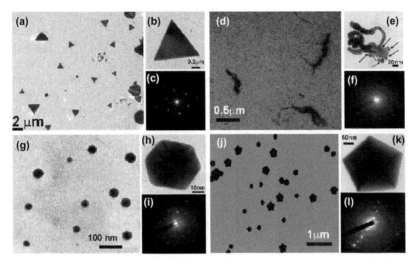

Figure 8.16 TEM images of directly synthesised nanoparticles (a, b in gel **63**; d, e in gel **64**; g, h in gel **65**; j, k in gel **66**) at MGC and the corresponding selected-area electron diffraction (SAED) pattern (c in gel **63**; f in gel **64**; i in gel **65**; and l in gel **66**).
Reprinted with permission from ref. 80. Copyright 2008 American Chemical Society.

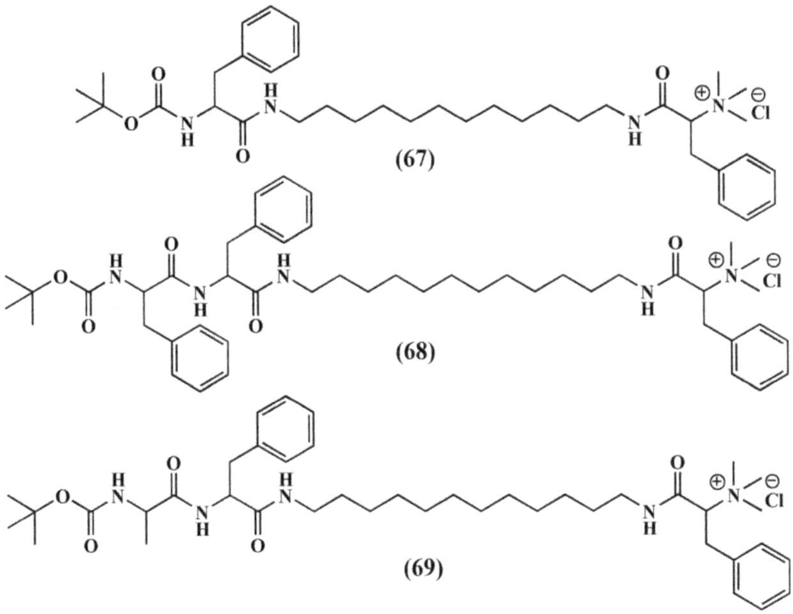

(67)

(68)

(69)

Chart 8.32

revealed the formation of triangular, decahedron, octahedron and some spherical AuNPs. However, these cationic gelators were found to be ineffective for the synthesis of AgNPs as these amphiphiles had chloride counterion, which readily reacts with $AgNO_3$ to precipitate AgCl that has a low solubility product. To this end Dutta and coworkers developed a library of efficient hydrogelators by changing the counterions of amphiphile (70, Chart 8.33) from chloride to aromatic carboxylates.[82] Interestingly, with the change of counterion of the L-tryptophan containing hydrogelating amphiphiles from chloride to benzoate and acetate, AgNPs were successfully synthesised within the gel matrix under mild conditions. AgNPs are well known for their intrinsic antimicrobial activity. In this context Das and coworkers exploited the *in situ* synthesised AgNPs to complement the antibacterial properties of cationic hydrogelators (70,71 Chart 8.33) resulting in the formation of a functional antibacterial gel-nanocomposite.[83–85] The TEM images of the nanocomposites revealed the formation of AgNPs within the self-assemblies of amphiphiles having diameters in the range of 10–25 nm. These nanocomposites exhibited excellent antibacterial activity against both Gram-positive and Gram-negative bacteria. These nanocomposites also showed considerable biocompatibility to the mammalian cell, NIH3T3.

So far, we have mostly discussed the integration of nanoparticles within SAFIN of hydrogels. However, reports on the development of organogel-based soft nanocomposites are very rare. In a recent study Kar and coworkers have shown transformation of *in situ* synthesised Au-hydrogel nanocomposites to an Au-organogel hybrid simply by altering the pH of the systems.[86] Amphiphilic dipeptides containing aromatic residue at the side chain and a free carboxylic acid group (72, 73 Chart 8.34) showed excellent

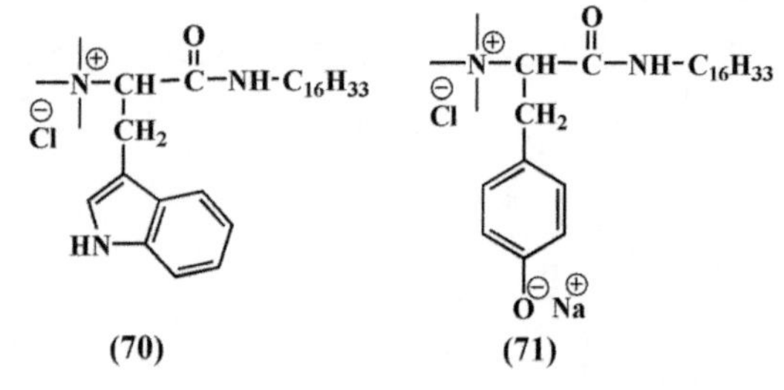

(70)   (71)

Chart 8.33

(72)   (73)

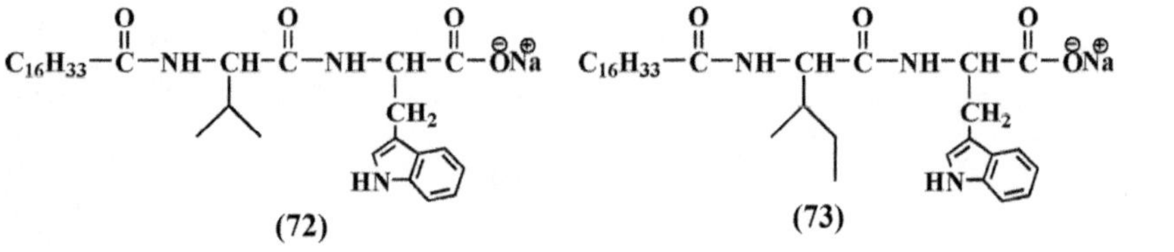

Chart 8.34

organogelation efficacy while their sodium salts were capable of gelating water.[87] AuNPs were synthesised *in situ* within the hydrogel of dipeptide-based amphiphilic carboxylates in the absence of exogenous reducing agent. Upon changing the pH of the hydrogel to acidic by adding the required amount of hydrochloric acid, amphiphilic carobxylates were converted to the corresponding carboxylic acids that gelated the organic solvent with simultaneous entrapment of the AuNP leading to the development of AuNP-organogel composite. The TEM images (Figures 8.17a and b) revealed the importance of supramolecular networks in the stabilisation of the nanoparticles in both hydro/organogel as the AuNPs were aligned on the surface of the amphiphilic nanofibres. Interestingly, the mechanical strength of the AuNP-gel composite improved as confirmed from rheology, probably through the formation of more compact and dense fibres (Figures 8.17c and d).

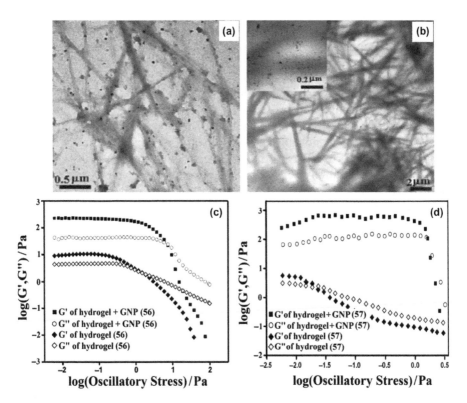

Figure 8.17 TEM images of *in situ* synthesised AuNPs embedded in (a) hydrogels of (**72**), (b) GNPs embedded in organogel after phase transfer, Plots of storage modulus (G') and loss modulus (G'') *versus* stress for hydrogel and AuNP-hydrogel composite of **72** (c) and **73** (d).
Reproduced with permission from ref. 86, The Royal Society of Chemistry.

8.5   Carbon Nanotube and Graphene-Based Gel-Nanocomposites

Carbon nanotubes (CNTs), a most fascinating class of nanomaterials have attracted much attention due to their unique properties like electrical conductivity, thermal stability, mechanical strength and so forth.[88–90] However, these materials suffer from an inherent insolubility in solvents of extreme polarities as they prefer to remain in the bundled state. The aggregated state of CNTs substantially limits their applications and hence debundling of CNTs is very important for their exploitation in different branches of science. In this context, attempts have been made to prepare soft nanocomposites by hybridising this pseudo-one-dimensional allotrope of carbon with LMW gels where both components complement their properties in a synergistic reciprocity.

To this end, Fukushima and coworkers for the first time demonstrated the dispersion of CNTs and formation of CNT-based composite gel at room temperature when the nanotube was ground in imidazolium-based ionic liquids (ILs).[91] The gelation occurs in a variety of imidazolium-ion-based ILs upon grinding with 0.5–1 wt% of single-walled carbon nanotubes (SWNT). This observation implied that the ionic liquid interact with the π-electronic surface of the SWNTs by means of cation–π and/or π–π interactions. The phase transition and rheological studies on this CNT/IL-based nanocomposite revealed that the gel is formed due to the molecular ordering of the ILs instead of the entangled networks of the CNTs. This serendipitous finding has opened up a new possibility for ILs as modifiers for CNTs. The authors also mentioned that the resultant nanocomposites consisting of highly electroconductive nanowires in the form of CNTs and fluid electrolytes in the form of ILs, can be utilised for fabrication of modified electrodes, sensors, capacitors, and actuators.[92]

Apart from ILs, SAFINs of LMW gels can also be successfully applied toward templating novel CNT composites. The presence of SAFIN in gels makes them ideal for the dispersion of carbon nanomaterials due to their ability to interact mutually *via* various modes of noncovalent interactions. Ajayaghosh and coworkers developed a stable CNT–gel composite by dispersing both SWNT and multiwalled carbon nanotube (MWNT) within oligo(p-phenylenevinylene) (OPV) gels in aromatic organic solvents like toluene (**74–76**, Chart 8.35).[93] Interestingly, inclusion of either kind of CNTs accelerated the

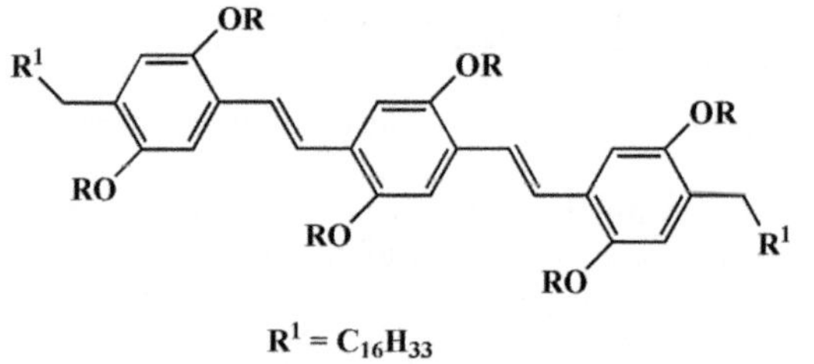

OPV1- R^1 = OH   (**74**)

OPV2- R^2 = OCOCH$_3$ (**75**)

OPV3- R^1 = OCH$_3$ (**76**)

R^1 = C$_{16}$H$_{33}$

Chart 8.35

gelation of the OPV derivatives even below the normal MGC. The composite gels were also found to be thermally more stable than the native gels. The rheological study revealed that the ratio of the G' to the G'' was higher for composite gels compared to the native gels, indicating enhanced stability and formation of a stronger solid-like network in the hybrid gels in the presence of CNT. These observations may be attributed to the efficient wrapping of SWNTs (diameter 1–2 nm) within the self-assembled tapes of OPV-based gelators through strong π–π stacking and van der Waals interactions in aromatic solvents and thereby reinforcing the OPV supramolecular gel (Figure 8.18). The TEM image of SWNTs showed the presence of aggregated nanotubes (Figure 8.19a) and OPV gels exhibited characteristic nanotape morphology in the TEM study (Figure 8.19b) with a width of 10–200 nm and length of several micrometres.

However, in OPV1-SWNT composite debundling of SWNT was observed and these were aligned in a side-wise fashion within the self-assembled OPV (Figure 8.19c). The magnified TEM images indicated the efficient wrapping of SWNT by self-assembled OPV molecules resulting in reinforced supramolecular tapes (Figure 8.19d).

Carbon-nanomaterials-based composite materials have attracted a lot of attention not only due to their ability to modulate the mechanical properties of the soft materials but also for their enhanced electrical conductivity, therapeutic applications and others. For instance, Rao and coworkers hybridised an existing organogel (**48**, Chart 8.23) with SWNT, MWNT and functionalised

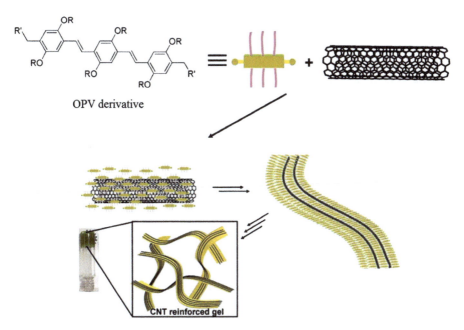

OPV derivative

Figure 8.18 Schematic representation of OPV-CNT gel formation. Reproduced with permission from ref. 93, Wiley-VCH.

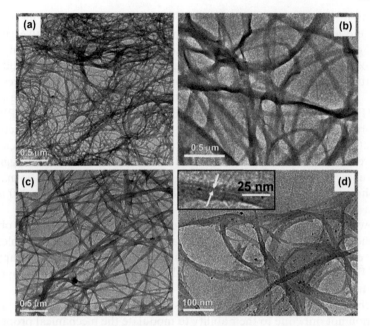

Figure 8.19   TEM images (unstained) of (a) SWNT (b) OPV1 (c) and (d) OPV1-
SWNT nanocomposite. All samples are drop cast from toluene on
carbon-coated TEM grids.
Reproduced with permission from ref. 93, Wiley-VCH.

SWNTs decorated with amides of different chain lengths and used the
nanotubes as a molecular heater for the gel–sol transition by irradiation with a
near-IR laser.[94] This elegant study has huge significance simply for the fact that
living systems are transparent to and are unharmed by near-infrared (NIR)
radiation (700–1100 nm). At the same time CNTs show NIR laser (1064 nm)
driven exothermicity that can be utilised as a stimulus for the disintegration of
the gels and possible release of cargo, and thereby would have implications in
stimuli-responsive drug delivery. However, pristine SWNTs exhibited a limited
dispersion in the organogels. To circumvent this problem, the amphiphobic
SWNTs were made solvophilic through functionalisation with different
aliphatic and aromatic chains such as hexadecyl, dodecyl, octyl and benzyl
and resulted in the formation of gel nanocomposites with higher SWNT
content. Interestingly, a selective gel–sol phase transition of the
nanocomposites at room temperature was accomplished by irradiating with a
near-IR laser at 1064 nm for only 1 min, while prolonged irradiation (30 min)
of the organogel under identical conditions did not cause gel melting.

A similar near-IR laser-induced gel–sol transition was also shown by Rao
and coworkers for the organogelator all-*trans* tri(p-phenylenevinylene) bis-
aldoxime (**77**, Chart 8.36) integrated with pristine and long-chain
functionalised SWNTs.[95] Here also, the incorporation of SWNTs improved
the mechanical properties of the nanocomposites suggesting reinforcement of

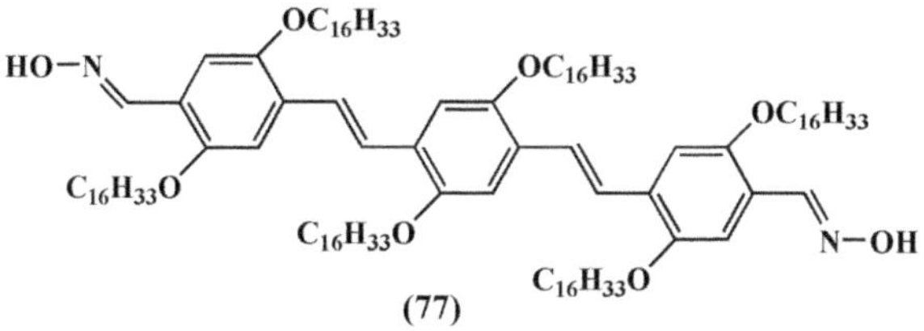

Chart 8.36

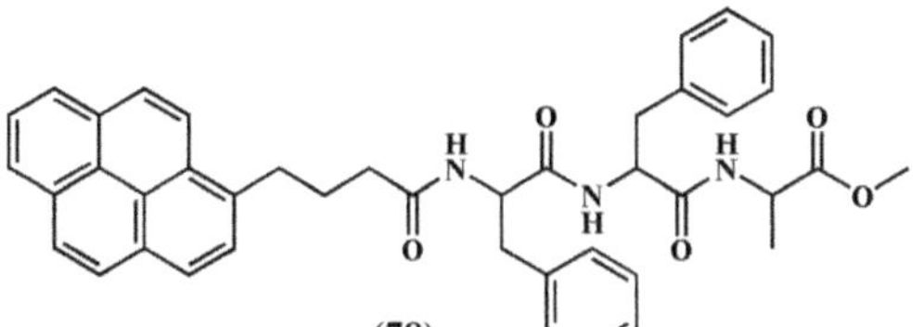

(78) 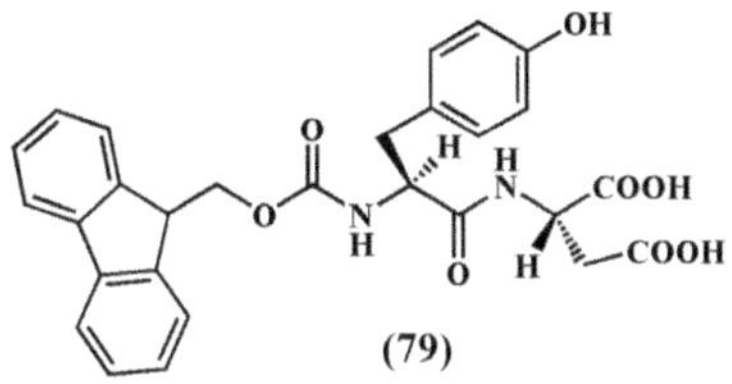   (79)

Chart 8.37

SAFIN with CNT. Interestingly, the nanocomposites showed enhanced electrical conductivity.

To this end, Adhikari and coworkers reported the successful incorporation of graphene within a pyrene-containing peptide-based fluorescent organogel (**78**) to form a stable hybrid organogel (Chart 8.37). The presence of an aromatic pyrene moiety in the gelator peptide facilitated the exfoliation of graphene sheets through noncovalent π–π stacking interaction.[96] Interestingly, here also the MGC of the hybrid organogel was lowered significantly in the presence of graphene. The rheological investigation suggested that the flow of the hybrid organogel had become more resistant towards the applied angular frequency upon the incorporation of graphene into the organogel. The hybrid gel was about seven times more rigid than the native gel.

Adhikari and Banerjee utilised peptide-based hydrogels **79** (Chart 8.37) for the successful incorporation of reduced graphene oxide (RGO) into the hydrogel to obtain a well-dispersed RGO containing stable hybrid hydrogel.[97] The TEM images of the xerogel obtained from the hybrid hydrogel confirmed the coexistence of both dispersed graphene sheets and gel nanofibres (Figure 8.20). The morphology of the hydrogel did not change significantly even after the incorporation of RGO.

Tan and coworkers demonstrated the formation of a supramolecular hydrogel based on SWNTs triggered by a bile salt surfactant, sodium deoxy-cholate (NaDC, **80**).[98] NaDC is an amphiphilic molecule with a hydrophilic part (including two hydroxyls and one carboxyl group) and a hydrophobic part (steroid group). This unique structure helps NaDC molecules to get adsorbed stably onto the surfaces of SWNTs through hydrophobic interactions, resulting in the solubilisation of SWNTs in water. It is known that NaDC forms a fibrous self-assembly in aqueous solution when the concentration is much higher than the CMC.

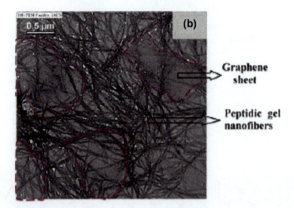

Figure 8.20   TEM image showing the presence of both peptide-based gel nanofibres and graphene nanosheets. The closed area in violet color indicates the presence of graphene sheets.
Reproduced with permission from ref. 97, The Royal Society of Chemistry.

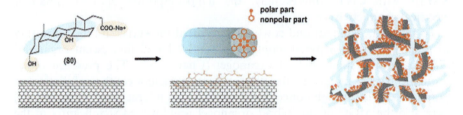

Figure 8.21   Schematic of the formation of a supramolecular hydrogel of SWNTs triggered by NaDC.
Reproduced with permission from ref. 98, Wiley-VCH.

Interestingly, NaDC self-assembled to form supramolecular hydrogel in the presence of dispersed SWNTs in the fibrous matrix (Figure 8.21). The supramolecular hydrogel exhibited excellent viscoelastic properties. Further nanowires and nanopatterns were prepared using the hydrogel as a "solid" ink that showed improved conductivity.

Similarly, Das and coworkers reported the remarkable improvement of organogelation efficiency (in toluene) of amphiphilic dipeptides with a free carboxylic acid group (**81,82** Chart 8.38) after the addition of a small amount of acid-functionalised SWNTs (SWNT-COOH/*f*-SWNT).[99] Addition of SWNT-COOH to a free flowing solution of a weak amino-acid-based gelator (**81**) at 17 fold lower concentration than its respective MGC, resulted in a self-supporting gel. Even more surprising was the formation of SAFIN in the presence of SWNT-COOH (FESEM images) in a solution of gelator that had few isolated/solitary fibres with no networks whatsoever (Figure 8.22). It was observed that the increased presence of planar aromatic residues in the small molecule maximises the gelation efficacy, possibly through enhanced π–π interaction

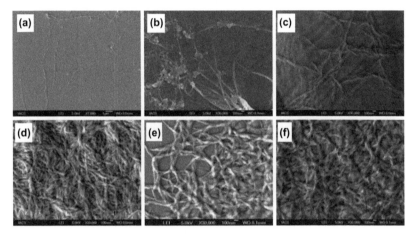

(81)　　　　　　　　　　　　　　　　(82)

Chart 8.38

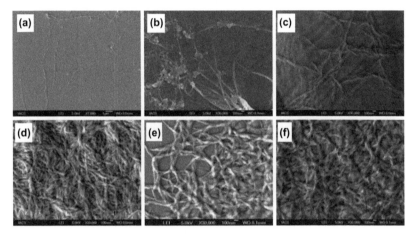

Figure 8.22　FESEM images of (a) **81**, (b) SWNT-COOH, (c) *f*-SWNT (0.005% w/v)–**81**, (d) f-SWNT (0.01% w/v)– **81**, (e) *f*-SWNT (0.005% w/v)– **82**, (f) *f*-SWNT (0.03% w/v)– **82** hybrids in toluene. [gelator] = 0.3% w/v. Reproduced with permission from ref. 99, The Royal Society of Chemistry.

with the backbone of SWNT-COOH. These dipeptide-based amphiphiles helped in the dispersion of SWNT-COOH and the dispersed nanotubes in return helped in the formation of the intertwined networks where the solvent gets arrested and yields a self supporting gel. This result has immense significance in the fact that an otherwise insoluble carbonaceous nanomaterial can remarkably impact gelation (an order of magnitude improvement) when present in such miniscule amounts.

They also reported the development of pristine SWNT-molecular hydrogel composites by dipeptide carboxylate amphiphiles (**83,84** Chart 8.39) having efficient hydrogelation ability along with SWNT dispersion capability.[100] Importantly, the dispersed SWNTs participated in supramolecular gelation as a physical crosslinker between SAFIN through complementary hydrophobic interactions and thus improving the gelation efficiency of the resultant composite by 2-fold (Figures 8.23a–c). The mechanical properties of the developed soft nanocomposites also showed manifold improvement compared to the native hydrogels without SWNTs. Importantly, the fitting fusion of

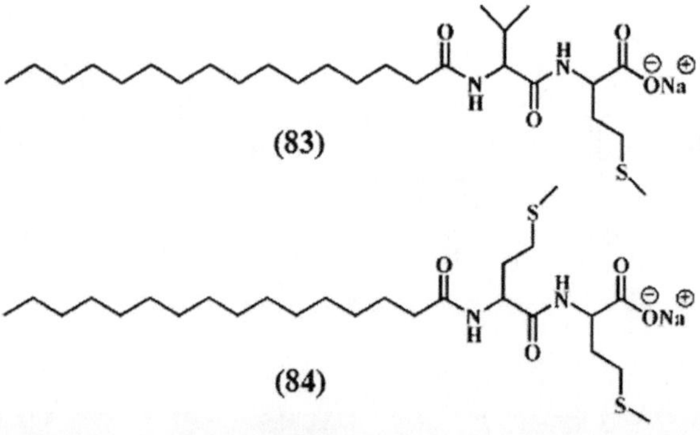

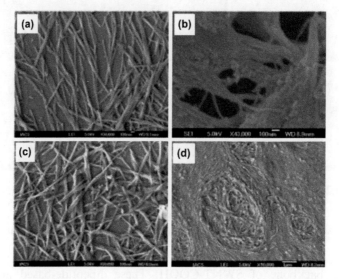

(83)

(84)

Chart 8.39

Figure 8.23 FESEM images of (a) hydrogel **83**, (b) pristine-SWNT, (c) SWNT-**83** composite and (d) SWNT-**83** hybrid gel containing cyt c.
Reproduced with permission from ref. 100, The Royal Society of Chemistry.

fibril-like dispersed SWNTs within the amphiphilic SAFIN of hydrogels re-markably improved the biocatalytic activity of immobilised cytochrome c (cyt c) in toluene by 120-fold compared to the activity of cyt c in water. Also, cyt c immobilised within soft nanocomposites exhibited 1.7-fold higher activity to that of native hydrogel immobilised enzyme.

8.6 Summary and Perspectives for the Future

In this chapter, we have discussed a wide range of examples of organic templates and methods for the formation of morphologically interesting inorganic materials. By using the self-assembled LMW gels as organic templates, well-defined functional inorganic nanomaterials were obtained by simple sol–gel chemistry. The self-assembled fibrillar networks of LMW gels function as templates not only for the tubular shapes, but also for helices, intertwined double helices, spirals, paper roll-like structures and others. In particular, helical structures have been shown to induce chirality in the corresponding transcribed inorganic material. If we consider the different structure scaffolds discussed in this chapter that have been employed as templates for transcription, it is noteworthy that they arise from a limited number of gelating scaffolds that are cholesterol-, sugar-, or cyclohexane-based. However, in spite of this limitation in the diversity of gelating scaffolds that has been used to prepare most of the transcribed materials thus far, a large diversity of inorganic shapes has been found. Clearly, use of other types of gelators will further broaden the synthesis of a variety of other interesting materials. On the other hand, many other inorganic compounds are now also being employed for transcription processes, thus widening the scope of applications of the resultant materials. Furthermore, the ability to precisely control tubular dimensions and to introduce different functional groups to the inner and outer surfaces of the silica nanotubes using the template synthesis method, makes these constructs promising candidates for biomedicinal and biotechnological applications. After successfully employing LMW gels for the transcription of inorganic materials, this knowledge was subsequently extended to prepare hybrid materials. Hybrid materials comprising metal nanoparticles are synthesised and/or stabilised/immobilised in the gel matrix by incorporating appropriate functionalities in the gelator's structure. Additionally, gel-stabilised nanoparticles like Au, Ag and CNTs synergistically helped in improving the physicochemical properties (such as mechanical strength, conductivity *etc.*) of the hybrid gels. The reversible gel–sol transition property of supramolecular gels has been utilised to make multiresponsive materials.

In regards to the synthesis of the small-molecule gelators and the subsequent formation of gel-nanocomposites, the protocols should be simple enough to be viable for their large-scale production. However, as mentioned above, the growing contributions from researchers working in multiple disciplines will surely aid the nanocomposite to overcome these challenges and form a truly advanced material for practical applications. Most importantly, the exponentially growing field of gel-nanocomposites has the potential of reaching out to widely diverse applications.

Acknowledgements

P.K.D. is thankful to the Department of Science and Technology, India for financial assistance to carry out research work on gel-nanocomposites. T.K. acknowledges the Council of Scientific and Industrial Research, India for his Research Fellowships.

References

1. P. Terech and R. G. Weiss, *Chem. Rev.*, 1997, **97**, 3133.
2. *Molecular Gels, Materials with Self-Assembled Fibrillar Networks*, ed. R. G. Weiss and P. Terech, Springer, New York, 2006.
3. M. George and R. G. Weiss, *Acc. Chem. Res.*, 2006, **39**, 489.
4. L. A. Estroff and A. D. Hamilton, *Chem. Rev.*, 2004, **104**, 1201.
5. M. de Loos, B. L. Feringa and J. H. van Esch, *Eur. J. Org. Chem.*, 2005, 3615.
6. A. Ajayaghosh and V. K. Praveen, *Acc. Chem. Res.*, 2007, **40**, 644.
7. N. M. Sangeetha and U. Maitra, *Chem. Soc. Rev.*, 2005, **34**, 821.
8. A. R. Hirst, B. Escuder, J. F. Miravet and D. K. Smith, *Angew. Chem. Int. Ed.*, 2008, **47**, 8002.
9. D. K. Smith, *Chem. Soc. Rev.*, 2009, **38**, 684.
10. F. Zhao, M. L. Ma and Bing Xu, *Chem. Soc. Rev.*, 2009, **38**, 883.
11. J. W. Steed, *Chem. Commun.*, 2011, **47**, 1379.
12. M. Suzuki and K. Hanabusa, *Chem. Soc. Rev.*, 2009, **38**, 967.
13. M. Zelzer and R. V. Ulijn, *Chem. Soc. Rev.*, 2010, **39**, 3351.
14. M. O. M. Piepenbrock, G. O. Lloyd, N. Clarke and J. W. Steed, *Chem. Rev.*, 2010, **110**, 1960.
15. L. Frkanec and M. Zinic, *Chem. Commun.*, 2010, **46**, 522.
16. R. V. Ulijna and D. N. Woolfson, *Chem. Soc. Rev.*, 2010, **39**, 3349.
17. R. V. Ulijn and A. M. Smith, *Chem. Soc. Rev.*, 2008, **37**, 664.
18. J. W. Steed, *Chem. Soc. Rev.*, 2010, **39**, 3686.
19. S. Banerjee, R. K. Das and U. Maitra, *J. Mater. Chem.*, 2009, **19**, 6649.
20. A. Dawn, T. Shiraki, S. Haraguchi, S.-i Tamaru and S. Shinkai, *Chem. Asian J.*, 2011, **6**, 266.
21. J. H. Jung, M. Park and S. Shinka, *Chem. Soc. Rev.*, 2010, **39**, 4286.
22. K. Sada, M. Takeuchi, N. Fujita, M. Namata and S. Shinkai, *Chem. Soc. Rev.*, 2007, **36**, 415.
23. R. G. Chapman and J. C. Sherman, *Tetrahedron*, 1997, **53**, 15911.
24. K. J. C. van Bommel, A. Friggeri and S. Shinkai, *Angew. Chem. Int. Ed.*, 2003, **42**, 980.
25. S. S. Kim, W. Zhang and T. J. Pinnavaia, *Science*, 1998, **282**, 1302.
26. D. H. W. Hubert, M. Jung, P. M. Frederik, P. H. H. Bomans, J. Meuldijk and A. L. German, *Adv. Mater.*, 2000, **12**, 1286.
27. H. Nakamura and Y. Matsui, *J. Am. Chem. Soc.*, 1995, **117**, 2651.
28. F. Miyaji, S. A. Davis, J. P. H. Charmant and S. Mann, *Chem. Mater.*, 1999, **11**, 3021.
29. N. Kawahashi and E. J. Matijevic, *Colloid Interface Sci.*, 1991, **143**, 103.
30. K. W. Gallis and C. C. Landry, *Adv. Mater.*, 2001, **13**, 23.
31. F. Caruso, *Chem. Eur. J.*, 2000, **6**, 413; Y. Lu, Y. Yin and Y. Xia, *Adv. Mater.*, 2001, **13**, 271.
32. F. J. M. Hoeben, P. Jonkheijm, E. W. Meijer and A. P. H. J. Schenning, *Chem. Rev.*, 2005, **105**, 1491; D. Das, T. Kar and P. K. Das, *Soft Matter*, 2012, **8**, 2348.

33. C. J. Brinker and G. W. Scherer, *Sol-Gel Science*, Academic Press, San Diego, 1990.
34. K. Murata, Ph.D. Thesis, Graduate School of Engineering, Kyushu University, 1997.
35. Y. Ono, K. Nakashima, M. Sano, Y. Kanekiyo, K. Inoue, J. Hojo and S. Shinkai, *Chem. Commun.*, 1998, 1477.
36. Y. Ono, K. Nakashima, M. Sano, J. Hojo and S. Shinkai, *Chem. Lett.*, 1999, 1119.
37. J. H. Jung, Y. Ono, K. Hanabusa and S. Shinkai, *J. Am. Chem. Soc.*, 2000, **122**, 5008.
38. K. Sugiyasu, S.-i. Tamaru, M. Takeuchi, D. Berthier, I. Huc, R. Oda and S. Shinkai, *Chem. Commun.*, 2002, 1212.
39. Y. Ono, Y. Kanekiyo, K. Inoue, J. Hojo and S. Shinkai, *Chem. Lett.*, 1999, 23.
40. J. H. Jung, Y. Ono and S. Shinkai, *J. Chem. Soc., Perkin Trans.*, 1999, **2**, 1289.
41. J. H. Jung, Y. Ono and S. Shinkai, *Angew. Chem. Int. Ed.*, 2000, **39**, 1862.
42. J. H. Jung, Y. Ono and S. Shinkai, *Langmuir*, 2000, **16**, 1643.
43. J. H. Jung, Y. Ono, K. Sakurai, M. Sano and S. Shinkai, *J. Am. Chem. Soc.*, 2000, **122**, 8648.
44. J. H. Jung, K. Nakashima and S. Shinkai, *Nano Lett.*, 2001, **1**, 145.
45. J. H. Jung and S. Shinkai, *J. Chem. Soc. Perkin Trans.*, 2000, **2**, 2393.
46. J. H. Jung, M. Amaike and S. Shinkai, *Chem. Commun.*, 2000, 2343.
47. H. Xu, Y. Wang, X. Ge, S. Han, S. Wang, P. Zhou, H. Shan, X. Zhao and J. R. Lu, *Chem. Mater.*, 2010, **22**, 5165.
48. J. Jiang, T. Wang and M. Liu, *Chem. Commun.*, 2010, **46**, 7178.
49. M. Yamanaka, Y. Miyake, S. Akita and K. Nakano, *Chem. Mater.*, 2008, **20**, 2072.
50. J. J. E. Moreau, L. Vellutini, M. W. C. Man and C. Bied, *J. Am. Chem. Soc.*, 2001, **123**, 1509.
51. J. J. E. Moreau, L. Vellutini, M. W. C. Man and C. Bied, *Chem. Eur. J.*, 2003, **9**, 1594.
52. J. J. E. Moreau, L. Vellutini, M. W. C. Man, C. Bied, J. L. Bantignies, P. Dieudonne and J. L. Sauvajol, *J. Am. Chem. Soc.*, 2001, **123**, 7957.
53. S. Kobayashi, K. Hanabusa, N. Hamasaki, M. Kimura, H. Shirai and S. Shinkai, *Chem. Mater.*, 2000, **12**, 1523.
54. S. Kobayashi, N. Hamasaki, M. Suzuki, M. Kimura, H. Shirai and K. Hanabusa, *J. Am. Chem. Soc.*, 2002, **124**, 6550.
55. M. Suzuki, Y. Nakajima, T. Sato, H. Shiraib and K. Hanabusa, *Chem. Commun.*, 2006, 377.
56. S. Dutta, D. Das, A. Dasgupta and P. K. Das, *Chem. Eur. J.*, 2010, **16**, 1493.
57. P. Xue, R. Lu, Y. Huang, M. Jin, C. Tan, C. Bao, Z. Wang and Y. Zhao, *Langmuir*, 2004, **20**, 6470.
58. P. Xue, R. Lu, D. Li, M. Jin, C. Tan, C. Bao, Z. Wang and Y. Zhao, *Langmuir*, 2004, **20**, 11234.

59. G. Gundiah, S. Mukhopadhyay, U. G. Tumkurkar, A. Govindaraj, U. Maitra and C. N. R. Rao, *J. Mater. Chem.*, 2003, **13**, 2118.
60. G. Palui, J. Nanda, S. Ray and A. Banerjee, *Chem. Eur. J.*, 2009, **15**, 6902.
61. J. D. Hartgerink, E. Beniash and S. I. Stupp, *Science*, 2001, **294**, 1684.
62. E. R. Zubarev, M. U. Pralle, E. D. Sone and S. I. Stupp, *J. Am. Chem. Soc.*, 2001, **123**, 4105.
63. M. Kimura, S. Kobayashi, T. Kuroda, K. Hanabusa and H. Shirai, *Adv. Mater.*, 2004, **16**, 335.
64. S. Bhat and U. Maitra, *Chem. Mater.*, 2006, **18**, 4224.
65. J. van Herrikhuyzen, S. J. George, M. R. J. Vos, N. A. J. M. Sommerdijk, A. Ajayaghosh, S. C. J. Meskers and A. P. H. J. Schenning, *Angew. Chem. Int. Ed.*, 2007, **46**, 1825.
66. N. M. Sangeetha, S. Bhat, G. Raffy, C. Belin, A. Loppinet-Serani, C. Aymonier, P. Terech, U. Maitra, J. P. Desvergne and A. D. Guerzo, *Chem. Mater.*, 2009, **21**, 3424.
67. S. Bhattacharya, A. Srivastava and A. Pal, *Angew. Chem. Int. Ed.*, 2006, **45**, 2934.
68. S. Bhattacharya and Y. K. Ghosh, *Chem. Commun.*, 2001, 185.
69. H. Basit, A. Pal, S. Sen and S. Bhattacharya, *Chem. Eur. J.*, 2008, **14**, 6534.
70. M. A. Firestone, M. L. Dietz, S. Seifert, S. Trasobares, D. J. Miller and N. J. Zaluzec, *Small*, 2005, **1**, 754.
71. C. S. Love, V. Chechik, D. K. Smith, K. Wilson, I. Ashworth and C. Brennan, *Chem. Commun.*, 2005, 1971.
72. B. Adhikari and A. Banerjee, *Chem. Eur. J.*, 2010, **16**, 13698.
73. J.-S. Shen and B. Xu, *Chem. Commun.*, 2011, **47**, 2577.
74. M. O. M. Piepenbrock, N. Clarke and J. W. Steed, *Soft Matter*, 2011, **7**, 2412.
75. P. K. Vemula and G. John, *Chem. Commun.*, 2006, 2218.
76. S. Ray, A. K. Das and A. Banerjee, *Chem. Commun.*, 2006, 2816.
77. P. K. Vemula, U. Aslam, V. A. Mallia and G. John, *Chem. Mater.*, 2007, **19**, 138.
78. C. Zhan, J. Wang, J. Yuan, H. Gong, Y. Liu and M. Liu, *Langmuir*, 2003, **19**, 9440.
79. Y. Li and M. Liu, *Chem. Commun.*, 2008, 5571.
80. R. N. Mitra and P. K. Das, *J. Phys. Chem. C*, 2008, **112**, 8159.
81. D. Das, S. Maiti, S. Brahmachari and P. K. Das, *Soft Matter*, 2011, **7**, 7291.
82. S. Dutta, A. Shome, S. Debnath and P. K. Das, *Soft Matter*, 2009, **5**, 1607.
83. S. Dutta, A. Shome, T. Kar and P. K. Das, *Langmuir*, 2011, **27**, 5000.
84. A. Shome, S. Dutta, S. Maiti and P. K. Das, *Soft Matter*, 2011, **7**, 3011.
85. S. Roy and P. K. Das, *Biotechnol. Bioeng.*, 2008, **100**, 756.
86. T. Kar, S. Dutta and P. K. Das, *Soft Matter*, 2010, **6**, 4777.
87. T. Kar, S. Debnath, D. Das, A. Shome and P. K. Das, *Langmuir*, 2009, **25**, 8639.

88. P. M. Ajayan, *Chem. Rev.*, 1999, **99**, 1787.
89. A. P. Goodwin, S. M. Tabakman, K. Welsher, S. P. Sherlock, G. Prencipe and H. Dai, *J. Am. Chem. Soc.*, 2009, **131**, 289.
90. M. Prato, K. Kostarelos and A. Bianco, *Acc. Chem. Res.*, 2008, **41**, 60.
91. T. Fukushima, A. Kosaka, Y. Ishimura, T. Yamamoto, T. Takigawa, N. Ishii and T. Aida, *Science*, 2003, **300**, 2072.
92. T. Fukushima and T. Aida, *Chem. Eur. J.*, 2007, **13**, 5048.
93. S. Srinivasan, S. S. Babu, V. K. Praveen and A. Ajayaghosh, *Angew. Chem. Int. Ed.*, 2008, **47**, 5746.
94. A. Pal, B. S. Chhikara, A. Govindaraj, S. Bhattacharya and C. N. R. Rao, *J. Mater. Chem.*, 2008, **18**, 2593.
95. S. K. Samanta, A. Pal, S. Bhattacharya and C. N. R. Rao, *J. Mater. Chem.*, 2010, **20**, 6881.
96. B. Adhikari, J. Nanda and A. Banerjee, *Chem. Eur. J.*, 2011, **17**, 11488.
97. B. Adhikari and A. Banerjee, *Soft Matter*, 2011, **7**, 9259.
98. Z. Tan, S. Ohara, M. Naito and H. Abe, *Adv. Mater.*, 2011, **23**, 4053.
99. S. K. Mandal, T. Kar, D. Das and P. K. Das, *Chem. Commun.*, 2012, **48**, 1814.
100. T. Kar, S. K. Mandal and P. K. Das, *Chem. Commun.*, 2012, **48**, 8389.

Subject Index

Page numbers in *italics* refer to figures or tables.